V. I. Babitsky (Ed.)

Dynamics of Vibro-Impact Systems

Springer

Berlin
Heidelberg
New York
Barcelona
Hong Kong
London
Milan
Paris
Singapore
Tokyo

V. I. Babitsky (Ed.)

Dynamics of Vibro-Impact Systems

Proceedings of the Euromech Colloquium 15-18 September 1998

With 217 Figures

Springer

V. I. Babitsky
Loughborough University
Department of Mechanical Engineering
LE11 3TU Loughborough, Leicestershire
United Kingdom

Cataloging-in-Publication Data applied for

Die Deutsche Bibliothek - CIP-Einheitsaufnahme
Dynamics of vibro-impact systems: proceedings of the Euromech Colloquium 15-18 September 1998 / V.I. Babitsky (ed.)
Berlin; Heidelberg; NewYork; Barcelona; Hong Kong; London; Milan; Paris; Singapore; Tokyo: Springer, 1999
ISBN-13: 978-3-642-64263-0

ISBN-13: 978-3-642-64263-0 e-ISBN-13: 978-3-642-60114-9
DOI:10.1007/978-3-642-60114-9

Softcover reprint of the hardcover 1st edition 1999

Typesetting: Camera-ready copies from authors
Cover-Design: MEDIO, Berlin
SPIN 10715178 62/3020 5 4 3 2 1 0 Printed on acid-free paper

Preface

The EUROMECH Colloquium "Dynamics of Vibro-Impact Systems" was held at Loughborough University on September 15th-18th, 1998. This was the first international meeting on this subject continuing the traditions of the series of Russian meetings held regularly since 1963.

Mechanical systems with multiple impact interactions have wide applications in engineering as the most intensive sources of mechanical influence on materials, structures and processes. Vibro-impact systems are used widely in machine dynamics, vibration engineering, and structural mechanics.

Analysis of vibro-impact systems involves the investigation of mathematical models with discontinuities and reveals their behaviour as strongly non-linear. Such systems exhibit complex resonances, synchronisation and pulling, bifurcations and chaos, excitation of space coherent structures, shock waves, and solitons.

The aim of the Colloquium was to facilitate the exchange of up-to-date information on the analysis and synthesis of vibro-impact systems as well as on the new developments in excitation, control and applications of vibro-impact processes.

The main topics were the following:

- excitation, synchronisation and stabilisation of vibro-impact processes
- dynamics of vibro-impact machines and technological processes
- non-linear phenomena due to vibro-impact interaction of solids
- synergistic effects of repeated impacts on solids and granular media
- control of vibration by impacts
- analytical, experimental and numerical methods for the analysis of vibro-impact systems and processes
- synthesis and optimisation of vibro-impact systems
- measurements of vibro-impact processes

The Colloquium attracted 51 participants from 19 countries. They presented 39 lectures of 25 minutes each. The important feature of the Colloquium was the participation of both academic and industrial specialists. Many well-known experts on the subject contributed to the programme as well as plenty of young participants. All the topics were well presented and actively discussed.

The articles included in the Proceedings were selected after refereeing of the discussed Colloquium papers by international experts. For the convenience of the readers we grouped them thematically.

We hope that the presented papers will give rich information to engineers and researchers involved in the analysis and development of dynamical systems as well as to the lecturers and graduate students interested in this developing area of non-linear dynamics and engineering.

We thank all the persons who contributed to the organisation of the Colloquium and preparation of the Proceedings as well as all participants for their collaboration. The organisers are very grateful to the European Mechanics Society and Faculty of Engineering of Loughborough University for patronage and sponsorship of the Colloquium. We thank the staff of Springer for support and accomplishing this publication.

Vladimir Babitsky

Chairman of the Colloquium — Loughborough, Leicestershire, UK

CONTENTS

Vocal folds as a vibro-impact system

Polina S. Landa *

A model for human vocal folds in the form of two spring-suspended plates, which can collide with one another, is considered. It is shown that due to air flow between the plates excitation of chaotic self-oscillations occurs. In this process the shape of oscillations of air flow volume velocity was found to be close to observed experimentally.

1 Introduction

Nowadays many different models of voice production are known. A review of some of these models is given in the book by Sorokin [1]. These models are of paramount importance both for understanding the mechanism of voice production and in certain practical applications, for example, in speech synthesis and recognition [2], in voice pathology [3], and so on. The heart of any such model is a model of vocal source, i.e., of vocal folds located in the human (or animal) larynx.

A variety of models of vocal folds, both complicated [4,5,6] and moderately simple [7,8,9,10] are known. The more used among these models is a two-mass model by Ishizaka and Flanagan [7]. In this model aerodynamic forces were calculated in the quasi-static approximation using the Bernoulli law. In so doing, the velocity of motion of the glottis' walls, the viscosity and inertia of air were ignored. Such a technique was also used in a simplified version of this model suggested by Herzel and Knudsen [3].

Another model of vocal folds, which by its nature is close to Ishizaka and Flanagan's model, was suggested in [11,12]. Contrary to the first, this model permits more correct calculation of aerodynamical forces. However, the study of this model with properly calculated aerodynamical forces, even numerically, is a very complicated problem. Therefore, in this work we present the results of numerical simulation of the model with aerodynamical forces calculated in quasi-static approximation with the use of the Bernoulli law.

2 The model and its equations

The model involves two absolutely rigid plates suspended by springs to the walls of a tube with right angled cross section (see Fig. 1). Air enters the tube from a reservoir of a sufficiently large volume V as a result of the pressure drop $\Delta P = P - P_a$ and can cause self-oscillations of the plates. It can be shown that the excitation of self-oscillations is possible if each plate has

* Moscow State University, Department of Physics

at least two degrees of freedom, i.e., it can both move progressively in the direction orthogonal to the air flow and turn about axis O, passing through its centre of mass. In this regard the mechanism of the excitation of self-oscillations of the plates is akin to that as for the bending-torsion flutter of an aeroplane wing [13,14,15].

Assuming the motion of the plates to be completely symmetric relative to the tube mid-plane we write the motion equations for each of the plates as

$$m\ddot{h}_a + \alpha\dot{h}_a + k(h_a - \tilde{h}_a) + \kappa(\varphi - \tilde{\varphi}) = F,$$
$$J\ddot{\varphi} + \beta\dot{\varphi} + K(\varphi - \tilde{\varphi}) + \kappa(h_a - \tilde{h}_a) = M, \quad (1)$$

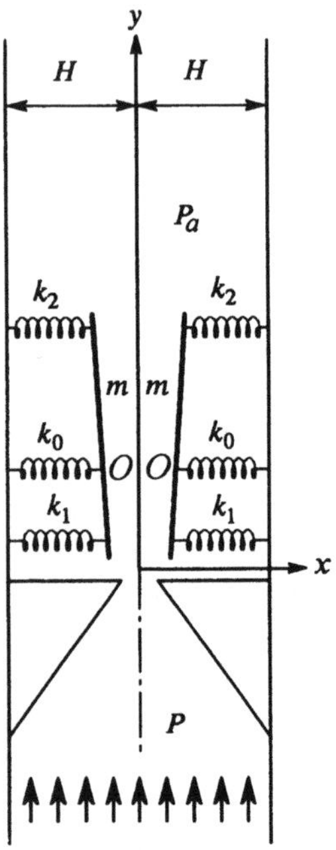

Figure 1. The model of the vocal folds.

where h_a is the displacement of plate's centre of mass, φ is the angle of rotation clockwise about the axis passing through the centre of mass, $\tilde{h}_a$ and $\tilde{\varphi}$ are the values of h_a and φ for undeformed springs, m is the plate mass, J is the plate moment of inertia about the axis passing through the centre of mass, $k = k_1 + k_2 + k_0$ is the total rigidity factor of the springs, $K = k_1a^2 + k_2(b - a)^2$ is the total rigidity factor with respect to the rotation, a is the distance between the lower edge of the plate and its centre of mass, b is the plate length along the flow, $\kappa = k_1a - k_2(b - a)$ is the coupling factor characterising the effect of the displacement of the plate centre of mass on the plate rotation, and vice versa, α and β are the damping factors,

$$F = l\int_0^b p(y)\,dy, \quad M = l\int_0^b (y - a)p(y)\,dy, \quad (2)$$

are the aerodynamic force and moment acting on the plate which are created by the air flow, $p(y)$ is the difference between the air pressure in the inter-plate slot and atmospheric pressure P_a, and l is the plate transverse dimension. It should be noted that Eqs. (1) can be obtained by the Galerkin method as a discrete model of bending-torsion oscillations of a beam lying on an elastic base.

Eqs. (1) are conveniently written in another form:

$$\ddot{S}_0 + 2\delta_{11}\dot{S}_0 + 2\delta_{12}\dot{S}_b + \omega_1^2(S_0 - \tilde{S}_0) + \kappa_1(S_b - \tilde{S}_b) = F_0,$$
$$\ddot{S}_b + 2\delta_{21}\dot{S}_0 + 2\delta_{22}\dot{S}_b + \omega_2^2(S_b - \tilde{S}_b) + \kappa_2(S_0 - \tilde{S}_0) = F_b,, \quad (3)$$

where

$$\delta_{11} = \frac{b-a}{b}\frac{\alpha}{2m} + \frac{a}{b}\frac{\beta}{2J}, \quad \delta_{12} = \frac{a}{b}\left(\frac{\alpha}{2m} - \frac{\beta}{2J}\right), \quad \delta_{21} = \frac{b-a}{a}\delta_{12},$$

$$\delta_{22} = \frac{a}{b}\frac{\alpha}{2m} + \frac{b-a}{b}\frac{\beta}{2J}, \quad \omega_1^2 = \frac{b-a}{b}\frac{k}{m} + \frac{a}{b}\frac{K}{J} + \frac{\kappa}{mJb}\Big(J + ma(b-a)\Big),$$

$$\omega_2^2 = \frac{a}{b}\frac{k}{m} + \frac{b-a}{b}\frac{K}{J} - \frac{\kappa}{mJb}\Big(J + ma(b-a)\Big),$$

$$\kappa_1 = \frac{a}{b}\left(\frac{k}{m} - \frac{K}{J}\right) - \frac{\kappa}{mJb}\Big(J - ma^2\Big),$$

$$\kappa_2 = \frac{b-a}{b}\left(\frac{k}{m} - \frac{K}{J}\right) + \frac{\kappa}{mJb}\Big(J - m(b-a)^2\Big),$$

$$\tilde{S}_0 = 2l(\tilde{h}_a + a\tilde{\varphi}), \quad \tilde{S}_b = 2l(\tilde{h}_a - (b-a)\tilde{\varphi}),$$

$$F_0 = 2l\left(\frac{F}{m} + \frac{aM}{J}\right), \quad F_b = 2l\left(\frac{F}{m} - \frac{(b-a)M}{J}\right). \tag{4}$$

It should be emphasised that the vocal folds, executing self-oscillations, collide with one another. This process plays a great role in voice production because it results in the generation of pulses containing a number of high harmonics which enrich speech. Therefore we have to add impact conditions to Eqs. (3). These conditions may be obtained as follows: because the plates are assumed to be absolutely rigid, then only the plate edges can collide. For definiteness, first we consider the collision of the plate edges corresponding to $y = 0$. Let an impulse of force $F\Delta t$, causing changes of the plate momentum $m\Delta\dot{h}_a$ and of the plate angular momentum $J\Delta\dot{\varphi}$, arise as a result of the collision. Thus, $F\Delta t = m\Delta\dot{h}_a$, $Fa\Delta t = J\Delta\dot{\varphi}$. From this it follows that

$$\Delta\dot{\varphi} = \frac{ma}{J}\Delta\dot{h}_a. \tag{5}$$

On the other hand, it is easily shown that

$$\Delta\dot{h}_a + a\Delta\dot{\varphi} = \Delta\dot{h}_0. \tag{6}$$

If the velocity restitution coefficient after impact is $R \le 1$ then

$$\Delta\dot{h}_0 = \dot{h}_0^+ - \dot{h}_0^- = -(1+R)\dot{h}_0^-, \tag{7}$$

where $\dot{h}_0^+$ is the value of $\dot{h}_0$ after impact, and $\dot{h}_0^-$ is the value of $\dot{h}_0$ before impact. From (5)–(7) we obtain the following conditions for the collision of the plate edges under consideration:

$$\Delta\dot{h}_a = -\frac{(1+R)\dot{h}_0^-}{1+ma^2/J}, \quad \Delta\dot{\varphi} = -\frac{ma}{J}\frac{(1+R)\dot{h}_0^-}{1+ma^2/J}. \tag{8}$$

In a like manner we obtain the conditions for the collision of the plate edges corresponding to $y = b$:

$$\Delta\dot{h}_a = -\frac{(1+R)\dot{h}_b^-}{1+m(b-a)^2/J}, \quad \Delta\dot{\varphi} = \frac{m(b-a)}{J}\frac{(1+R)\dot{h}_b^-}{1+m(b-a)^2/J}. \tag{9}$$

From (8) and (9) we can find the conditions for the collision of the plate edges corresponding to $y = 0$ and $y = b$ in terms of S_0 and S_b: for $y = 0$

$$\Delta\dot{S}_0 = \dot{S}_0^+ - \dot{S}_0^- = -(1+R)\dot{S}_0^-,$$

$$\Delta\dot{S}_b = \dot{S}_b^+ - \dot{S}_b^- = -(1+R)\dot{S}_0^- \frac{J - ma(b-a)}{J + ma^2}, \tag{10}$$

and for $y = b$

$$\Delta\dot{S}_b = -(1+R)\dot{S}_b^-, \quad \Delta\dot{S}_0 = -(1+R)\dot{S}_b^- \frac{J - ma(b-a)}{J + m(b-a)^2}. \tag{11}$$

We note that for sufficiently small R the impact may be quasi-plastic [16], i.e. the duration of the contact of the plate edges during the impact may be finite.

It can be shown that the sound pressure above the slot is

$$p_s(t) = \frac{\rho c_0}{Hl}\left(u - u_{st}\right), \tag{12}$$

where c_0 is the velocity of sound, u is the volume velocity, u_{st} is the steady state value of u, and H is the width of the tube. Thus, the sound pressure is determined by oscillations of the volume velocity of air flow.

As opposed to flowing an aeroplane wing, we are dealing here with one-sided flow. Therefore we cannot use the expressions for aerodynamic forces which are known for a bending-torsion flatter problem. Let us calculate these forces in quasi-static approximation using the Bernoulli theorem.

First of all, we use the formula for a dynamical pressure drop as a fluid flow is constricted gradually [17]. In a zero approximation with respect to h/H we obtain

$$p(0) = \Delta P - \zeta_1 \frac{\rho u^2}{2S_0^2}, \tag{13}$$

where $1 < \zeta_1 < 2$ is the coefficient depending on the bevel shape at the slot input and, in general, on the Reynolds number.

In agreement with the Bernoulli theorem we find $p(y)$:

$$p(y) = p(0) - \frac{\rho u^2}{2S_0^2}\left(\frac{S_0^2}{(S_0 - (S_0 - S_b)y/b)^2} - 1\right). \tag{14}$$

Unknown volume velocity u can be found from the formula for a dynamic pressure drop as the flow diverges abruptly at the slot output [17]

$$p(b) = \zeta_2 \frac{\rho u^2}{2S_b^2}, \tag{15}$$

where ζ_2 is the coefficient depending on the velocity profile at the slot's output and on the Reynolds number **Re** (for **Re** $> 10^3$, if the velocity profile is uniform then $\zeta_2 = 0$, and if the velocity profile is Poiseuille's then $\zeta_2 \approx 0.6$). Using further the expression (14), in view of (13), for $y = b$ we obtain the following equation for the volume velocity u:

$$\frac{\rho u^2}{2} = S_0^2 S_b^2 \frac{\Delta P}{(\zeta_1 - 1)S_b^2 + (1 + \zeta_2)S_0^2}, \tag{16}$$

Substituting (14) in (2) and taking into account (13) we find

$$F = bl\,\frac{\rho u^2}{2S_b^2}\left(1 + \zeta_2 - \frac{S_b}{S_0}\right),$$

$$M = rF - \frac{\rho b^2 l u^2}{2(S_0 - S_b)^2}\left(\ln\frac{S_b}{S_0} + \frac{S_0^2 - S_b^2}{2S_0 S_b}\right),$$

where $r = b/2 - a$.

3 Results of numerical simulation of the model equations

The parameters which are necessary for the simulation of self-oscillations of the human vocal folds can be estimated on the basis of the information presented in the book by Sorokin [1]. We set the following values of the parameters: $\zeta_1 = 1.37$, $\zeta_2 = 0.2$, $\rho = 1.3 \times 10^{-3}\,\mathrm{g/cm^3}$, $m = 0.15\,\mathrm{g}$, $J = 0.004\,\mathrm{g\,cm^2}$, $k = 8 \times 10^4\,\mathrm{g/s^2}$, $K = 2400\,\mathrm{g\,cm^2/s^2}$, $\kappa = 10^3\,\mathrm{g\,cm/s^2}$, $\alpha = 40\,\mathrm{g/s}$, $\beta = 1\,\mathrm{g/s}$, $a = 0.15\,\mathrm{cm}$, $b = 0.5\,\mathrm{cm}$, $l = 1.6\,\mathrm{cm}$, $\tilde{h}_0 = 0.07\,\mathrm{cm}$, $\tilde{h}_b = 0.06\,\mathrm{cm}$. The parameters ΔP and R are varying.

A steady state solution of Eqs. (3) is determined from the following algebraic equations:

$$\omega_1^2(S_0 - \tilde{S}_0) + \kappa_1(S_b - \tilde{S}_b) = F_0, \quad \omega_2^2(S_b - \tilde{S}_b) + \kappa_2(S_0 - \tilde{S}_0) = F_b. \tag{17}$$

(Here the subscripts 'st' are omitted for brevity.)

Taking into account (4), (16) and (17) we can find the dependencies of steady-state values of the slot cross-section area at the slot input and output and of the volume flow velocity u on the pressure drop ΔP. These dependencies are given in Fig. 2.

Writing linearized equations for small deviations from the steady state values found, one can find a condition for the self-excitation of the system and the self-oscillation frequency in the neighbourhood of the boundary of self-excitation. It can be shown that the condition of self-excitation is satisfied for $\Delta P \geq \Delta P_{cr} \approx 5650\,\mathrm{g/cm\,s^2} \approx 5.7$ cm of water. The critical value of u associated with this value of ΔP is $u_{cr} \approx 527\,\mathrm{cm^3/s}$. The self-oscillation

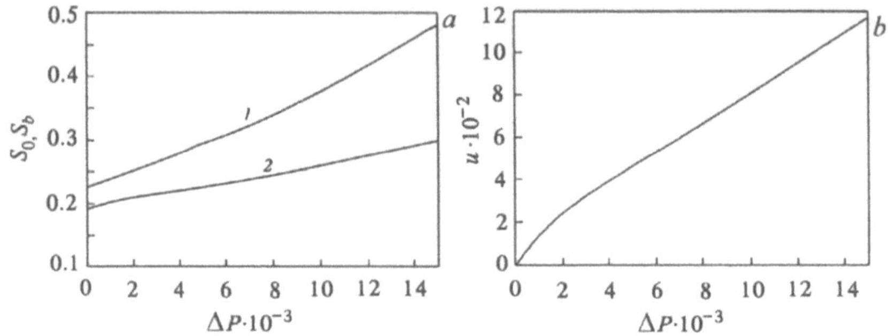

Figure 2. (a) The plots of S_0^{st} (the curve labelled *1*) and S_b^{st} (*2*) versus ΔP, and (b) the plot of u^{st} versus ΔP for the values of the parameters given above.

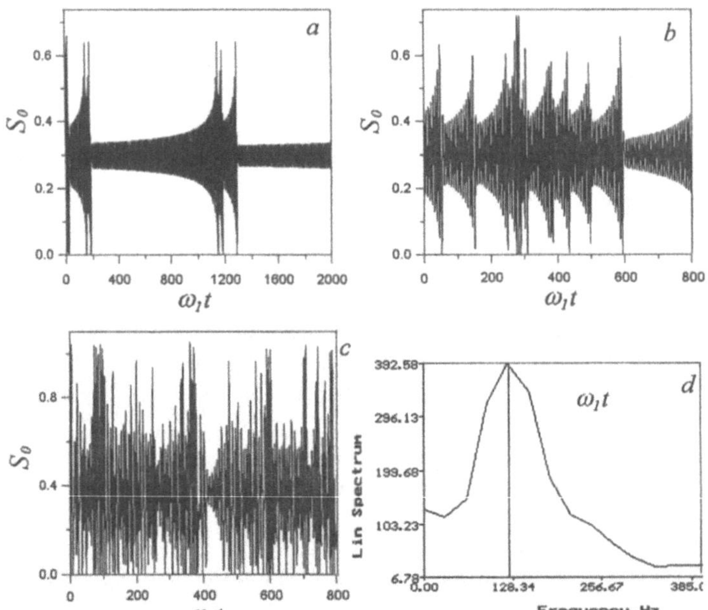

Figure 3. The dependencies of S_0 on $\omega_1 t$ for $R = 1$, $\Delta P = 5710$ (a), $\Delta P = 6000$ (b) and $\Delta P = 10000$ (c); (d) the power spectrum of the volume velocity u for $\Delta P = 10000$.

frequency in the neighbourhood of the self-excitation boundary is $f = \omega/2\pi \approx 120$ Hz. The results obtained correspond to known experimental data.

Numerical simulation of Eqs. (3), with using (17) and the impact conditions (10), (11), shows that excitation of self-oscillations is hard. The character of the transition through the self-excitation threshold essentially depends on the restitution coefficient R. For small R a chaotic attractor exists on each side of the transition point, but for $\Delta P^* < \Delta P < \Delta P_{cr}$, where $\Delta P^* < 3700\,\mathrm{g/cm\,s^2}$, in addition to this attractor the steady state is also an attractor. For R close to unity and $\Delta P < \Delta P_{cr}$ no chaotic attractor has been found. In this case the transition through the self-excitation threshold occurs via intermittency [18].

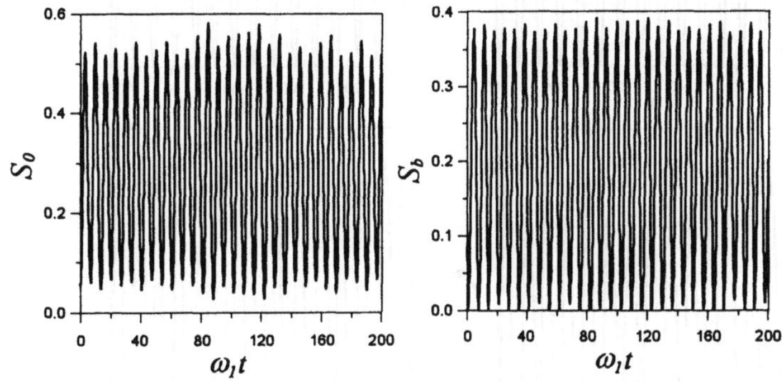

Figure 4. The dependence of S_0 on $\omega_1 t$ for $\Delta P = 4000$, $R = 0$.

The shape of oscillations of S_0 for $R = 1$ and different values of ΔP is given in Fig. 3 *a*, *b* and *c*. The shape of oscillations of S_b and u is similar to that for S_0. For $\Delta P = 6000$ the oscillations resemble in their shape stochastic oscillations in Neimark's pendulum that is an example of stochastic generator [18,19]. Power spectra of the volume velocity u for all indicated values of ΔP contains only a single line, but this line is moderately wide. An example of the power spectrum is shown in Fig. 3 *d*.

For small values of the restitution coefficient R stable self-oscillations are possible down to ΔP^*. For $\Delta P^* < \Delta P < \Delta P_{cr}$ their existence depends on initial conditions. In this range of ΔP the oscillations excited in their shape are close to periodic with the fundamental frequency about 120 Hz. An example of such oscillations is illustrated in Fig. 4.

As ΔP increases the oscillations become more chaotic and the duration of the contact of the plate edges during the impact increases. This is illustrated in Fig. 5. Power spectra of u contain a number of moderately narrow lines at the frequencies divisible by the fundamental frequency. An example of such spectrum for $\Delta P = 10000$ is given in Fig. 5, bottom centre.

It is interesting that the phase shift between oscillations of the lower and upper edges of the plates is closed to that was observed in experiments [20]. This is illustrated in Fig. 6, where oscillations of S_0 and S_b both numerical (for $\Delta P = 6000$, $R = 0$) and experimental are shown.

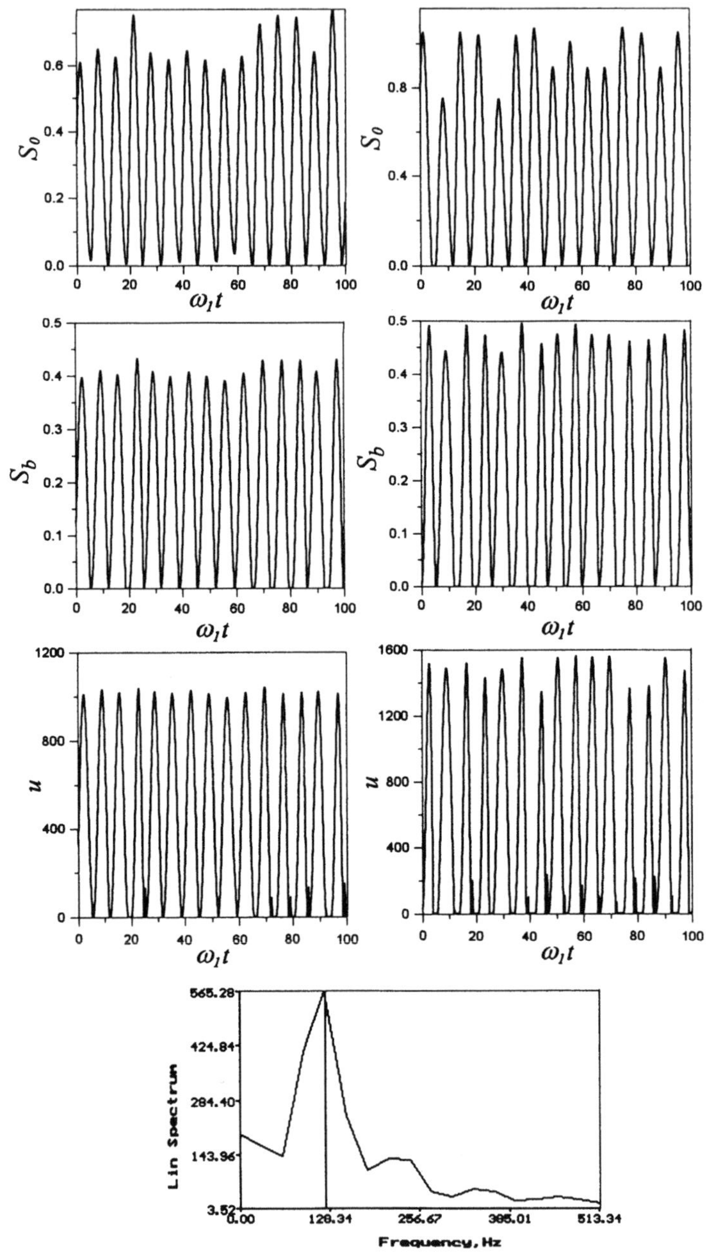

Figure 5. Dependencies of S_0, S_b, u on $\omega_1 t$ and the power spectrum of u (bottom centre) for $R = 0$, $\Delta P = 6000$ (at the left) and $\Delta P = 10000$ (at the right).

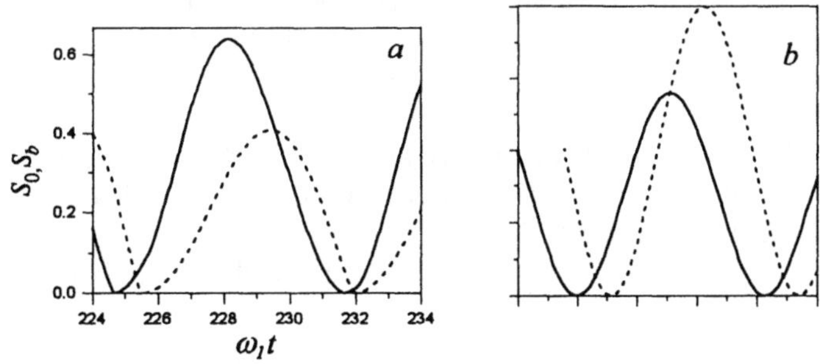

Figure 6. Plots of S_0 (solid lines) and S_b (dashed lines) versus time for numerical simulation (a) and an experiment (b).

4 Conclusion

We have shown that the model considered allows us to simulate chaotic self-oscillations of human vocal folds taking into account collisions between them. Experiments with speech synthesiser, in which have used periodic vocal sources, show that for better sounding it is necessary to add small noise [1]. It seems likely that in nature slight chaos plays a role of this noise.

References

1 Sorokin V N, The Theory of Speech Production, Radio i svyaz, Moscow, 1985 (in Russian).
2 Flanagan J L, Speech Analysis, Synthesis and Perception, Springer, New York, 1972.
3 Herzel H, Bifurcations and chaos in voice signals, Appl. Mech. Rev. 46(7): 399–413, 1993.
4 Titze I R and Talkin D T, A theoretical study of the effects of various laryngeal configurations on the acoustics of phonation, J. Acoust. Soc. Amer. 6:60–74, 1979.
5 Vocal Fold Physiology: New Frontiers in Basic Science, ed. by Titze I.R. Singular Publ. Group, San Diego, 1992.
6 Liljencrants J, Numerical simulations of glottal flow, in: Proc. EUROSPEECH, Geneva, 1991.
7 Ishizaka K and Flanagan J L, Synthesis of voiced sounds from a two-mass model of the vocal cords, Bell Syst. Techn. J. 51:1233–1268, 1972.
8 Stevens K N, Physics of laryngeal behavior and larynx modes, Phonetica 34:264–279, 1977.
9 Broad D, The new theories of vocal fold vibration, in: Speech and Language: Advances in Basic Research and Practice, ed. by Lass N., Academic Press, New York, 1979.

10 Story B H and Titze I R, Voice simulation with a body-cover model of the vocal folds, J. Acoust. Soc. Amer. 97(2): 1249–1260, 1995.
11 Landa P S and Rudenko O V, Two sound generation mechanism, Soviet Physics – Acoustics 35(5): 497-501, 1989.
12 Landa P S, Rudenko O V, and Bakscis B P, Excitation of body self-oscillations and sound generation in gas or fluid flows, Optical& Acoustical Rev. 1(3):277–290, 1990–1991.
13 Strelkov S P, Introduction to the Theory of Oscillations, Nauka, Moscow, 1964 (in Russian).
14 Panovko Ya G and Gubanova I I, Stability and Oscillations of Elastic Systems, Nauka, Moscow, 1979 (in Russian).
15 Panovko Ya G, Introduction to the Theory of Mechanical Oscillations, Nauka, Moscow, 1991 (in Russian).
16 Nagaev R F, Mechanical Processes with Repeating Damped Impacts, Nauka, Moscow, 1985 (in Russian).
17 Idel'chik I E, Handbook of Hydraulic Resistances, Mashinostroenie, Moscow, 1975 (in Russian); English Translation: Hemisphere Publ. Corp., Washington, New York, London, 1986.
18 Neimark Yu I and Landa P S, Stochastic and Chaotic Oscillations, Nauka, Moscow, 1987 (in Russian); English Translation: Kluwer Academic Publ., Dordrecht, 1991.
19 Landa P S, Nonlinear Oscillations and Waves in Dynamical Systems, Kluwer Academic Publ., Dordrecht, 1996.
20 Hiroto J, Introductory remarks, Vocal Fold Physiology Conference, Kurume, 1980, pp. IR2-1–IR2-6.

Behaviour of granular materials in the field of vibration

M. Saeki[*1] *and E. Takano*[*2]

Vibration has been widely used in various industrial operations dealing with granular materials. Classification, conveying and packing provide typical examples of the vibration technique. In spite of these actual uses, the behaviour of granular materials is poorly understood. Most of the knowledge related to handling of particulates is empirical and no generalized approach for the analysis of the response of granular materials to external forces exists. This paper presents an investigation into the development of a numerical method for size segregation, which is the typical behavior of granular materials in the field of vibration. This numerical method is modified on the basis of the discrete element method proposed by Cundall. In order to consider the shape anisotropy of particles, they are first treated as elliptic models. Then mechanical units such as spring, dashpots and friction sliders are used to express the contact forces. The effect of the size ratio and the shape of each particle on the response of granular materials is considered.

1 Introduction

A vibration technique such as packing, conveying and smooth discharge of particles from the hopper is often utilized in various industries handling granular materials. When a mixture containing particles of various sizes and specific gravities is subjected to vibrations, the phenomenon of segregation occurs. This phenomenon is usually undesirable. Rosato et al. [(1)] introduced a new method by simulating the size segregation process based on the Monte Carlo method and presented the typical dynamical pictures. Much of the current interest [(2), (3)] in size segregation has been stimulated by this simulation. However, previous numerical simulation attempts to model the size segregation process do not consider the interactive effects due to interparticle friction and collisions. Since most of these investigations have used discs or spheres as particles, the effect of shape has not been clarified. Recently, the discrete element method (DEM) was used in the investigation of soil mechanics and powder technology [(4), (5)]. In this method, the dynamics of granular materials is determined by analyzing the motion of the individual particles. In addition, this method concerns with how to consider the dependence of interparticle friction and collisions. In the present work, this method is adapted for the size segregation in mixtures of a large particle within an

*1 Department of Mechanical and Control Engineering, Niigata Institute of Technology, Fujihashi, Kashiwazaki, Niigata, 945-1195, Japan

*2 Department of Mechanical and Production Engineering, Niigata University, Igarashi, Niigata, 950-2181, Japan

assembly of small particles when subjected to vibrations. We discuss the effect of size ratio on the segregation process. Particles are treated as elliptic models and particle-shape effects are also studied.

2 Size segregation model

Figure 1 shows a size segregation model. A large particle is placed within an assembly of small particles in the container. When the container undergoes translational harmonic vibrations parallel to the y-axis, the large particle rises up to the top of the bed. We assume that the motion of the container is independent of the behaviour of granular materials. As shown in Fig. 1, x_i and y_i denote the coordinates of any particle i. The displacement of the container V at time t is given by Eq. (1), using the x-y coordinate system.

$$\left.\begin{aligned} X &= 0 \\ Y &= a\sin(2\pi f t) \end{aligned}\right\} \quad (1)$$

where a and f are the amplitude and frequency of the vibrating container, respectively.

3 Numerical method

3.1 The equation of motion for particles

In the present work, the two-dimensional motion of cylindrical bodies with elliptic cross sections (It is termed 'particles' as follows) is considered, taking into account the effect of shape anisotropy on the size segregation of granular materials. Considering the external force acting on a particle i, the equation of motion for the particle is given by

$$\left.\begin{aligned} m_i\ddot{x}_i &= F_{xi} \\ m_i\ddot{y}_i &= F_{yi} - m_i g \\ I_i\ddot{\theta}_i &= T_i \end{aligned}\right\} \quad (2)$$

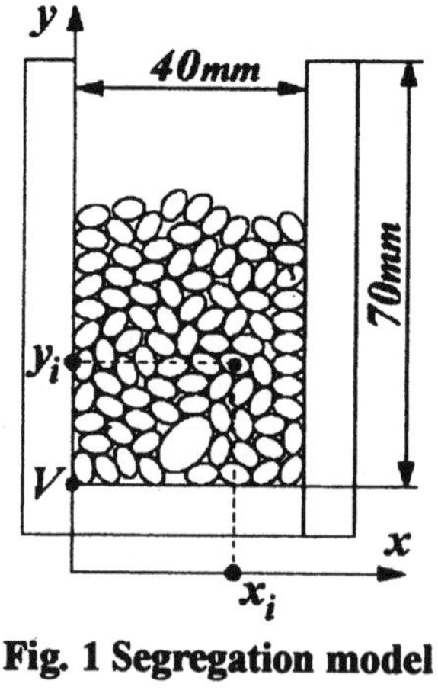

Fig. 1 Segregation model

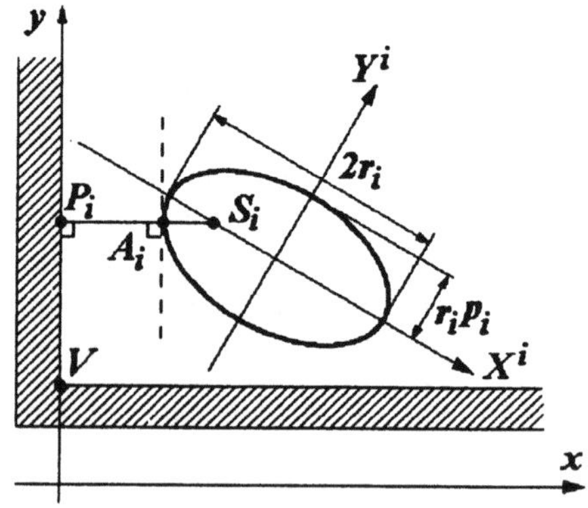

Fig. 2 Distance between particle and container

where m_i is the mass of the particle i, I_i is the moment of inertia of the particle, g is the acceleration due to gravity, θ_i is the rotation angle of the particle, F_{xi} and F_{yi} are the x and y directional components of the total force acting on the particle, T_i is the total torque on the particle about an axis normal to the paper and the dot denotes a time derivative.

3.2 Contact judgement

Let us consider the contact judgement between a particle and the y-axis, which means an inside wall of container. Figure 2 shows a particle i in the vibrating container. The major axis and flat ratio (=minor axis / major axis) of the particle are $2r_i$ and p_i, respectively. Initially, the line normal to the ellipse of the particle i at point A_i crosses the X^i-axis at point S_i. A contact between the particle and the wall of the container is realized if the line S_iA_i is perpendicular to the y-axis at point P_i and S_iA_i is longer than S_iP_i. In the case of a contact between a particle and the other wall of the container, a similar procedure is performed.

Next, consider the interparticle contact. Figure 3 shows two particles i and j in the vibrating container. The line normal to the ellipse of the particle i at point A_i crosses the major axis at point S_i. Points A_j and S_j of particle j are determined as follows. An interparticle contact is realized if the line A_iS_i coincides with the line A_jS_j and S_iS_j is shorter than the sum of S_iA_i and S_jA_j.

3.3 Contact force

Figure 4 shows the contact state between two particles i and j. As shown in this figure, consider the normal component f_N and tangential component f_T of the contact force acting on the particle i. In order to determine the contact force, the angle ϕ_{ij} between the line S_iS_j and x-axis is introduced. The normal component f_N of the contact force is given by the sum of the spring force and the viscous damping force, as shown in Eq. (3). The tangential component f_T of the contact force is given by Eq. (4), considering the Coulomb law of friction.

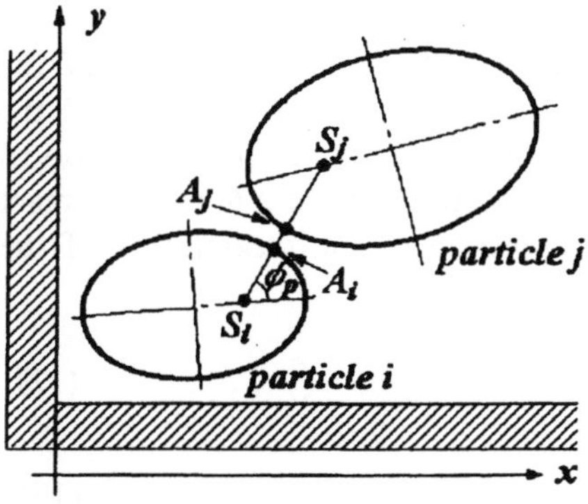

Fig. 3 Distance between two particles

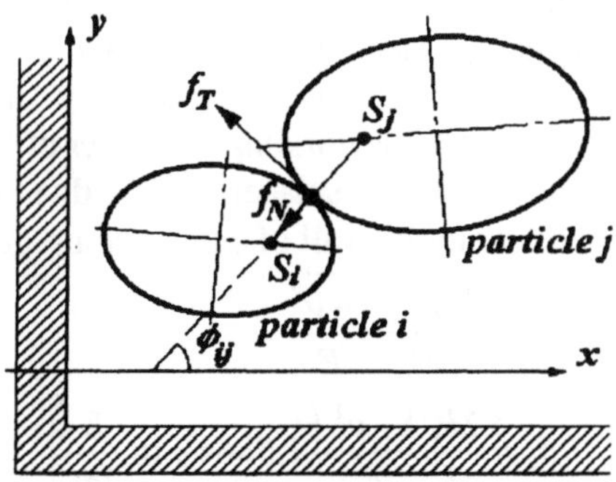

Fig. 4 Contact force

$$f_N = P_N + c_N \dot{\delta}_N = k_N \delta_N + c_N \dot{\delta}_N \tag{3}$$

$$f_T = -\mu f_N \frac{\dot{\delta}_T}{\left|\dot{\delta}_T\right|} \tag{4}$$

$$\delta_N = \sqrt{(x_{Aj} - x_{Ai})^2 + (y_{Aj} - y_{Ai})^2} \tag{5}$$

$$\dot{\delta}_N = u_{rx} \cos\phi_{ij} + u_{ry} \sin\phi_{ij} \tag{6}$$

$$\dot{\delta}_T = -u_{rx} \sin\phi_{ij} + u_{ry} \cos\phi_{ij} \tag{7}$$

$$u_{rx} = u_{xi} - u_{xj} \tag{8}$$

$$u_{ry} = u_{yi} - u_{yj} \tag{9}$$

where δ is the displacement, k_N is the spring constant, c_N is the damping coefficient and μ is the coefficient of friction. Subscripts N and T refer to the normal and the tangential directions of the contact force, respectively, u_{xi} and u_{xj} are the horizontal components of the velocities at the center of particles i and j, respectively and u_{yi} and u_{yj} are the vertical components of the velocities at the center of particles i and j, respectively. In the present work, the adhesive force between two particles is neglected.

3.4 Spring constant and damping coefficient

By assuming that the contact problem for cylindrical bodies with elliptic cross sections is the same as that for cylinders in the vicinity of contact point, the spring constant k_N in Eq. (3) is determined by means of Herzian theory of elastic contact for cylinders. In the case of the contact between two cylinders of uniform size and density, the width b of the contact area and the displacement δ due to the approach of the central axis of one cylinder to the other are given by

$$b = \sqrt{\frac{8P}{\pi L} \frac{r_1 r_2}{r_1 + r_2} \frac{1-\nu^2}{E}} \tag{10}$$

$$\delta = \frac{2P}{\pi L} \frac{1-\nu^2}{E} \left(1.8864 + \ln \frac{L}{2b}\right) \tag{11}$$

where E and ν are the modulus of elasticity and the Poisson's ratio of the cylinder, respectively and P and L are the compressive applied force and the length of the cylinder, respectively. In the case of a contact between the cylindrical bodies and the vibrating container of the same material, the width b_1 of the contact area and the displacement δ_1 due to the approach of the central axis of the cylinder to the wall of container are given by

$$b_1 = \sqrt{\frac{8 r_1 P_1}{\pi L} \frac{1-\nu^2}{E}} \tag{12}$$

$$\delta_1 = \frac{2P_1}{\pi L} \frac{1-\nu^2}{E} \left(1.8864 + \ln \frac{L}{2b_1}\right) \tag{13}$$

where P_1 is the compressive applied force. By assuming that the statical relations

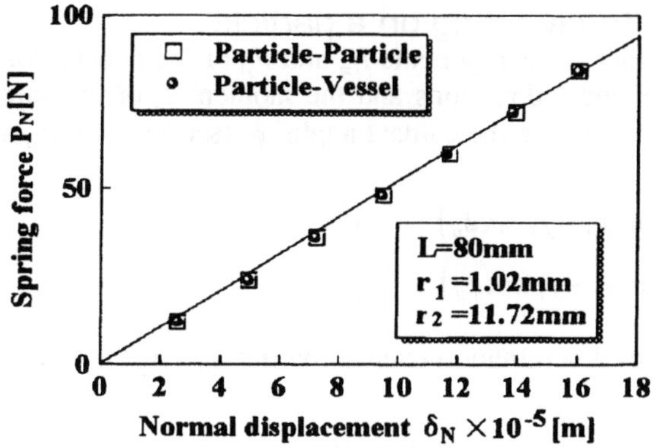

Fig. 5 Relation between spring force and normal displacement

given in Eqs. (10)~(13) hold for the interparticle collision and the collision with the vibrating container and by setting P_N equal to P or P_I and δ_N equal to δ or δ_I, the relation between the normal displacement δ_N and P_N can be obtained as shown in Fig. 5. An acrylic resin particle is used in the calculation. In this figure, the solid lines are obtained by an approximate fit to the relation between δ_N and P_N. Therefore, the spring constant k_N in Eq. (3) is determined from the slope. The spring constant of the particle used in this work was seen to be almost constant.

Next, consider the damping force $c_N \dot{\delta}_N$ that is referred to as collision energy dissipation. If the collision duration is assumed to be equal to the half period of the damped harmonic oscillation and by using the coefficient of restitution e, the damping coefficient c_N is expressed as [5]

$$c_N = \frac{2\sqrt{m \cdot k_N}}{\sqrt{1+(-\pi/\ln e)^2}} \tag{14}$$

Table 1 Calculation conditions

Number of particles	
Large particles	**1**
Small particles	**103**
Spring constant	
Spring constant	**5.2×10^6N/m**
Coefficient of friction	
Particle-Particle	**0.46**
Particle-container	**0.52**

3.5 External force acting on a particle

Next, consider the components F_{xi} and F_{yi} of the contact force acting on the particle i along x and y directions and the moment T_i of the force about an axis normal to the paper. Using the contact angle ϕ_{ij} (see Fig. 4.), F_{xi} and F_{yi} are given by

$$\left.\begin{aligned} F_{xi} &= \sum_j \left(-f_N \cos\phi_{ij} - f_T \sin\phi_{ij}\right) \\ F_{yi} &= \sum_j \left(-f_N \sin\phi_{ij} + f_T \cos\phi_{ij}\right) \end{aligned}\right\} \tag{15}$$

where Σ indicates a summation over all particles j in contact with particle i. T_i is expressed as

$$T_i = \sum_j \left\{X_A{}^i\left(f_T \cos\phi_p - f_N \sin\phi_p\right) + Y_A{}^i\left(f_N \cos\phi_p + f_T \sin\phi_p\right)\right\} \tag{16}$$

where $X_A{}^i$ and $Y_A{}^i$ are the coordinates of point A_i in the local coordinates $X^i - Y^i$ and ϕ_p is the angle between the line S_iS_j and the major axis of particle i (see Fig. 3).

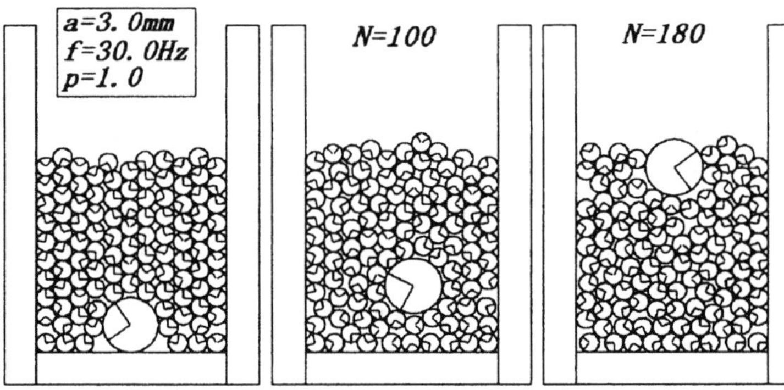

Fig. 6 Typical configuration of a large cylinder within an assembly of small cylinders during vibration

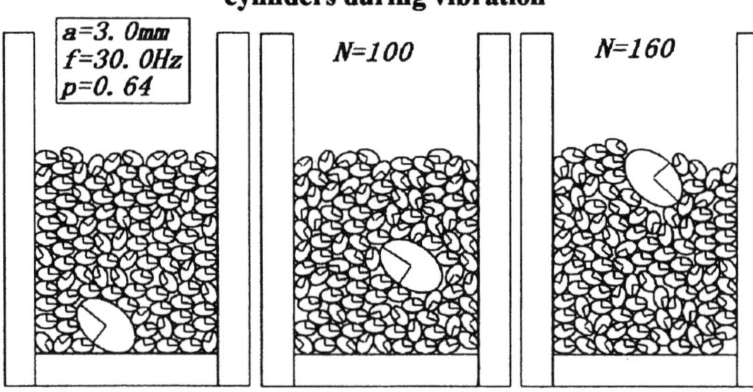

Fig. 7 Typical configuration of a large elliptic cylinder within an assembly of small elliptic cylinders during vibration

4 Analytical results

The computational conditions used for the simulation are shown in Table 1. The coefficient of static friction was measured using a cylinder and plate of acrylic resin. Figure 6 shows the typical configurations during the vibration of cylindrical granular materials. We first place one large particle ($p_{104}=1.0, r_{104}=6.0mm$) at the center at the bottom of the container and next pour an assembly of small particles ($p_1 \sim p_{103}=1.0$, $r_1 \sim r_{103}=2mm$) on it by a random packing. In this figure, N denotes the number of vibrations of the container. Since individual particles are in free flight during the vibration, it is difficult to observe how the large particle moves up within the assembly of small particles. Therefore, for the case of $N \neq 0$, the container is kept stationary after N number of vibration cycles. It is clear that, as the number of vibration cycles N increases, the large particle rises up to the top of the bed. Figure 7 shows how a large elliptic particle ($p_{104}=0.64, r_{104}=6.0mm$) rises up within an assembly of small elliptic particles($p_1 \sim p_{103}=0.64$, $r_1 \sim r_{103}=2mm$) in the vibrating environment. In the case of Fig. 7, the large particle moves horizontally. Figure 8 shows a plot of the height of a large particle versus the number of vibration cycles.
This figure also shows the effect of the size ratio Φ (=major axis of the large particle / major axis of the small particle). The greater the height of the large

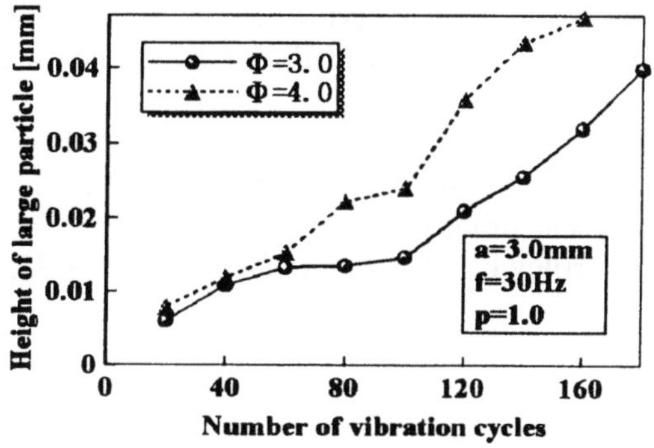

Fig. 8 Effect of size ratio Φ

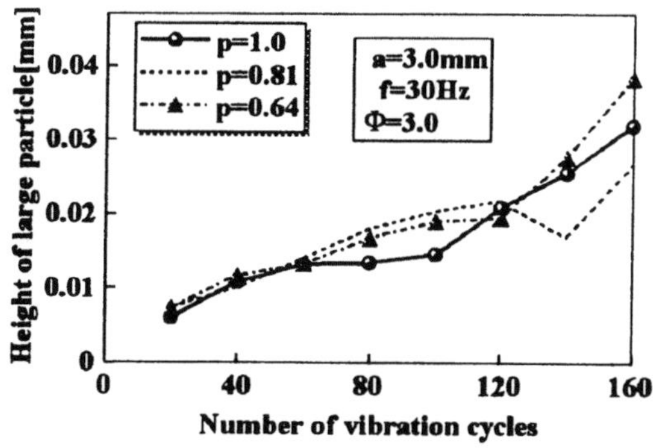

Fig. 9 Effect of flat ratio p

particle, the greater the tendency of the system to segregate. It is clear that the height of the large particle increases as the number of vibration cycles increases or the size ratio increases. Figure 9 shows the effect of flat ratio p on size segregation in the same plane as in Fig. 8. As shown in this figure, the flat ratio of the particle (of equal mass) does not produce any significant effect on segregation. These trends are in qualitative agreement with the experimental results presented by Ahmad and Smalley [(6)].

5 Conclusions

In this study, we have investigated the size segregation in a vibrating container filled with monodispersed small particles and one large particle by means of a modified discrete element method, in which granular materials are treated as elliptic models by considering shape anisotropy. Segregation increases with the increase in size of the large particle or number of vibration cycles. In addition, the flat ratio of a particle (of equal mass) may not have any significant effect on segregation.

References

1 Rosato A., Prinz F., Standburg K. J. and Swenson R., Powder Technology, **49**, 59-69, 1986.

2 Jullien R. and Meakin P., Physical Review Letters, **69**, 640-642, 1992.

3 Duran J., Rajchenbach J. and Clement E., Physical Review Letters, **70**, 2431-2434, 1993.

4 Cundall P. A. and Strack O. D., Geotechnique, **29**, No.1, 47-65, 1979.

5 Kawaguchi T., Tanaka T., and Tsuji Y., Transactions of the Japan Society of Mechanical Engineering, 58-551B, 2119-2125, 1992.

6 Ahmad B. H. and Smalley I. J., Powder Technology, **8**, 69-75, 1973.

To the general theory of hand-held percussion machines

*V. I. Babitsky**

The generic approach for the analysis and synthesis of a hand-held percussion machine as a discrete nonlinear converter is presented. The optimisation of the interaction of the machine with the operator and load is formulated and solved as a problem of optimal control. For the simplification of design, the sequence of estimations is developed for the series of possible quasi-optimal excitations.

1 Introduction

Percussion machines are complex dynamic systems demonstrating strongly nonlinear behaviour. Used in machines, vibro-impact dynamic processes have a severe influence on the operator and the environment. The main feature of vibro-impact systems is a strong dependence of the process on operational parameters. This complicates the control of these processes and applies strict limitations to the parameters of hand-held machines.

The complexity of hand-held percussion machine dynamics limited their analysis to the use of some special models [1-4]. Such models did not permit the discovery of common dynamical characteristics of the machines as a whole. The variety of mechanisms of vibration excitation in hand-held percussion machines, in conditions of absence of the general theory, makes difficult their comparative analysis, choice of optimal design concepts and estimation of their improvement potential.

The range of constructions for vibro-impact mechanisms and systems requires development of methods of synthesis of dynamical structures which best fulfil the optimal form of motion. Also, the problem of decreasing the vibration-induced load for all types of hand-held percussion machines demands the development of a general optimal approach for estimation and optimisation of their excitation [5,6]. The ideal is to find the connection between the main output and resulting unfavourable vibrations with optimal forms of the excitation.

An approach to the general theory of hand-held percussion machines formulated in the paper as a problem of optimisation of a discrete nonlinear converter which modulates the operator effort by sequences of impact impulses. The optimal generation of impulses is specified as a problem of periodic optimal control. This formulation is transformed to a mathematical problem of moments in the proper normed function linear space.

---

*Department of Mechanical Engineering, Loughborough University,Leicestershire LE11 3TU, UK. Fax: +44 (0)1509 223934, E-mail: V.I.Babitsky@lboro.ac.uk

An exact solution of the problem gives the opportunity to receive absolute estimations of the improvement potential for all modifications of similar machines. For the simplification of design, the sequence of exact estimations is developed for the series of possible quasi-optimal excitations.

2 Hand-held percussion machine as pulse modulator

The general structure of every percussion machine consists of a source of energy and a vibro-impact converter which transforms the energy of source into a vibro-impact process. In order to transfer the energy of the source to the object or medium being treated, an operator has to apply an effort P to the machine body by pressing it to the object (Fig. 1). This effort, which can be called *feed*, reveals the main channel of permanent processing control by the operator.

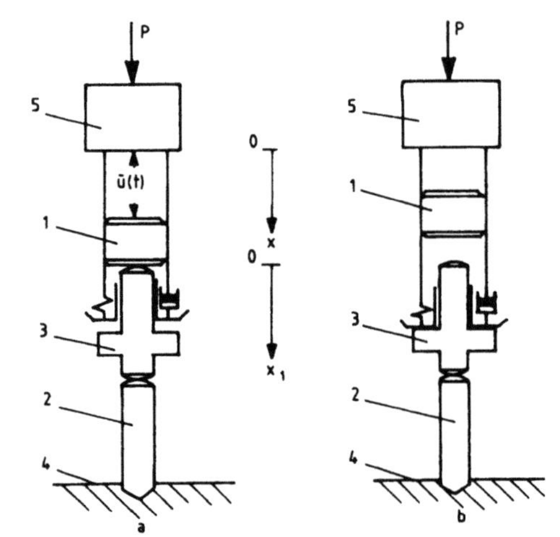

Fig. 1

The feed P, as an external force, depends on the interaction of the machine as a whole with the object of treatment through the impact unit (Fig. 1). This includes an interaction of the striker 1 with a tool 2 or intermediate transformer 3 for processing of the object or media 4 (Fig. 1,a), a setting of the machine 5 and the tool and transformer against the object 4 in a position of impact (Fig. 1,b) and additional interactions with impact unit due to rebound.

Depending on the vibro-impact dynamics, the striker with mass M_1 can accomplish mainly either single impact periodic motions [3] or periodic motions with repeated attenuated impacts (chatter) [7].

The regimes with single-impact interaction have the most intensive (resonant) nature [3]. They are used preferably for processing and demolition. During such processing, the momentum of impact interaction per period T equals $M_1(1+R)\,\dot{x}_{1-}$, where $\dot{x}_{1-}$ is the velocity of striker before the impact and R is the coefficient of striker reflection after the impact. The coefficient R indicates the change of absolute velocity of the striker. It has to be measured preferably for each specific impact pair in conditions of collision close to reality.

The impact momentum has to be equalised by the impulse of feed P_{ef} from the operator according to the principle of impulse and momentum: $P_{ef}T=M_1(1+R)\;\dot{x}_{1-}$ which defines the value of minimum feed for the accomplishment of vibro-impact periodic motion with the prescribed energy (velocity $\dot{x}_{1-}$) and frequency f of impacts:

$$P_{ef} = M_1(1+R)\ \dot{x}_{1-} f. \qquad (1)$$

It follows from (1) that an increase of the processing throughput, either by the rise of impact energy or frequency leads to the intensification of P_{ef}.

Any additional interactions between the machine and the impact unit lead to P being exceeded (1). If, for example, the interaction represents an unlimited series of attenuated impacts during the period T, then for the same velocity $\dot{x}_{1-}$ of the first impact, it follows:

$$P = M_1(1+R)\ \dot{x}_{1-} f(1+R+R^2+...) = \frac{M_1(1+R)\dot{x}_{1-} f}{1-R} \geq P_{ef}\ . \qquad (2)$$

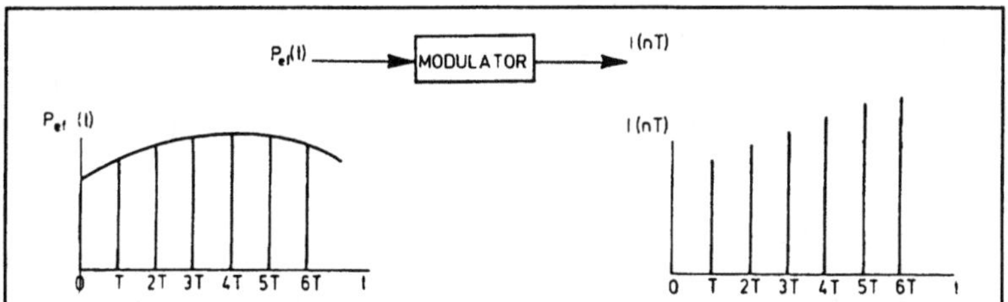

Fig. 2

Further increase of feed occurs if this series of impacts dies out after time $\tau < T$ or through the interaction of the machine body with the tool or intermediate striker during the setting before impact or due to rebound of the tool after impact against the object of treatment. If real feed P exceeds P_{ef}, it imposes an additional stress upon the operator. The validity of a hand held machine, from this point of view, can be estimated with *the coefficient of feed effectiveness:*

$$\lambda = \frac{P_{ef}}{P} \qquad (0 < \lambda \leq 1). \qquad (3)$$

The ideal hand-held percussion machine, which capable of producing a periodic vibro-impact process under arbitrary feed with a single impact during the period, can be treated as a discrete pulse modulator (Fig. 2) which converts the permanent feed P_{ef} into a series of impulses with period T:

$$I(nT) \cong \left(\int_{(n-1)T}^{nT} P_{ef}(t)dt \right) \sum_n \delta(t - nT) \qquad (4)$$

where $\delta(t)$ is Dirac's function, n is the current number of impacts. Small movements of the centre of machine mass have been neglected here. The values of the impulses equal the integral in a discrete form of the feed of operator.

The output of the machine is defined usually as the energy of the striker before the impact $E_- = 0.5M_1\ \dot{x}_{1-}^2$. With the help of expression (1) this yields the characteristic for modulator conversion:

$$E_- f^2 = \frac{1}{2(1+R)^2 M_1} P_{ef}^2 . \tag{5}$$

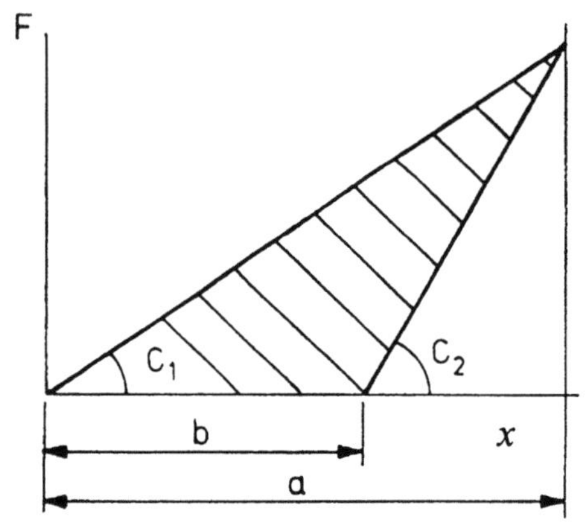

Fig. 3

It is seen that the intensification of chiselling by an increase of impact energy or frequency leads to the increase of P_{ef} in the different degree.

Let the machine, for example, process concrete with displacement (x) - force (F) diagram of demolition during one impact shown in Fig. 3. The shaded area represents the work of demolition. Supposing that the whole energy of impact $E_i = 0.5(1-R^2)M_1\dot{x}_{1-}^2$ loads the concrete $E_i = 0.5c_1a^2$, it is possible to calculate the efficiency of demolition with the formula $\eta = 1 - c_1/c_2$ and the capacity of chiselling:

$$v = (a-b)f = 2f\left(1-\frac{c_1}{c_2}\right)\sqrt{\frac{E_-(1-R^2)}{2c_1}} = \left(1-\frac{c_1}{c_2}\right)\sqrt{\frac{1-R}{(1+R)c_1M_1}}P_{ef} . \tag{6}$$

The machine will be able to accomplish this capacity if under any feed P_{ef} its exciting mechanism develops a steady state vibro-impact process with single impact during the period. It can be achieved either by the development of robust dynamical exciting systems or with the use of control facilities. In the last case it is preferably to realise an autoresonant vibro-impact structure [8].

3 Excitation of a hand-held percussion machine as a problem of optimal control

Let us investigate the optimal conditions of machine excitation with single impact during the period. It is supposed that under the influence of excitation the periodic force acting on the striker is described by an unknown periodic function $\tilde{u}(t)$ (Fig. 1). This can be due to the variation of pressure in the gas chamber of the hammer drill, hydraulic pressure in the cylinder of the breaker, compression of some spring elements of the striker suspension or electromagnetic interaction of the striker as an element of a linear motor. In all cases, this force has a simultaneous opposite action on the body of the machine and through it on the operator. Below, the problem of optimisation of excitation of $\tilde{u}(t)$ will be formulated.

The differential equation of motion of the striker can be written as follows

$$\ddot{x}_1 = u(t) \tag{7}$$

where x_1 is the coordinate of the striker measured from the point of its impact, when $x_1=0$, $u(t)= \tilde{u}(t)/M_1$. The positive direction of x_1 is shown in Fig. 1. In order to realise the periodic motion of the striker with the prescribed energy E_- or velocity of impact $\dot{x}_{1-}$ and frequency f, the solution of equation (7) has to satisfy the periodic conditions:

$$\begin{array}{lll} t=0 & x_1=0, & \dot{x}_1=\dot{x}_{1+} \\ t=T & x_1=0, & \dot{x}_1=\dot{x}_{1-} \end{array} \tag{8}$$

where $T=2\pi/\omega=1/f$ is the period of motion, $\dot{x}_{1+}=-R\dot{x}_{1-}$.

Rewriting the equation (7) in Cauchy form [9], gives the following:

$$\begin{aligned} \dot{x}_1(t) &= \dot{x}_{1+} + \int_0^t u(\tau)d\tau \\ x_1(t) &= x_1(0) + t\dot{x}_{1+} + \int_0^t (t-\tau)u(\tau)d\tau \end{aligned} . \tag{9}$$

Applying the periodic conditions (8) to (9), the results will be as follows

$$\begin{aligned} (1+R)\dot{x}_{1-} &= \int_0^T u(\tau)d\tau \\ RT\dot{x}_{1-} &= \int_0^T (T-\tau)u(\tau)d\tau \end{aligned} . \tag{10}$$

In the left hand parts of equations (10) there are known numbers. The right hand parts of (10) represent the scalar products of the unknown function $u(t)$ with prescribed linear independent basic functions $h_i(t)$ $(i=1,2)$, where $h_1(t)=1$, $h_2(t)=T-t$ and all functions are defined in segment $t\in[0,T]$. The problem of finding the unknown function $u(t)$ under such conditions is known in mathematics as *the moment problem* [10]. For reduction of influence on the operator, the excitation $u(t)$ has to be limited in some technical sense. We will use $|u(t)|$ which represent the important characteristics influencing vibration of the machine body. As a result, *the optimal solution* of the moment problem for periodic excitation with the prescribed period T has to be minimised as

$$\min_u \max_{t\in[0,T]} |\, u(t)\,| . \tag{11}$$

4 Optimal excitation

The basic mathematical foundations leading to the solution of formulated optimal problem are given in [3, 11]. Applying them to our case we receive that the optimal excitation $u^0(t)$ satisfying (10) and minimising (11) can be found in the form:

$$u^0(t) = \frac{\mathrm{sgn}[n_1^0(T-t)+n_2^0]}{\int_0^T |n_1^0(T-t)+n_2^0|dt} = U_0\,\mathrm{sgn}[n_1^0(T-t)+n_2^0] \tag{12}$$

where numbers n_1^0 and n_2^0 are the solution of the problem: to find

$$\min_{n_1, n_2} \int_0^T |n_1(T-t) + n_2| dt = \frac{1}{U_0} \qquad \text{when} \qquad n_1 RT\dot{x}_{1-} + n_2(1+R)\dot{x}_{1-} = 1 .$$

In this case it is not necessary to follow the entire regular procedure, because the solution can be found by a simpler approach.

According to (12), the optimal excitation $u^0(t)$ is a piecewise constant function with no more than one switch during the interval of $t \in (0,T)$. It is because of the linearity of the function under the signum sign. In our case it can only be a single possible version:

$$u^0(t) = \begin{cases} -U_0, & \text{if} \quad t \in [0, t_1) \\ +U_0, & \text{if} \quad t \in [t_1, T) \end{cases} \tag{13}$$

where t_1 is an unknown value. Another combination of signs contradicts the nature of the mechanical problem.

Substituting (13) into (10) yields, after integrating the system of two equations with two unknown values U_0 and t_1:

$$\begin{aligned} (1+R)\dot{x}_{1-} &= (T - 2t_1)U_0 \\ RT\dot{x}_{1-} &= [(T-2t_1)T + t_1^2 - T^2/2]U_0 \end{aligned} . \tag{14}$$

After solution of these equations, it follows finally

$$\begin{aligned} \frac{t_1}{T} &= \frac{1}{1+R}\left(1 - \sqrt{1 - \frac{1-R^2}{2}}\right), \\ U_0 &= \frac{(1+R)^2 \dot{x}_{1-}}{2T\left(\sqrt{1-0.5(1-R^2)} - 0.5(1-R)\right)}. \end{aligned} \tag{15}$$

It is important to mention that t_1 does not depend on the "power" characteristics of the machines. The amplitude of excitation U_0 is proportional to the feed P_{ef}, taking into account (1). This characterises the dependence of the excitation energy on the intensity of the chiselling process. After the calculation of optimal excitation, it is possible to estimate the proper variable component of body acceleration $\tilde{a}(t)$ perceived by the operator:

$$\tilde{a}(t) = \frac{M_1}{M} u^0(t), \tag{16}$$

where M is the mass of the body.

5 Optimal motion of the striker

The optimal laws of striker motion can be received by the integration of equations (9) with respect to (13), (15). From the first of the equations (9) it follows that

$$\dot{x}_1(t) = \dot{x}_{1+} + \int_0^t u(\tau) d\tau = -R\dot{x}_{1-} - 2U_0 t_1 + U_0 t . \tag{17}$$

A maximum of striker deflection corresponding to when the velocity equals zero: $\dot{x}_1(t_2)=0$, where t_2 is the moment of maximum deflection. Then it follows from (17): $t_2 = 2t_1 + R\dot{x}_{1-}/U_0$ or by taking into account equation (15)

$$\frac{t_2}{T} = \frac{R+2t_1/T}{1+R} = \frac{2\left[1-\sqrt{1-0.5\left(1-R^2\right)}\right]}{\left(1+R\right)^2} + \frac{R}{1+R}. \tag{18}$$

The integration of the second equation (9) gives

$$x_1(t) = \dot{x}_{1+}t + \int_0^t (t-\tau)\,u(\tau)d\tau = -R\dot{x}_{1-}t + U_0t_1^2 - 2U_0t_1t + U_0t^2/2 \ .$$

Then, after transformations

$$x_{1\max} = x_1(t_2) = \frac{U_0T^2}{(1+R)^2}\left[\left(\frac{t_1}{T}\right)^2(1+R)^2 - \frac{1}{2}\left(R+2\frac{t_1}{T}\right)^2\right],$$

where t_1/T and U_0 are defined by (15).

This value characterises the length of the working stroke which influences the size of the machine. The optimal laws of striker motion are shown in Fig. 4. They are calculated for the typical parameters of a heavy hammer drill: $\dot{x}_{1-}$ = 8.8 m/s (E_-~7 Nm), M_1 = 0,2 kg, M = 8.0 kg, f = 45 Hz, R = 0.25. These parameters will be used in all the following examples.

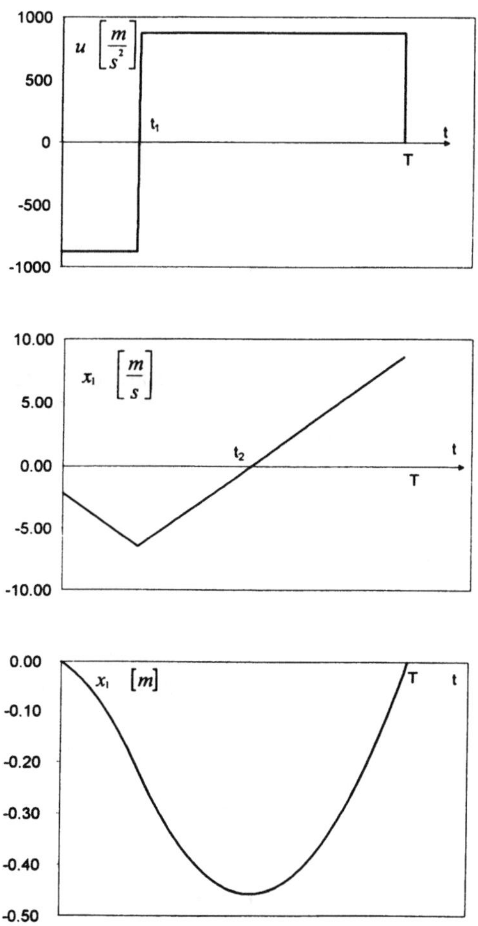

Fig.4

6 Quasi-optimal excitations by condensed impulses

In real mechanisms it is difficult to accomplish an optimal excitation. For example, in typical hammer drills with an air chamber and crank-slide mechanism, it is possible to achieve only limited expansion of the gas during the initial stage of excitation just after an impact. In these conditions an attempt to realise the optimal excitations (13) looks impractical. In order to bring the improvement potential to realisation, it is useful to find some quasi-optimal practical excitations and their estimations.

The reduction of initial negative impulse of excitation within the framework of limited amplitude can be achieved with the help of the following quasi-optimal excitations:

$$u(t)=\begin{cases} -U, & \text{if} \quad t\in[0,t) \\ +nU, & \text{if} \quad t\in[t_1,T) \end{cases}, \tag{19}$$

where U, t_1 are unknown values, n - an approximating coefficient.

Substituting (19) into (10), gives after integrating the system of two equations for unknown values U and t_1:

$$\begin{aligned} (1+R)\dot{x}_{1-} &= -Ut_1(1+n)+nUT \\ RT\dot{x}_{1-} &= -UTt_1(1+n)+t_1^2\frac{U(1+n)}{2}+\frac{nU}{2}T^2 \end{aligned} \tag{20}$$

The solution of these equations gives finally:

$$\begin{aligned} \frac{t_1}{T} &= \frac{1}{1+R}\left[1-\sqrt{1-\frac{n(1-R^2)}{1+n}}\right] \\ U &= \frac{(1+R)\dot{x}_{1-}f}{n-\frac{1+n}{1+R}\left[1-\sqrt{1-\frac{n(1-R^2)}{1+n}}\right]} \end{aligned} . \tag{21}$$

The quasi-optimal laws of excitation (19), (21) for different n are presented in Fig. 5. They demonstrate the tendency to reduce the initial negative acceleration and to condense positive acceleration impulse. It is seen that if n=100 the exciting force (acceleration) acts practically in the positive direction only. Further estimations of possibility of these condensed positive impulses of excitation can be calculated with the help of the following approximations of the equation

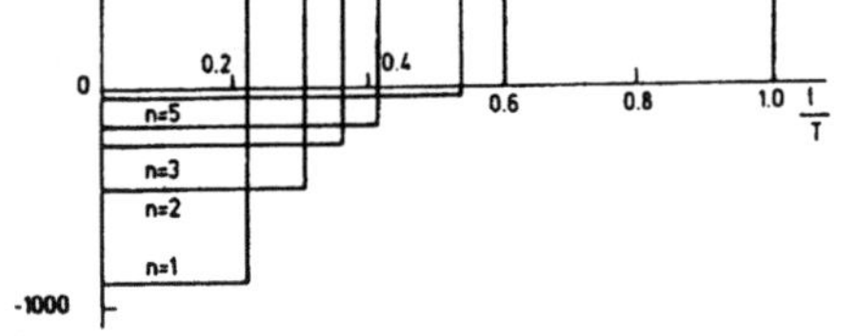

Fig. 5

$$u(t)=\begin{cases} 0, \text{if } t\in[0,t_1) \\ U, \text{if } t\in[t_1,t_1+\Delta) \\ 0, \text{if } t\in[t_1+\Delta,T) \end{cases} \tag{22}$$

where U, t_l are unknown values, Δ is the permissible duration of positive acceleration, $(t_1 + \Delta)/T \le 1$.

Substituting (22) into (10) then integrating brings us once again to a system of two equations for unknown values U and t_l. The solution of these equations gives finally:

$$\frac{t_1}{T} = 1 - \frac{1}{2}\left(\frac{\Delta}{T}\right) - \frac{R}{1+R} \, . \qquad U = \frac{(1+R)\dot{x}_{1-}f}{\Delta / T} \tag{23}$$

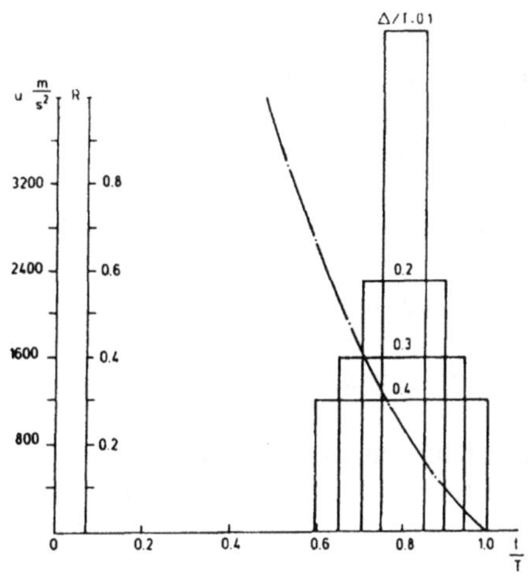

Fig. 6

Figure 6 demonstrates the different possible excitations of the striker with the aid of positive impulses of different duration Δ/T. This continues the sequence of quasi-optimal excitations of Fig. 6. The dash dotted line shows the position of the middle of the impulse interval dependent upon *R*. The shortening of interval Δ leads naturally to the essential increase of the exciting force amplitude. The worst case when Δ/T=0.1 corresponds approximately to the excitation of modern hammer drills.

7 Conclusions

The reorganisation of hand-held percussion machine excitation in the direction of optimal excitation is an effective and conformable way of excitation improvement and reduction of unfavourable vibration. It also relieves the load on the drive. This can be achieved mainly through the extension of excitation impulse and stabilisation of resonant vibro-impact excitation.

The developed technique gives the possibility of estimating an improvement potential at every stage of reorganisation. The optimal cycle of excitation permits also the comparison of percussion machines with different designs by calculation of their improvement potential from the point of their capacity and unfavourable vibration activity.

References

1. Fu CC, Paul B, Stability of motion of impact tools, Int. Journal of Solids and Structures, **4**:897-905, 1968.
2. Sikarskie DL, Paul B, Periodic motions of a two-body system subjected to

repetitive impact. Journal of Engineering for Industry, Trans. ASME, **91**: 931-938, 1969.
3. Babitsky VI, Theory of Vibro-Impact Systems and Applications. Springer-Verlag, Berlin, 1998. Revised translation from Russian, Nauka, Moscow, 1978.
4. Astashev VK, Babitsky VI, Dynamics of resonant machines. In handbook "Dynamics and Control of Machines", Ed. G.V.Kreinin, Machinostroenie, Moscow, 1988 (in Russian).
5. Byhovsky I I, Goldstein BG, Goppen AA Vibration protection of hand-held machine operators. In handbook "Vibration in Engineering", **6**: Vibration and Impact Protection, Ed. K.V.Frolov, Machinostroenie, Moscow, 1995 (in Russian).
6. Kyriakidis K, Survey of exposure to hand-arm vibration in Great Britain. Health and Safety Executive, 1988.
7. Nagaev RF Mechanical Processes with Repeated Attenuated Impacts. Nauka, Moscow, 1985 (in Russian).
8. Babitsky VI Autoresonant mechatronic systems. Mechatronics, **5**: 483-495, 1995.
9. Lefschetz S Differential Equations: Geometric theory, 2nd ed., Wiley, Interscience, New York, 1963.
10. Krasovsky NN, Theory of Motion Control. Nauka, Moscow, 1968 (in Russian).
11. Babitsky VI, Hand-held percussion machine as a discrete nonlinear converter. Journal of Sound and Vibration, 214(1):165-182, 1998.

Free manifolds of dynamical billiards

V V Beletsky, E I Kugushev, E L Starostin[1]

A mechanical system with impacts is called a mathematical billiard if it may be reduced to a mass point moving in a plane and reflecting from a bounding curve [1]. A billiard is called dynamical if an external force field is imposed [2–6]. The objective of this work is to suggest a new method of representation of dynamical billiards as generalized mathematical billiards on curved surfaces, or free manifolds embedded in $\mathbb{R}^3$. *The trajectories of the mass point on a free manifold consist of segments of geodesics. The mass point moves on a free manifold with no external forces acting on it. The free manifolds of dynamical billiards in the constant and Newtonian force fields are shown to be surfaces of revolution in the three-dimensional Euclidean space. The parabolic case is of particular interest, because it is equivalent to the classical planar mathematical billiard.*

1 Free manifolds of dynamical billiards

Let $\mathbb{R}^2$ be the plane with Cartesian coordinates (x, y). We define a dynamical billiard B on this plane as a triple $\{G, U, h\}$, where G is a closed connected domain in $\mathbb{R}^2$ (an action domain of the billiard) and $U = U(x, y)$ a smooth function in this domain, which is a force function of the forces, acting on a billiard point, which moves inside G. The constant h is an energy level for which the motion of the billiard point proceeds. If $U \equiv 0$, then we have a classical mathematical billiard.

We assume that the billiard point has a unit mass. Its kinetic energy T in the Cartesian coordinates (x, y) is a homogeneous quadratic form $T = \frac{1}{2}(\dot{x}^2 + \dot{y}^2)$. Therefore, the system may be treated with the help of the least action principle in the Jacobi form. Now we formulate it.

Let w_0 be a collision-free interval of the point's trajectory, with its beginning in some point p_1 and its end in point p_2. For every point of trajectory, the energy integral is valid: $T = U + h$. Let $W = \{w\}$ be the family of smooth curves w with the beginning in the point p_1 and the end in the point p_2. We define the functional Φ_0 on the family W:

$$\Phi_0 = \int_{p_1}^{p_2} \sqrt{(U + h)}ds,$$

where the curve's arc differential ds is determined by the quadratic form of the kinetic energy T: $ds^2 = dx^2 + dy^2$. The least action principle in the Jacobi form suggests that the curve w_0 is an extremal of the functional Φ_0 on the set W and vice versa, if the curve $w_1 \in W$ is an extremal of the functional Φ_0

[1]Keldysh Institute of Applied Mathematics, Miusskaya sq. 4, 125047 Moscow, Russia.

on the set W, then it may be parameterized in such a way that it becomes a trajectory of the system under consideration.

Instead of the functional Φ_0, we further use the functional $\Phi = \sqrt{\lambda}\Phi_0$, where $\lambda > 0$ is a positive constant normalizing factor, the value of which will be adjusted below. The extremals of the functionals Φ and Φ_0 are the same.

Let M^2 be a two-dimensional surface (manifold), parameterized with (u, v) and such that its linear element takes the form

$$ds^2 = \lambda(U(u,v) + h)(du^2 + dv^2).$$

The geodesics on this surface are extremals of the functional Φ (expressed in coordinates (u, v)), and vice versa. We map the plane of the dynamical billiard onto M^2 by the simple rule

$$u = x, \quad v = y. \tag{1.1}$$

Therefore, the billiard domain G maps onto some domain $\tilde{G}$ in M^2, and the collision-free intervals of the billiard trajectories map in a one-to-one manner onto the geodesics in M^2, not intersecting the boundary of the region $\tilde{G}$. In addition, since the coordinates (u, v) on M^2 are isothermal (i.e., the metric tensor is spherical), then the above mapping preserves the angles between curves in the billiard's plane and, in particular, the "reflection law", i.e., the correspondence between the angles at which the trajectory approaches and leaves the boundary of the domain.

Consider now a mathematical billiard on M^2 in the domain $\tilde{G}$. The point moves freely in this system, without action of external forces. In this case the trajectories of motion coincide with the corresponding geodesics on M^2. This means that the initial dynamical billiard and the free billiard on M^2 coincide geometrically, i.e., the mapping (1.1) maps the trajectories of the first billiard onto the trajectories of the second one and vice versa. However, notice that the temporary law of motion along the trajectories may not retain.

The manifold M^2 is called the *free manifold* of the original dynamical billiard. In what follows, we consider the billiards with two particular potentials: constant gravitational and Newtonian. In each case, we shall show that a planar billiard with the imposed force field is trajectorically equivalent to the force-free billiard on the corresponding surface of revolution. The procedure of construction of that surface will be developed.

2 Free manifold of a billiard in the constant force field

As a first example, we examine the free manifold for a planar dynamical billiard in the constant force field $U = -gy$. According to the Jacobi principle, the linear element of the free manifold M^2 then takes the form

$$ds^2 = \lambda(h - gv)(du^2 + dv^2), \tag{2.1}$$

where h is the energy constant. We shall show that M^2 may be embedded in $\boldsymbol{R}^3$ isometrically as a many-sheeted surface of revolution. The image of the domain G may be located entirely within one sheet unless it is too large.

Let us introduce new coordinates (f, l) in M^2 such that

$$f = \sqrt{\lambda}u, \qquad l = \sqrt{\lambda}\int_{v}^{h/g}\sqrt{(h - g\tau)}d\tau.$$

It is an easy matter to obtain

$$l = \sqrt{\lambda}\frac{2}{3g}(h - gv)^{3/2} \tag{2.2}$$

and

$$h - gv(l) = \left(\frac{3g}{2\sqrt{\lambda}}l\right)^{2/3}.$$

In the new coordinates, the linear element is

$$ds^2 = \left(\frac{3g}{2\sqrt{\lambda}}l\right)^{2/3} df^2 + dl^2. \tag{2.3}$$

Now introduce three variables (φ, η, ζ), related to the variables (f, l) as follows:

$$\varphi = f, \qquad \eta = \sqrt{h - gv(l)}, \qquad \left(\frac{d\eta}{dl}\right)^2 + \left(\frac{d\zeta}{dl}\right)^2 = 1. \tag{2.4}$$

By using Eq. (2.2), we obtain from the two last formulas in (2.4)

$$\frac{d\eta}{dl} = \frac{g}{2\eta^2\sqrt{\lambda}}, \qquad \frac{d\zeta}{dl} = \sqrt{1 - \left(\frac{d\eta}{dl}\right)^2} = \frac{1}{\eta^2}\sqrt{\eta^4 - \frac{g^2}{4\lambda}}. \tag{2.5}$$

Eqs. (2.5) parametrically define a curve $\Gamma(l) = (\eta(l), \zeta(l))$ in the plane (η, ζ). Further, we obtain

$$\eta = 3^{1/3}2^{-1/3}g^{1/3}l^{1/3}\lambda^{-1/6}, \qquad \frac{d\zeta}{dl} = \sqrt{1 - 3^{-4/3}2^{-2/3}l^{-4/3}\lambda^{-1/3}g^{2/3}}.$$

By virtue of the last condition (2.4), the parameter l is an intrinsical parameter along this curve. We have from Eq. (2.3) $ds^2 = \eta^2 d\varphi^2 + dl^2$, or

$$ds^2 = \eta^2 d\varphi^2 + \left(\left(\frac{d\eta}{dl}\right)^2 + \left(\frac{d\zeta}{dl}\right)^2\right) dl^2. \tag{2.6}$$

Let (x_1, x_2, x_3) be a three-dimensional Cartesian space. We introduce in it cylindrical coordinates (φ, η, ζ), such that (η, φ) are the radius and the polar angle in the plane (x_1, x_2), and ζ is a displacement along the axis x_3. Then the linear element of the manifold M^2, written in the form (2.6), coincides with the linear element of the surface of revolution M_ρ^2, formed by rotation of the curve $\Gamma(l)$ around the axis x_3 in the Cartesian space (x_1, x_2, x_3). Therefore, the map $(\varphi(u,v), \eta(u,v), \zeta(u,v)) : M^2 \to M^2_\rho$ is an isometric embedding of M^2 in $\boldsymbol{R}^3$.

From Eqs. (2.5), one can find $\dfrac{d\zeta}{d\eta}$ to give:

$$\varphi = \sqrt{\lambda} u, \qquad \eta = \sqrt{h - gv}, \qquad \frac{d\zeta}{d\eta} = \sqrt{\frac{4\lambda}{g^2}\eta^4 - 1}. \tag{2.7}$$

Having found from Eqs. (2.7) the variables (φ, η, ζ) as functions of (u, v), we obtain the following isometrical map of the space (u, v) with metric (2.1) into the space $\boldsymbol{R}^3 : (x_1, x_2, x_3)$ with metric $ds^2 = dx_1^2 + dx_2^2 + dx_3^2$:

$$x_1 = \eta \cos\varphi, \qquad x_2 = \eta \sin\varphi, \qquad x_3 = \zeta. \tag{2.8}$$

Note that the right-hand side of the last equation (2.7) has the domain of definition $\eta^2 \geq \dfrac{g}{2\sqrt{\lambda}}$, or

$$v \leq \frac{h}{g} - \frac{1}{2\sqrt{\lambda}}. \tag{2.9}$$

This means that the map of the isometric embedding (2.8) exists only in the domain (2.9), where the coordinate ζ rises monotonically as η increases. It is interesting to note that the rest of the manifold M^2 may be embedded isometrically in the three-dimensional pseudoeuclidean space $\boldsymbol{R}^3_1$ with pseudometric $ds^2 = dx_1^2 + dx_2^2 - dx_3^2$.

The last equation (2.7) may be integrated to give $\zeta(\eta) = \frac{1}{3}\left[\eta\sqrt{\frac{4\lambda}{g^2}\eta^4 - 1} - \frac{\sqrt{g}}{\lambda^{1/4}} F\left(\alpha, \frac{1}{\sqrt{2}}\right)\right] + C$, where $\alpha = \arccos\sqrt{\dfrac{g}{2}}\dfrac{1}{\lambda^{1/4}\eta}$, and F is the elliptic integral of the first kind. If we put $\zeta = 0$ for $\eta = \sqrt{\dfrac{g}{2}}\dfrac{1}{\lambda^{1/4}}$, then the constant $C = 0$.

Now consider a sufficiently fast billiard, for which the inequality $h - gv > \varepsilon$ is valid everywhere in the billiard domain G for some $\varepsilon > 0$. This means that the trajectory of this billiard can reach every point of the boundary of its domain with a uniformly positive velocity $(\dot{x}^2 + \dot{y}^2 > 2\varepsilon)$. Then, having taken $\lambda = \dfrac{g^2}{4\varepsilon^2}$, we have the situation when the whole billiard domain G may be isometrically embedded by the map (2.8) in the three-dimensional Euclidean space as a part of the surface of revolution.

3 Free manifold of a billiard in the Newtonian force field

Let us now consider the form of the free manifold for a planar dynamical billiard in the Newtonian force field $U = \gamma/r$, where $r = \sqrt{x^2 + y^2}$. We introduce in the plane of the billiard the polar coordinates (θ, r), such that $x = r\cos\theta$, $y = r\sin\theta$. (The angle θ is assumed to lie in the range

$(-\infty, \infty)$, i.e., the billiard plane is treated as many-sheeted.) In these coordinates, the metric takes the form:

$$ds^2 = r^2 d\theta^2 + dr^2, \tag{3.1}$$

and the kinetic energy of the unit mass point is $T = \frac{1}{2}(r^2\dot{\theta}^2 + \dot{r}^2)$. According to the Jacobi principle, the linear element of the free manifold M^2 will look like:

$$ds^2 = \lambda\left(h + \frac{\gamma}{v}\right)(v^2 du^2 + dv^2), \tag{3.2}$$

where h is the energy constant and $\lambda > 0$ the positive normalizing factor. Notice that, unlike the above sections where the variable u was introduced as a linear coordinate which then became angular, u is considered such in this section.

Since the metric (3.2) in the free manifold differs from the original one (3.1) by a factor only, the mapping $B : (\theta, r) \to M^2 : (u, v)$ preserves angles (i.e., it is conformal).

We shall show that, in this case, M^2 may be also embedded isometrically in $\mathbb{R}^3$ as a many-sheeted surface of revolution. We introduce new coordinates (f, l) in M^2, such that

$$f = \sqrt{\lambda}u, \qquad l = \sqrt{\lambda}\int_0^v \sqrt{h + \frac{\gamma}{\tau}}\, d\tau. \tag{3.3}$$

It is easy to check that for $h > 0$

$$l(v) = \sqrt{\lambda}\left(\sqrt{v(hv+\gamma)} + \frac{\gamma}{\sqrt{h}}\ln(\sqrt{hv} + \sqrt{hv+\gamma}) - \frac{\gamma}{\sqrt{h}}\ln(\sqrt{\gamma})\right),$$

for $h = 0$

$$l(v) = 2\sqrt{\lambda\gamma v}$$

and

$$l(v) = 2\sqrt{-\frac{\lambda}{h}}\left(\arctan\sqrt{\frac{-hv}{hv+\gamma}} - \arctan\frac{1}{\sqrt{\gamma}}\right)$$

for $h < 0$.

Denote

$$\eta(l) = v(l)\sqrt{h + \frac{\gamma}{v(l)}} = \sqrt{hv^2(l) + \gamma v(l)}. \tag{3.4}$$

Then the linear element takes the form $ds^2 = \eta^2 df^2 + dl^2$.

Similarly to the previous section, we introduce the variable $\varphi = f$ and the variable ζ, related to the variables (f, l) by the following formula:

$$\left(\frac{d\eta}{dl}\right)^2 + \left(\frac{d\zeta}{dl}\right)^2 = 1. \tag{3.5}$$

Eqs. (3.3) – (3.5) define parametrically a curve $\Gamma(l) = (\eta(l), \zeta(l))$ in the plane (η, ζ), and Eqs. (3.3) – (3.4) imply

$$\frac{dv}{dl} = \frac{\sqrt{v}}{\sqrt{\lambda(hv+\gamma)}}, \qquad \frac{d\eta}{dl} = \frac{d\eta}{dv}\frac{dv}{dl} = \frac{1}{\sqrt{\lambda}}\left(1 - \frac{\gamma}{2(hv+\gamma)}\right),$$
$$\frac{d\zeta}{dl} = \sqrt{1 - \frac{1}{\lambda}\left(1 - \frac{\gamma}{2(hv+\gamma)}\right)^2}. \tag{3.6}$$

The derivative $\dfrac{d\zeta}{dv}$ may be expressed from Eq. (3.6) as a function of v and the general solution of the resulting differential equation may be shown to be obtained in terms of elliptic integrals.

In view of Eq. (3.5), the parameter l is an intrinsical parameter along this curve. We have in the new coordinates

$$ds^2 = \eta^2 d\varphi^2 + \left(\left(\frac{d\eta}{dl}\right)^2 + \left(\frac{d\zeta}{dl}\right)^2\right) dl^2. \tag{3.7}$$

Let us consider the three-dimensional Cartesian space (x_1, x_2, x_3) with the cylindrical coordinates (φ, η, ζ) in it, such that (η, φ) are the radius and the polar angle in the plane (x_1, x_2), and ζ is a displacement along the axis x_3. Then the linear element of the manifold M^2, written in the form (3.7), coincides with the linear element of the surface of revolution M^2_ρ, formed by rotation of the curve $\Gamma(l)$ around the axis x_3 in the Cartesian space (x_1, x_2, x_3). Therefore, the map $(\varphi(u,v), \eta(u,v), \zeta(u,v)) : M^2 \to M^2_\rho$ is an isometric embedding of M^2 in $\boldsymbol{R}^3$.

Having found from Eqs. (3.3), (3.4), (3.6) the variables (φ, η, ζ) as functions of (u, v), we obtain the following isometric map of the space (u, v) with metric (2.1) into the space $\boldsymbol{R}^3 : (x_1, x_2, x_3)$ with metric $ds^2 = dx_1^2 + dx_2^2 + dx_3^2$:

$$x_1 = \eta\cos\varphi, \qquad x_2 = \eta\sin\varphi, \qquad x_3 = \zeta. \tag{3.8}$$

Note that the right-hand side of the last equation (3.6) has the domain of definition

$$2(1 - \sqrt{\lambda}) \le \frac{\gamma}{(hv+\gamma)} \le 2(1 + \sqrt{\lambda}). \tag{3.9}$$

This means that the map of the isometric embedding (3.8) exists only in the domain (3.9). It is interesting to note that the rest of the manifold M^2 may be embedded isometrically in the three-dimensional pseudoeuclidean space $\boldsymbol{R}^3_1$ with pseudometric $ds^2 = dx_1^2 + dx_2^2 - dx_3^2$.

Now consider a billiard, for which the following inequality is valid everywhere in the billiard domain G:

$$|\, hv + \gamma \,| > \varepsilon \tag{3.10}$$

for some $\varepsilon > 0$. Then, having taken $\lambda > \left(1 + \frac{|\gamma|}{2\varepsilon}\right)^2$, we come to the situation when the whole billiard domain G may be isometrically embedded by the map (3.8) in the three-dimensional Euclidean space as a part of the surface of revolution.

Let the potential be repulsive, i.e., $\gamma < 0$. Then $h = T - \frac{\gamma}{v} > 0$ (where T is the kinetic energy). The accessibility domain of this mechanical system is an exterior of the circle $v = -\frac{\gamma}{h-T} \geq -\frac{\gamma}{h}$. If the billiard domain lies strictly inside the accessibility domain, i.e., if there exists R such that $-\frac{\gamma}{h} < R \leq v$, then the condition (3.10) is valid for $\varepsilon = \frac{hR+\gamma}{2}$.

Let the potential be attracting, i.e., $\gamma > 0$. Three cases are possible depending on the sign of the energy integral constant h.

Elliptic motion for $h < 0$. All collision-free segments of the billiard trajectory are elliptic arcs. In order to satisfy the condition (3.10), it is necessary and sufficient that the whole billiard domain lies in the accessibility domain of this mechanical system, i.e., there exists R such that $v \leq R < -\frac{\gamma}{h}$. This means that the billiard is fast and that its trajectory may approach any point of its boundary with a uniformly positive velocity.

For $\lambda = 1$, the shape of the ruling curve in the plane (η, ζ) is determined by Eq. (3.4) for the coordinate η and by the following solution for ζ:

$$\zeta = \frac{\sqrt{3}\gamma}{\sqrt{-h}} E\left(\beta, \frac{2}{\sqrt{3}}\right) + const,$$

where $\beta = \arcsin\sqrt{-\frac{hv}{\gamma}}$, and E is the elliptic integral of the second kind.

Parabolic motion for $h = 0$. All collision-free segments of the billiard trajectory are parabolic arcs. The condition (3.10) is satisfied, e.g., for any billiard domain for $\lambda = 4$, i.e., $\varepsilon = \frac{\gamma}{2}$. The parabolic billiard is of particular interest, because, in some sense, it is equivalent to the classical mathematical billiard in a plane. This case is considered in the next section in more detail.

Hyperbolic motion for $h > 0$. All collision-free segments of the billiard trajectory are hyperbolic arcs. The condition (3.10) is satisfied, e.g., for any billiard domain for $\lambda = 4$, i.e., $\varepsilon = \frac{\gamma}{2}$.

To sum up all the listed cases, we conclude that both for the attracting and repulsive Newtonian potential, if the billiard domain belongs strictly to the interior of the accessibility domain of this mechanical system, then the free manifold of the billiard may be embedded isometrically in the three-dimensional Euclidean space as a part of a surface of revolution. The condition formulated means that the billiard is sufficiently "energetic" and the absolute value of the billiard point velocity on the boundary of the billiard domain is uniformly bounded below by a positive value.

4 The parabolic billiard

Let us consider the form of the free manifold of the planar dynamical billiard in the attracting Newtonian force field in the parabolic case. Then $h = 0$, hence, Eq. (3.6) implies

$$\frac{dv}{dl} = \sqrt{\frac{v}{\lambda\gamma}}, \qquad \frac{d\eta}{dl} = \frac{1}{2\sqrt{\lambda}}, \qquad \frac{d\zeta}{dl} = \sqrt{1 - \frac{1}{4\lambda}},$$

or, after integration,

$$l = 2\sqrt{\gamma\lambda v}, \qquad \eta = \frac{l}{2\sqrt{\lambda}}, \qquad \zeta = \sqrt{1 - \frac{1}{4\lambda}}\,l. \tag{4.1}$$

According to Eq. (3.8), these equations determine a many-sheeted conic surface of revolution. This surface may be developed into the plane $\tilde{R}^2(\tilde{x}, \tilde{y})$ with polar coordinates l and φ in it

$$\tilde{x} = l\cos\varphi, \qquad \tilde{y} = l\sin\varphi.$$

The original billiard lies in the plane $R^2(x, y)$ with the Newtonian force field. In this plane, the full circle $(r, \theta), \theta \in [0, 2\pi)$ maps into the arc $(\eta, f), f \in [0, 2\pi\sqrt{\lambda})$. n such arcs should make up one full circle of radius l: $n2\pi\sqrt{\lambda}\eta = 2\pi l$. Substituting η from (4.1) then immediately gives $n = 2$, i.e., a pair of sheets of the conic surface may be developed into the plane. Therefore, we take the two-sheeted part of the representation of the plane $\boldsymbol{R}^2(x, y)$, which is defined in the polar coordinates by $x = r\cos\theta, y = r\sin\theta, 0 \le \theta \le 4\pi$.

From the above it follows that the original plane $\boldsymbol{R}^2(r, \theta)$ is equivalent "isobilliardly" to the plane $\tilde{\boldsymbol{R}}^2(l, \varphi)$ with the following map:

$$l = 2\sqrt{\lambda\gamma r}, \qquad \varphi = \frac{\theta}{2}. \tag{4.2}$$

Now it is clear that in the parabolic case the map may be considered as the map $z \to \sqrt{z}$ from the two-fold Riemann surface rooted at zero to the ordinary complex plane. For this map, it is known that a parabola with focus in the origin becomes a straight line. More precisely, one can easily check that a parabola $r = \dfrac{p}{1 + \cos(\theta - \theta_0)}$, p, θ_0 parameters, in $\boldsymbol{R}^2(r, \theta)$ maps into the straight line $l = \dfrac{\sqrt{2\lambda\gamma p}}{|\cos(\varphi - \theta_0/2)|}$ in $\tilde{\boldsymbol{R}}^2(l, \varphi)$.

By way of illustration let us consider the dynamical billiard in the unit circle D^2 in $I\!R^2$. Let the centre of the circle be offset by r_0 from the gravitating centre. Without loss of generality, we assume that the circle is shifted along the axis Ox. Then the billiard domain is given by the inequality $r^2 - 2rr_0\cos\theta \le 1 - r_0^2$. In this case, the two-sheeted representation of the parabolic billiard is equivalent trajectorically to the one-sheeted planar mathematical billiard in the domain $\tilde{D}^2$, determined by the inequality

$$\frac{l^4}{16\gamma^2\lambda^2} - \frac{l^2 r_0 \cos 2\varphi}{2\gamma\lambda} \le 1 - r_0^2,$$

or

$$(\tilde{x}^2 + \tilde{y}^2)^2 - 8\gamma r_0 \lambda(\tilde{x}^2 - \tilde{y}^2) \le 16\gamma^2\lambda^2(1 - r_0^2). \tag{4.3}$$

According to Eq. (4.3), the bounding curve of the domain $\tilde{D}^2$ is a cassinoid.

If the centre of the billiard circle coincides with the gravitating centre, i.e., $r_0 = 0$, then the domain $\tilde{D}^2$ degenerates into a circle of radius $2\sqrt{\lambda\gamma}$. The Cassini oval is simply connected for $r_0 < 1$, when the billiard domain contains the gravitating centre. If the last one lies on the boundary ($r_0 = 1$), then the cassinoid becomes a lemniscate with an intersection point in the origin. If $r_0 > 1$ (the billiard domain does not contain the gravitating centre), then the whole circle D^2 is located in one sheet of $\mathbb{R}^2$. Therefore, the usual (one-sheeted) dynamical billiard is equivalent trajectorically to the planar mathematical billiard in one of the two simply connected domains (4.3).

Now look more closely at the special case $r_0 = 0$, when the parabolic billiard is equivalent to the classical mathematical billiard in the circle.

Let us construct periodic trajectories of the original parabolic billiard. Let them be centrally symmetrical and consist of n parabolic arcs ($n \ge 2$). Each arc should meet the bounding circle at points with the circular arc distance between them being equal to $2\pi\frac{m}{n}$; m, n coprime integers ($m \ge 1$). Then, in order for the parabola $r = \frac{p}{1+\cos\theta}$ to be a segment of the periodic trajectory, the following relation should be true: $1 = \dfrac{p}{1+\cos\frac{m}{n}\pi}$, from which we get $p = 1 + \cos\frac{m}{n}\pi$. It is easy to compute the angle μ between the tangent to the parabolic arc and the radius direction in the collision point $\theta = \frac{m}{n}\pi$: $\tan\mu = \dfrac{r}{\frac{dr}{d\theta}} = \dfrac{1+\cos\frac{m}{n}\pi}{\sin\frac{m}{n}\pi}$, that implies $\mu = \frac{\pi}{2} - \frac{m}{2n}\pi$.

The mapping (4.2) is conformal, hence, the angle of reflection is the same for the image periodic polygonal trajectory of the mathematical billiard. If $m = 1$, then we have $\mu = \frac{\pi}{2} - \frac{\pi}{2n}$, which corresponds to a regular $2n$-gon. Thus, the centrally symmetrical periodic n-segment trajectory of the parabolic billiard maps into the regular $2n$-gon. The simplest case is a trajectory consisting of two equal parabolic arcs. It should be passed twice to complete the corresponding square in the force-free billiard. Accordingly, the parabolic triangle maps into the regular hexagon and so on. When $m = 1$, all trajectories are convex and various star-like trajectories arise, if $m > 1$.

The constructed map allows us to extend the properties of the Birkhoff billiard to the parabolic one. For example, it is well known, that each periodic trajectory of the Birkhoff billiard in a circle consists of straight line segments of equal length. Therefore, we may conclude that the parabolic billiard in the circle with the non-shifted centre also possesses periodic trajectories containing only identical arcs. Of course, this particular statement may be also derived by means of direct considerations.

In conclusion, we consider the parabolic billiard in the ellipse, one focal point of which coincides with the attracting centre. The mapping (4.2) transforms such ellipses into the ellipses with the centre in the origin [7]. Therefore, the parabolic billiard in the ellipse $r = \frac{p_e}{1+e_e\cos\theta}, 0 < e_e < 1$, is found to be equivalent to the Birkhoff (force-free) billiard in the ellipse $\frac{x^2}{a^2} + \frac{y^2}{b^2} = 1, a = 2\sqrt{\frac{\lambda\gamma p_e}{1+e_e}}, b = 2\sqrt{\frac{\lambda\gamma p_e}{1-e_e}}$.

It is interesting to note that the inverse map $z \to z^2$ of the complex plane was applied to analysis of the billiards in the ellipse with the imposed Newtonian potential. A trajectory equivalence of the billiard in the ellipse with the Newtonian attracting force directed to the focal point and the billiard in the ellipse with the Hooke linear elastic force attracting the particle to the centre of the ellipse was proved [8]. In this context, the parabolic motion may be treated as a limiting case, corresponding to zero elastic constant in the Hooke law [7].

The authors acknowledge the financial support of Russian Foundation for Basic Research under grants N 95-01-00308 and N 98-01-00940.

References

1 Kozlov V V, Treschev D V, Billiards: A Generic Introduction to the Dynamics of Systems with Impacts (Translations of Mathematical Monographs, Vol 89), Amer. Mathematical Society, 1991.

2 Lehtihet H E, Miller B N, Numerical study of a billiard in a gravitational field, *Physica*, **D21**(1): 93–104, 1986.

3 Beletsky V V, Kasatkin G V, Starostin E L, Billard im Gravitationsfeld: Reguläre und chaotische Bewegungen. In: V.V.Beletsky, Reguläre und chaotische Bewegung starrer Körper, Teubner, Stuttgart, 1995, 89–120.

4 Beletsky V V, Kasatkin G V, Starostin E L, The pendulum as a dynamical billiard, *Chaos, Solitons & Fractals*, **7**(8): 1145–1178, 1996.

5 Starostin E L, Weak gravitation destroys non-symmetric periodic trajectories of a billiard in a circle. In: Proceedings of the 2nd European Nonlinear Oscillations Conference, Prague, 9–13 Sept. 1996, v. **2**, 235–238.

6 Beletsky V V, Kugushev E I, Starostin E L, Free manifolds of dynamic billiards, *Regular and Chaotic Dynamics*, **2**(3/4): 62–71, 1997.

7 Arnol'd V I, Huygens and Barrow, Newton and Hooke: Pioneers in Mathematical Analysis and Catastrophe Theory from Evolvements to Quasicrystals, Birkhauser, 1990, 118 p.

8 Panov A A, The elliptic billiard with the Newtonian potential, *Matematicheskie Zametki (Math. Notes)*, **55**(3): 139–140, 1994.

Vibro-impact processes in systems with multiple impact pairs and distributed impact elements

V L Krupenin*

This paper is a brief review of some interesting, and perhaps surprising, phenomena often encounterd in the discrete and distributed systems which vibrate with impacts. We examine such systems in some cases where we have multiple collisions of subsystems or the sizes of objects require the consideration of wave processes. Such systems are:

- *-Mechanical objects, which elements are subjected to impacts;*
- *-Various extended constructions, vibrating near pointwise, gridwork and continuous movement obstacles;*
- *-Vibro-conductive and vibro-arresting constructions equipped with facilities with solitary and multiple breaks, etc.*

Original calculation methods have been presented which are based on the frequency-time analysis and other methods of the modern nonlinear mechanics. Multiple nonlinear effects are theoretically described, which are related to the formation of specific nonlinear waves of trapezoidal profiles; advent of localization of intensive impacts in some areas of constructions; generation of higher harmonic components; generation of non-synchronic and chaotic movements, etc.

1 Systems with parallel impact pairs

1.1 ***Basic models.*** Systems with parallel impact pairs are the complex vibro-impact systems, in which some of the elements of basic sub-system interact with corresponding elements of other sub-system. Examples of such systems are shown in *Fig.1.*

Fig.1.a shows a system where basic subsystem is represented by a string with *N* balls fixed on it. In this case, the balls collide with a rigid one-side obstacle, representing the second subsystem. Obviously, the restriction may be also symmetrical. Besides that, the balls may constitute impact pairs with elements of more complicated form. Fig.1.b displays a transversally oscillating string or beam interacting with pointwise obstacle. Here the subsystem with distributed parameters is a basic one.

---

* Blagonravov Mechanical Engineering Research Institute RAS, M. Haritonievsky per. 4, Moskow 101830, Russia, e-mail: krupenin@glasnet.ru.

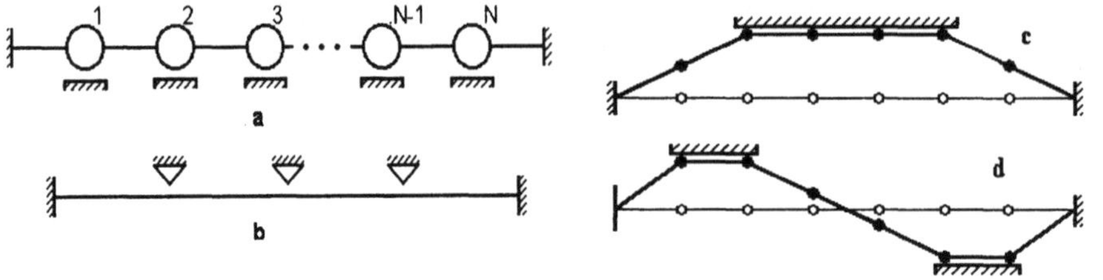

Fig.1

1.2 ***Motion equations.*** In general case for one-dimensional system with N parallel impact pairs (*Fig.1.a*) we can write equation of motion as

$$\rho u_{tt} - C^{*}[u] + B^{*}[u] = \sum_{j=1}^{N} \{-m_k u_{tt}(x,t) + P_k(t) - [u,u_t]\}\,\delta(x-x_j), \tag{1}$$

where:
$u(x,t)$ denotes displacement; C^{*} and B^{*} are linear "elastic" and "dissipative" operators of the system; ρ is linear density of the material, m_k are masses of the bodies located in points x_k ; **T**-periodic driving forces $P_k(t)$ acting, for instance, in points of location of impact pairs x_k; $\delta(t)$ is Dirac's delta-function, $[u,u_t]$ is the force of impact interaction, defined by relations following from Newton's or other hypothesis and assumptions concerning the nature of the process.
For example, we can write [1], in terms of

$$[u,u_t]\delta(x-x_j) = J_j \delta^{T}(t-\varphi_j),$$

where $\delta^{T}(t)$ is **T**-periodic delta-function; J_k - the impact impulse in the k-*th* pair; φ_K – is the phase of impact.
We do not have possibility to consider the formal features of the notation for a force of impact interaction. So we recommend, in particular, works [1-9].

Independent of the system structure Equation (1) may be written uniformly in the operator form. For the field of displacements $u(x,t)$:

$$u(x,t) = \sum_{j=1}^{N} L(x, x_j; p)[P_j(t) - [u(x,t); pu(x,t)]\delta(x-x_j)] \tag{2}$$

where the operators $L(x,y;p)$ of dynamic compliance are determined by the structures of the interacting subsystems; $p=d/dt$. For description of the system with lumped parameters in the operator of Equation (2) we should set $x=x_k$.

1.3 ***Analysis methods.*** The study of systems with parallel impact pairs is carried out numerically or by means of frequency-time methods [1,6]. In case of a periodic excitation in order to find the T-periodic modes, we can obtain the integral equation of periodic motion [6]:

$$u(x,t)=u_0(x,t)+\sum_{j=1}^{N}\int_0^T \chi(x,x_j;t\text{-}s)\,(u_j,u_{jt})ds \tag{3}$$

Here $u_0(x,t)$ is impactless stationary movement under driving forces in the absence of colliding; $u_j=u(x_j,t)$, and also

$$\chi(x,x_j;t\text{-}s)=T^1\sum_{k=-\infty}^{\infty} L(x,x_j;ik\omega)\exp(ik\omega t) \tag{4}$$

is a periodic Green's function (*PGF*), compliant to the operator $L(x,x_j;p)$ ([1]). If assumed that the impact is momentary and acted once during the period of the motion, so function (u_j,u_{jt}) can be represented as a combination of periodic delta-functions. Equation (3) can be reduced to the following representation of the vibro-impact process:

$$u(x,t)=u_0(x,t)+\sum_{j=1}^{N} J_j\chi(x,x_j;t\text{-}t_j) \tag{5}$$

where J_j is the impulse of impact forces in the *j-th* impact pair; t_j is moment of the impact in this pair.

The representation (5) is known as a "*2N-parametric*". The unknown parameters of the motion can be sought from the impact conditions

$$u(x,t_j)=\Delta_j,\quad J_j=m_j(R_j+1)u_t(x_j,t_j\text{-}0) \tag{6}$$

where Δ_j, R_j - clearance and restitution ratio in the *j-th* impact pair.

The solutions obtained should be analyzed regarding the stability and feasibility of geometric conditions of the type $u(x,tj)\geq\Delta_j$. The final solution of the problem in a visible analytical general form can be obtained for a limited number of models of such kind. However, for particular parameters it is usually possible to find a corresponding numerical-analytical solution. Besides that, based on the representation of Equation (5) and (6) it is possible to build up some approximate solutions.

1.4 ***The essential dynamic effects.*** Further, we will discuss some effects revealed as a result of analyzing the model *Fig.1.a* with a periodic structure: for each j all m_j, b_j and c_j values are the same. The external excitation was chosen in a sinusoidal form.

The main result (see also Clause *2.5* and [5, 6, 10]) is the discovery of existence of periodic modes with synchronous impacts in different impact pairs. Such modes are known as "claps". At excitation of these modes, the string with the balls fixed on it generates one, two or multi-trapezoidal forms. They are quite similar to the corresponding eigenforms of a linear system in terms of alternation of nodes and antinodes of waves, and feasible in frequency areas situated to the right from the eigenfrequencies of the corresponding linear system. *Fig.1.c,d* shows an example of claps for two lowest forms in the system with six symmetrical impact pairs.

The trapezoidal form appears even with two impact elements. This is not surprising, since in this case that is exactly the first eigenform of linear system. The same form with two elements in the antinodes of the wave is one of the higher eigenforms for systems containing 6n-1 elements (where **n=1,2,...**) and therefore for the realization of such form in vibro-impact processes is quite natural.

The surprising thing is that trapezoidal form becomes dominating as the number of elements increases. At realization of one and multi-trapezoidal standing clap-waves, the dynamic effects that are typical for the systems with a single impact pair arise: phenomena of frequency and amplitude pulling, and also hard initiation (see [2] and clause 2.4). Thus, the behavior of clap -waves is in many respects similar to that of a system with a single impact pair and, particularly, of a simple "impact oscillator".

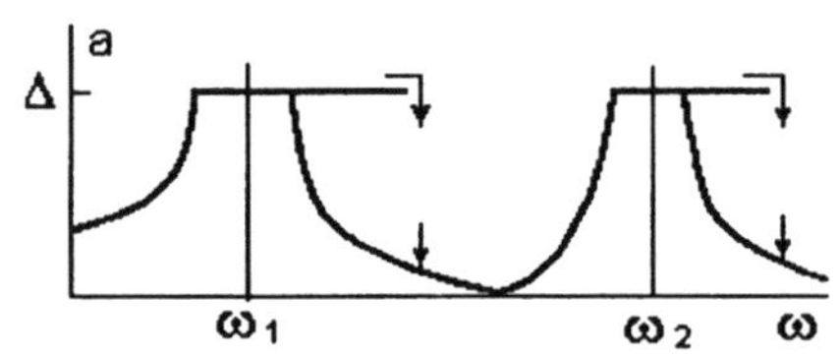

Fig.2

Fig.3 displays the amplitude-frequency characteristic of a system with parallel symmetrically situated obstacles near two lower forms of the linear system. At amplitudes which are less than the clearance ($a<\Delta$) the branches of resonance curves of the linear system are being realized. After reaching of obstacles a vibro-impact process of the clap type takes place, number of bodies involved in the clap depends on the excitation level. The figure illustrates the above-mentioned analogy between dynamics of claps and vibro-impact systems with a single impact pair.

Fig. 3

Besides, the above mentioned, these "similar" effects include, in particular, appearance of ambiguity areas of amplitude-frequency characteristics and jumping of oscillations away from one stable branch of resonance curve to another (arrows in *Fig.2*).

It should be noted that distributed systems with multiple parallel impact pairs like shown in *Fig.1,b* have similar characteristics.

The numerical analysis made it possible to find out more complicated modes, which can be considered as chaotic.

It would be necessary to notice, that in some cases in these systems we can detect the modes of motion of a solitons type [5, 10, and 11].

1.5 ***Remark: systems with serial impact pairs.*** Systems with serial impact pairs include vibro-impact systems with large number of degrees of freedom in which all elements with exception of outer ones are involved in two or more impact pairs. Many of the typical models of such systems are shown in [12,13].

2 Systems with distributed impact elements

2.1 ***About these systems:*** Now we'll briefly examine the problems concerned with the models of systems with distributed impact elements, that can appear while studying discrete systems of the type (1) as the corresponding long-wave approximations, or while using axiomatic approaches.

It is reasonable to consider distributed impact elements, particularly, when studying vibro-conductive continuous media with complicated structures, and also when studying various extended objects (strings, threads, flexible bars, cables, ropes, beams and other systems), moving near some solid or gridwork obstacles, partitions, walls, gridworks etc.

Reference [14] considered models of distributed linear media of complicated structure. The most specific feature of mentioned structures is the presence of two main "medium parts", namely: "carrier" and "attached" parts. Dynamic description of such systems, in general, consists of two groups of motion equations - reflecting "carrier" and "attached" subjects constrained behavior respectively. Likewise all the models in the multi-polar mechanics, the concept of a point is a subject of significant revision: its state can be defined by unspecified number of kinematics parameters.

Utilization of such models arise to be productive for solution of some practical problems, such as dynamical analysis of vibro-states of complicated mechanical structures, consisting of distributed single dimensioned "carrier" and of gross number flexurally "attached" solid devices.

It seems that mentioned considerations could be useful by the dynamic models creating for systems consisting of "carrier" and "attached" parts with multiple breaks in it. Due to the system nature the possibility of different type collisions arises in the "attached" subsystems.

Then we consider a completely distributed model. Thus, assuming the impact pairs to be "spreaded" within a certain space, we can use the concept of *distributed impact elements.*

Now, let's overview briefly some papers concerning the distributed impact elements. We can obtain the model of distributed impact element by at least two ways:

Firstly, in some cases it appears impossible to disregard wave process arising in the impact pairs itself. Impacting bodies can't be considered as the solid bodies since the lengths of the waves generated by collisions are comparable with the dimensions of impacting surface.

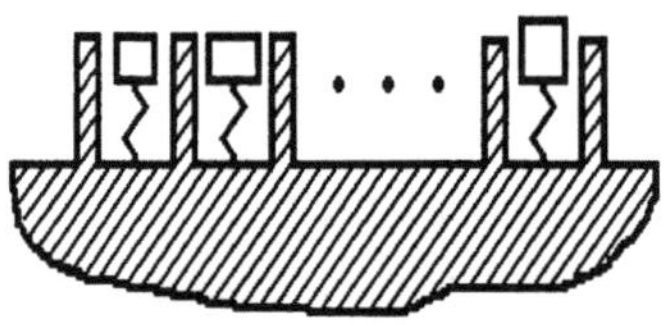

Fig.4

Secondly, considering the dynamic system with amount of convenient impact pairs are to be large enough, so we can perform long-wave approximation and transit to the distributed model with distributed impact element.

In this way, starting with a abstract model of "regularly concentrated beds on the strained thread" colliding with the solid stationary obstacle [15], we can obtain basic physical model of the string colliding with the solid stationary obstacle. In the *Fig.2.c* and *Fig.3* - the double-side obstacle is shown.

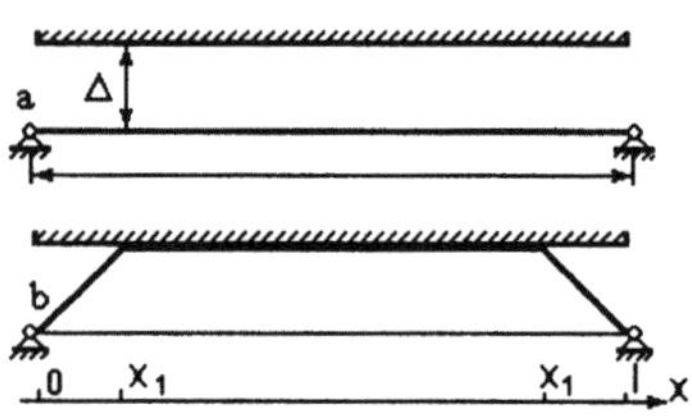

Fig.5

The number of technical applications for such a model is huge. The mentioned above basic model, initially considered in the paper [7] is now examined deeply enough both theoretically [8, 9, 15, 16] and experimentally [15-17] (further references may be found in the above papers).

The idea to use the concept of distributed impact elements seems to be also productive for dynamic analysis of large number of lightly damped vibro-isolated devices attached over the flexible carrier with the motion limiters and interacting dynamically via carrier (see *Fig.4*). This model is an exact model of strongly nonlinear vibroconductive media with complicated structure.

Evidently, for the first time, this problem was considered in [19]. The problem of vibration propagation through a strongly nonlinear vibro-conductor (bar with the distributed impact element, modeling the two degrees of freedom attached equipment) was solved in this paper. Main definitions and solution method are based on so-called non-linear forms of oscillations (periodical motion conditions having certain symmetric properties) and generalized methods of frequency-time analysis of vibro-impact processes [1] was suggested.

2.2 Models.

2.2.1 Let's postulate the existence of some elastic carrier medium, which motion described by function $u(x,t)\in R^3$ $(x\in R^3,\ t\in R)$ is subjected to the classical ***Lame*** equation

$$\rho u_{tt}=(\lambda+\mu)\,grad\,div\,u+\Delta u+F \qquad (7)$$

where ρ - density of the carrier medium and ***Lame*** factors, λ, μ - characterize the carrier elasticity properties.
Let the intensity of the volumetric forces has the following structure: $F=F_0+F_1$, where F_1 -dictated vector, and $F_0(X,t)=-C_1(u-Y_k^{(1)})-C_2(u-Y_k^{(2)})$, where it is assumed that each point of the medium carries an elemental impact pair, consisting of two interacting linear stationary subsystems $A^{(1)}(x)$ and $A^{(2)}(x)$, described by dynamic compliance operators: $L_{qj}^{(1)}(p)=O(p^{-2})$ and $L_{ls}^{(1)}(p)=O(p^{-2})$; subscript indexes q, j, l, s are varying within some sets, depending upon the dimensionalities of interacting subsystems, which parameters can be determined, in general, by the **x** value. In order to close the system of motions equations, let's add the following relationships:

$$Y_k^{(1,2)}=L_{nk}^{(1,2)}(p)C_{1,2}u\pm L_{kk}^{(1,2)}(p)\Phi_0(Y^0,Y^0_t)+f_k^{(1,2)} \qquad (8)$$

$$Y_n^{(1,2)}=L_{nn}^{(1,2)}(p)C_{1,2}u\pm L_{kn}^{(1,2)}(p)\Phi_0(Y^0,Y^0_t)+f_n^{(1,2)} \qquad (9)$$

where $Y_n^{(1,2)}(x,t)$ and $Y_k^{(1,2)}(x,t)$ - displacements of the hanger and contact points; $Y^0=Y_k^{(2)}-Y_k^{(1)}$ -relative convergence of impact elements of the interacting subsystems to which linear densities $m_1(x)$ and $m_2(x)$ are referred; $\Phi\{Y_n^{(1,2)}, Y_k^{(1,2)}, \partial/\partial t\,[Y_n^{(1,2)}], \partial/\partial t\,[Y_k^{(1,2)}]\}$ - density of impact force (for the system $A^{(1)}(x)$ in (8) and (9) we choose the "plus" sign and for $A^{(2)}(x)$ - "minus"); these equations can also include some functions **f**, describing an additional external effects.

The boundary conditions for (1) which models the carrier construction (namely, physical and geometrical characteristics) are the same as in the classical case. The frequency properties of the vibration-isolated equipment generating vibro-impact processes can be obtained from the model of the added medium part, containing the impact element. The connection mechanism of the carrier and added parts determines the structure of global vibration field.

This approach, adapted for analysis of the collective effect of impact pairs, possibly sacrifices information on "individual" features of certain concrete system elements, as well as effects, which can only be clarified if considering the model's discreteness.

2.2.2 Let's consider the second example [17]: supported flexible beam with total length L (see Fig.5.a) vibrates near the solid wall. Beam parameters are denoted as following: ρ - linear density, E,G - elasticity modules of the first and second order, F, Γ - cross sectional area and moment of inertia, k'- factor describing the stress distribution non-uniformity over the beam cross section, $P(x,t)$ - external force distribution, Δ - clearance between the beam equilibrium position and wall. Using the *Timoshenko* beam model and denoting by $u(x,t)$ and $[Y(x,t)-u_x(x,t)]$ the beam instantaneous linear and angular deflection shape, $\Phi(u)$ impact force distribution, the motion equations will be written as following:

$$\rho u_{tt} - k'FGu_{2x} + k'FGY_{2x} + \Phi(u) = P, \quad E\Gamma Y_{2x} - k'FGu_x - \Gamma\rho F^{-1}Y_{2t} = 0 \qquad \mathbf{(10)}$$

with the boundary conditions $u(0,t) = u(L,t) = 0,\ t \in]-\infty,\infty[,\ x \in [0,L]$.

2.3 ***On the analysis methods.*** In spite of the complexity of the motion equations, in some cases [19,20] it appears possible to perform not only the corresponding numerical analysis of mentioned models, but also to obtain an approximate analytical representation of the required motion distributions. These representations can be obtained by means of the modified frequency-time analysis of vibro-impact processes and other methods of the modern non-linear mechanics.

In many cases the equations of motion can be reduced to the following "two-functions" representation of the vibro-impact process $[u(x,t)]$:

$$u(x,t) = u_0(x,t) - \int_X J(z)\,\chi[x, z; \varphi[t-(z)]]dz \qquad \mathbf{(11)}$$

where $u_0(x,t)$ - steady-state movement field under inducing forces in the absence of impact interactions; X - some integration areas; $\chi(x,z;z)$- PGF of system.

The unknown functions $J(x)$ (the impact force impulse distribution) and $\varphi(x)$ (the impact force phase distribution) can be sought from the impact conditions.

2.4 Two principal examples:.

2.4.1 ***The first example*** - the longitudinal wave propagation in the rod with flexurally attached in its every point impact pair (see Fig.6). Applying the above mentioned approach we can obtain the following system of equations:

$$\rho u_{tt} - Eau_{xx} + C(u - y) = 0$$
$$my_{tt} + C(Y - u) - \Phi(y^0) = 0$$
$$m_1(y_{tt} + y^0_{tt}) + \Phi(y^0) = 0 \qquad (12)$$

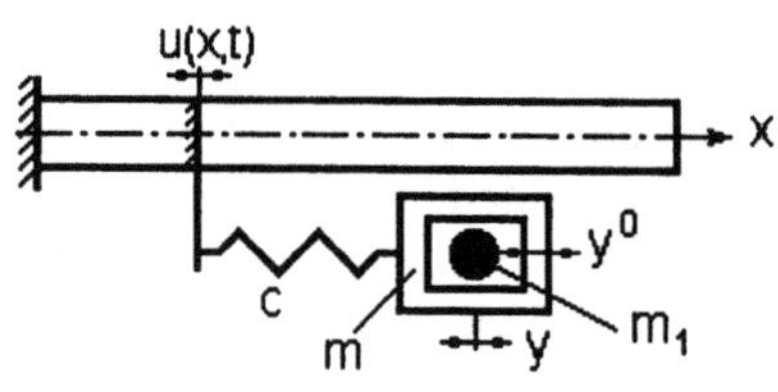

Fig.6

where u, y, y^0 - displacements of system structural elements with instant coordinate X; $u=u(x,t)$ is carrier displacements; $y=y(x,t)$, and $y^0=y^0(x,t)$ is displacements of attached elements, ρ, **E, A** is- linear density, cross sectional area and Young modules of rod material respectively;

$\Phi(y^0)$ is impact force density; m, m_1 are linear densities of attached parts. Boundary conditions for (12) looks like: $u(0,t)=0$, $u(L,t)=\mu\cos\omega t$, where L - the length of the rod.
Conditions of Newton's impact are as follows:

$$y^0[x, \Psi(x)]=\Delta,\ J(x)=M(1+R)y_t^0[x, \Psi(x)-0]>0.$$

where $\Psi(x)$ is impact force phase distribution, $J(x)$ is impact force impulse distribution, $M = mm_1/(m+m_1)$ is reduced mass distribution, R - restitution ratio. In general case values M, R and Δ are varied with **x**.

2.4.2 Let's consider again our second example - see clause *2.2.2*. Supposing the existence of periodical single-impact motion, the impact force distribution may be considered as follows:

$$\Phi(u)=J(X)\delta^T[t-\varphi(x)]$$

where $J(x)>0$ impact force impulse distribution, $\varphi(x)$ -impact phase distribution, T - period of oscillation.
In order to describe the "clap" shape (see *Fig.5,b.*), $\varphi(x)=\varphi=const$, the Newton impact conditions may be formulated as follows:

$$u(x,0)_-=u(x,0)_+=\ u(x,0)=(\Delta/x_1)x\ \{if\ x\in[0,L-x_1]\},$$
$$u(x,0)=\Delta\ \{if\ x\in[x_1,L-x_1]\},$$
$$u(x,0)=(\Delta/x_1)(L-x)\ \{if\ x\in[L-x_1,L]\},$$

and

$$u_t(x,0)=V_0\ (if\ x\in[x_1,L-x_1]),$$
$$u_t(x,0)=0\ (if\ x\notin[x_1,L-x_1]);$$
$$u_t(x,0)_+=-Ru_t(X,0);\ L_0=L-2\Delta\tan^{-1}[u_x(0,0)]$$

(*Fig.5.b*); subscript indexes **(-, +)** mean the function values directly before and after impact. Under such suppositions:

$$J(x)=(1+R)\rho V_0=const;$$
$$\varphi(X)=const=\varphi\ (x\in[x_1,L-x_1])$$

In addition, conditions of impact have to be considered for the beam midpoint, which is obviously involved in vibro-impact process.

3 Acknowledgements

This research was supported by RFBR (project 98-01-00925) and INTAS (project 96-2138).

References

1. Babitsky V I, Krupenin V L, Oscillations in Strongly Nonlinear Systems. Nauka, Moscow, 1985 (in Russian).
2. Babitsky V I, Theory of Vibro-Impact systems. Nauka, Moscow, 1978 (in Russian, Revised English edition: Springer, 1998) .
3. Astashev V K, Babitsky V I, The resonant oscillations of the visco-elastic bar with an obstacle. Mechanics of Solids, 1972, no.3 (in Russian).
4. Krupenin V L, In the development of the laws of motion for the bar vibro- impact systems, Machine Sciences, 1981, no.2 (in Russian).

5. Vedenova E G, Manevich L I, Periodical and located waves in vibro-impact systems with regular structures, Machine Sciences, 1984, no.4, (in Russian).
6. Krupenin V L, On study of the vibro-impact processes in the systems with many impact pairs, J. of Machine Manufacture and Reliability, 1994, no.2 (in Russian).
7. Amerio L, Prouse G, Study of the motion of a string vibrating against an obstacle, Rend. di Mat. Ser. 6(8), 1975, no. 2.
8. Cabannes H, Cordes vibrantes avec obstacles, Acustica, 1984, V.55.
9. Krupenin V L, To the theory of vibro-impact systems with distributed impact elements, Mechanics of Solids, 1986, no.1 (in Russian).
10. Babitsky V I, Veprik A M, Krupenin V L, Vibro-impact effects in the chain of elastically linked particles, DAN, 1988, V.300, no.3 (in Russian).
11. Babitsky V I, Vedenova E G, Goihman L V, Manevich L I, - DAN, 1985, v.281, 2 pp. 288-291 (in Russian).
12. Kobrinsky A E, Kobrinsky A A, Vibro-Impact Systems. Nauka, Moscow, 1973 (in Russian).
13. Nagaev R F, Mechanics Processes with Repeated Attenuated Impacts Nauka, Moscow, 1985 (in Russian).
14. Palmov V A, Oscillations of elastic-plastic bodies. Nauka, Moscow, 1976, (in Russian, Revised English edition Springer, 1998).
15. Veprik A M, Krupenin V L, On the solution of systems containing distributed impact elements, Machine Sciences, 1985, no. 6 (in Russian).
16. Krupenin V L, Transformation of oscillation forms of a string interacting with two extended obstacles. DAN USSR, 1990, V.313, no. 6 (in Russian).
17. Veprik A M, Krupenin V L, On the resonance oscillations of a system with a distributed impact element, Machine Sciences, 1988, no.6, (in Russian).
18. Astashev V K, Krupenin V L, Waves in discrete and distributed strongly nonlinear systems: the theory and experiments. - In the book «Nonlinear Oscillations in Mechanical Systems», St.Peterburg, 1997.
19. Krupenin V L, 1987. A contribution to the theory of strongly non-linear vibration guides. Machine Sciences, 1987, no.1, (in Russian).
20. Krupenin V L, Vibration fields in the systems with complicated structures and multiple discontinuities, DAN USSR, 1995, V.343, no. 6 (in Russian).

The impact based transportation process in a vibratory feeder

Peter Wolfsteiner and Friedrich Pfeiffer**

Vibratory feeders are used in automatic assembly to feed small parts. They are capable to store, transport, orient, and isolate the parts. *An oscillating track with frequencies up to 100 Hz excites the transportation process, that is mainly based on impact and friction phenomena between the parts and the track. This paper presents a complete mechanical model of part feeding dynamics, based on unilateral constraints with Coulomb friction. This model enables a theoretical investigation and consequently an improvement of the transportation process. The developed impact model was verified by measurements, that were carried out using a special experimental setup with laser distance measurement.*

1 Introduction

Vibratory feeders are applied for a wide variety of parts and for a lot of different tasks [1]. In the majority of cases, the parts are available as a sort of bulk material that is stored in a container. The transportation process, starting in this reservoir, is often combined with orienting devices that orient parts, or select only these parts having already the right orientation (Figure 1 shows an example of a vibratory bowl feeder with an orienting device). Each kind of parts, with its special geometry and mechanical properties, requires an individual adaptation of the feeder. This individual tuning comprises the development of suitable track and orienting device geometries and the adjustment of the excitation parameters frequency and amplitude. Due to the complex mechanics of the feeding process this design is usually done by trial and error without any theoretical background. A complete dynamical model of the transportation process allows a theoretical investigation and consequently an improvement in the properties of the feeder.

Friction and impact phenomena between the parts and the track are the most important mechanical properties of transportation processes. Consequently, the required dynamical model has to deal with unilateral constraints, dry friction, and multiple impacts. A general theory for multibody systems with unilateral contacts was developed by [2, 4]. The presented formulation is

*Institute B of Mechanics, Technical University, Munich, D-85747 Garching, Germany

capable to deal with unilateral contacts with Coulomb friction and impacts with friction in planar systems. Multiple dependent constraints and discontinuous transitions are included in these methods. The transformation of these methods for three dimensional problems is described in [5], the occurring nonlinearity in the contact equations demands a special formulation. This approach is based on a method developed by [3]. [6] gives a detailed description of the application of this theory for the parts transportation process in vibratory feeders.

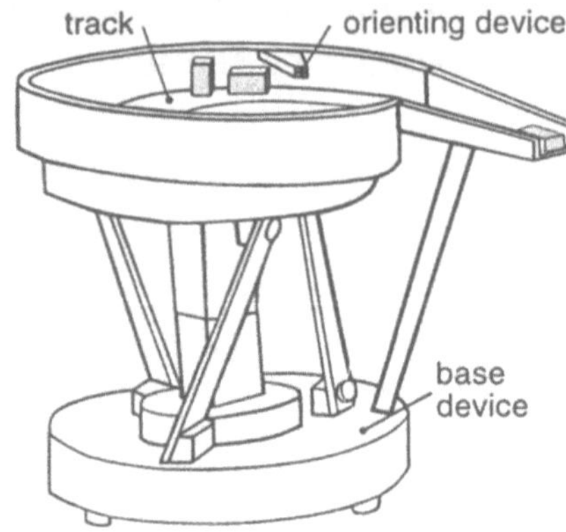

Figure 1: Vibratory Bowl Feeder

2 Mechanical model

The mechanical model of the vibratory feeder can be split in two parts: the transportation process and the base device. This paper focuses on modeling and simulation of the transportation process. The modeling of the base device can be done with well known standard technics for multibody systems, not described in this paper. Friction and impact effects have a fundamental importance for the transportation of parts. Changing contact configurations between the parts and the track and also between the parts itself are characteristic for the feeding process. The contacts appear either continuously for a certain time interval, or for a discrete time (impact). A structure varying multibody system with unilateral constraints with friction is an ideal technique for modeling the feeding process. Its formulation results in a set of differential equations with inequality constraints requiring special mathematical and numerical methods [2, 4, 5]. The spatial (three dimensional) geometry evokes a two dimensional model for the contacts. This proposed formulation is capable of dealing with time discrete multiple impacts with Coulomb friction. The assumption of rigid bodies is appropriate for a wide variety of potential parts (figure 2 shows the mechanical model).

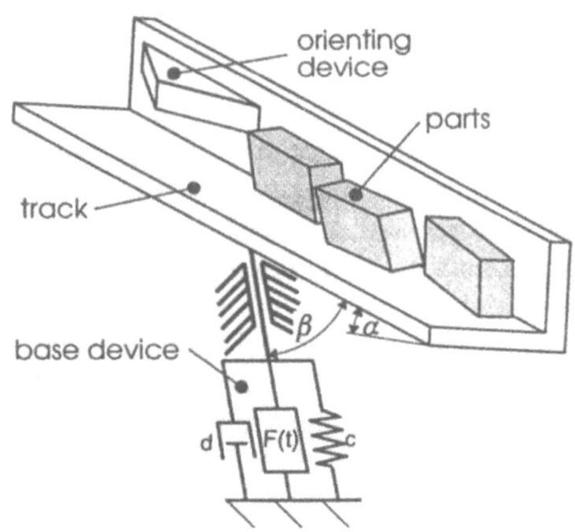

Figure 2: Mechanical Model of a Vibratory Feeder, with a one Degree of Freedom Model for the Base Device and a Track with Orienting Device

Equations of motion with superimposed constraints

The dynamic description of a structure varying system is based on the equation of motion of the unconstrained system: $\boldsymbol{M}(\boldsymbol{q},t)\ddot{\boldsymbol{q}}(t) - \boldsymbol{h}(\dot{\boldsymbol{q}},\boldsymbol{q},t) = \boldsymbol{0}$. t is the time variable, the vector of generalized coordinates $\boldsymbol{q}$ includes six degrees of freedom for every part and the coordinates describing the dynamics of the base device. $\boldsymbol{M}(\boldsymbol{q},t)$ is the mass matrix, the vector $\boldsymbol{h}(\dot{\boldsymbol{q}},\boldsymbol{q},t)$ contains the gyroscopic accelerations and all active forces. $\boldsymbol{M}$ and $\boldsymbol{h}$ can be derived with the Lagrange II formalism [6]. If contact between two bodies occurs, the degrees of freedom of the dynamical system decreases. The couplings are modeled with constraint forces $\boldsymbol{\lambda}$, inserted into the equations of motion as Lagrange Multipliers:

$$\boldsymbol{M}\ddot{\boldsymbol{q}} - \boldsymbol{h} - (\boldsymbol{W}_N, \boldsymbol{W}_T)\,(\boldsymbol{\lambda}_N^T \boldsymbol{\lambda}_T^T)^T = \boldsymbol{0} \tag{1}$$

The vector $(\boldsymbol{\lambda}_N^T, \boldsymbol{\lambda}_T^T)^T$ contains the constrained forces in normal (N) and tangential (T) direction for every active coupling i, with $\boldsymbol{\lambda}_N = (\cdots, \lambda_{Ni}, \cdots)^T$ and $\boldsymbol{\lambda}_T = (\cdots, \boldsymbol{\lambda}_{Ti}^T, \cdots)^T$, the matrices $\boldsymbol{W}$ project the constrained forces $\boldsymbol{\lambda}$ from the local contact coordinates of every contact point in the directions of the generalized coordinates $\boldsymbol{q}$. The vector of tangential forces is split into two directions of the tangential plane: $\boldsymbol{\lambda}_{Ti} = (\lambda_{Tix}, \lambda_{Tiy})^T$. These forces can result from passive (sticking–) or active (sliding–) forces. The active tangential friction forces in the sliding case are uniquely defined by the Coulomb friction law (coefficient of friction: μ_i, relative tangential velocity: $\dot{\boldsymbol{g}}_{Ti} = (\dot{g}_{Tix}, \dot{g}_{Tiy})^T$):

$$\boldsymbol{\lambda}_{Ti} = -\mu_i \cdot \dot{\boldsymbol{g}}_{Ti} / |\dot{\boldsymbol{g}}_{Ti}| \cdot \lambda_{Ni} \tag{2}$$

The passive normal contact forces and the passive tangential friction forces for the sticking case are defined by secondary conditions, that are formulated on an acceleration level.

Relative kinematics

The outline of the contact surfaces is modeled by flat planes, allowing the approximation of any geometric object. Potential point–contacts are therefore between *point–plane* and *line–line*. Contacts along a line and planar contacts are composed by single point–contacts. A disappearing distance in normal direction g_{Ni} indicates a new contact point: $g_{Ni}(\boldsymbol{q},t) = 0$. This normal direction is rectangular to both contacting surfaces. The two independent tangential directions are lying in the tangential plane, rectangular to the normal direction. For a contact point i we can write the normal and tangential accelerations in local contact coordinates $\boldsymbol{g}$ as: $\ddot{g}_{Ni} = \boldsymbol{w}_{Ni}^T\ddot{\boldsymbol{q}} + \overline{w}_{Ni}$, $\ddot{\boldsymbol{g}}_{Ti} = \boldsymbol{W}_{Ti}^T\ddot{\boldsymbol{q}} + \overline{\boldsymbol{w}}_{Ti}$ ($\boldsymbol{w}_{Ni}$ and $\boldsymbol{W}_{Ti}$ are submatrices of $\boldsymbol{W}_N$ and $\boldsymbol{W}_T$, $\overline{w}_{Ni}$ and $\overline{\boldsymbol{w}}_{Ti}$ represent relative magnitudes independent from $\ddot{\boldsymbol{q}}$, [2, 4]).

Contact laws

In an active contact we distinguish the impenetrability condition in normal direction and the Coulomb Friction law in tangential direction. Due to the structure varying nature of the problem, with a formulation of the constraint forces as secondary conditions together with the equations of motion, a formulation of the contact laws on the acceleration level is necessary (see eq. 7).

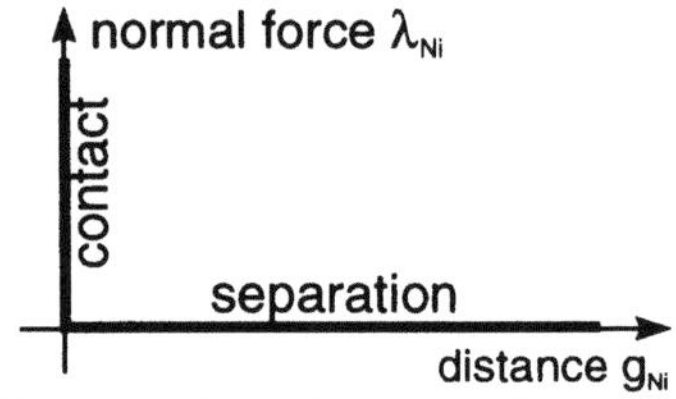

Figure 3: Contact Law in Normal Direction

Normal direction: Assuming rigid bodies, figure 3 shows the characteristic properties of a normal contact. The quantities g_{Ni} and λ_{Ni} describe the state of the constraint. The behavior is complementary, only two states are possible: either contact is maintained with a force $\lambda_{Ni} \geq 0$ and a disappearing distance $g_{Ni} = 0$ or the contact is separated with $\lambda_{Ni} = 0$ and $g_{Ni} \geq 0$. To describe the transitions from contact to separation with the relative accelerations we assume an active contact with $g_{Ni} = \dot{g}_{Ni} = 0$. The contact keeps active if: $\ddot{g}_{Ni} = 0 \wedge \lambda_{Ni} \geq 0$. $\ddot{g}_{Ni} \geq 0 \wedge \lambda_{Ni} = 0$ indicates a separation. Eq. 3 summarizes the complementary behavior of these conditions for multiple contacts. An equivalent formulation of these complementary conditions represents the variational inequality 4, with the convex set $C_N = \{\boldsymbol{\lambda}_N^* : \boldsymbol{\lambda}_N^* \geq \mathbf{0}\}$, containing all admissible normal forces λ_{Ni}^*, [4, 5]:

$$\ddot{\boldsymbol{g}}_N \geq 0 \,; \quad \boldsymbol{\lambda}_N \geq 0 \,; \quad \ddot{\boldsymbol{g}}_N^T \boldsymbol{\lambda}_N = 0 \tag{3}$$

$$-\ddot{\boldsymbol{g}}_N^T (\boldsymbol{\lambda}_N^* - \boldsymbol{\lambda}_N) \leq 0 \,; \; \boldsymbol{\lambda}_N \in C_N \,; \; \forall \, \boldsymbol{\lambda}_N^* \in C_N \tag{4}$$

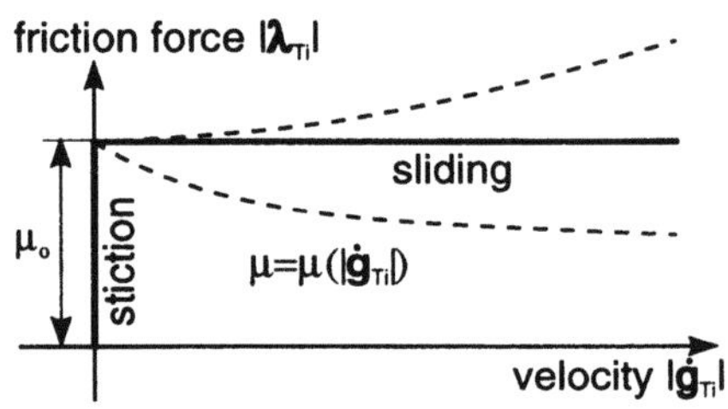

Figure 4: Coulomb Friction Law in Tangential Direction

Tangential direction: Figure 4 shows a typical Coulomb Friction characteristic. In the sliding case ($|\dot{\boldsymbol{g}}_{Ti}| \neq 0$), the friction force is proportional to the normal force. The sliding friction has the opposite direction of the relative sliding velocity. A velocity dependent friction coefficient ($\mu_i = \mu_i(|\dot{\boldsymbol{g}}_{Ti}|)$) can be considered (figure 4, dotted lines). In the case of sticktion ($|\dot{\boldsymbol{g}}_{Ti}| = 0$) the friction force lies in the interval between zero and the maximum $\mu\lambda_{Ni}$, and can take any direction in the tangential plane:

$$\begin{array}{llll} \text{sticktion:} & |\boldsymbol{\lambda}_{Ti}| < \mu_i \lambda_{Ni} & \Rightarrow & |\dot{\boldsymbol{g}}_{Ti}| = 0 \\ \text{sliding:} & |\boldsymbol{\lambda}_{Ti}| = \mu_i \lambda_{Ni} & \Rightarrow & \dot{\boldsymbol{g}}_{Ti} = -\rho \boldsymbol{\lambda}_{Ti} \,; \; \rho > 0 \end{array} \tag{5}$$

In order to formulate the transitions from sticktion to sliding, the conditions from equation 5 have to be formulated with relative accelerations $\ddot{\boldsymbol{g}}_{Ti}$ instead of the relative velocities $\dot{\boldsymbol{g}}_{Ti}$. If we assume a relative velocity $\dot{\boldsymbol{g}}_{Ti} = 0$, this transformation is admissible, because in the instant of transition from sticktion to sliding $\ddot{\boldsymbol{g}}_{Ti}$ and $\dot{\boldsymbol{g}}_{Ti}$ have the same direction. This behavior again can be expressed by the variational inequality 6. $C_{Ti}(\lambda_{Ni}) = \{\boldsymbol{\lambda}_{Ti}^* : \boldsymbol{\lambda}_{Ti}^* \leq \mu_i \lambda_{Ni}\}$ is the convex set, containing all friction forces λ_{Ti}^* that fulfill Coulomb's friction law.

$$-\ddot{\boldsymbol{g}}_{Ti}^T(\boldsymbol{\lambda}_{Ti}^* - \boldsymbol{\lambda}_{Ti}) \leq 0\,;\; \boldsymbol{\lambda}_{Ti} \in C_{Ti}\,;\; \forall\, \boldsymbol{\lambda}_{Ti}^* \in C_{Ti} \tag{6}$$

The equations of motion with secondary conditions

The complete formulation for the problem consists now of the equations of motion (eq. 1), the relative kinematics, and the contact mechanics (eq. 2 for the sliding contacts, eq. 4 for the normal constraints and eq. 6 for the tangential constraints during sticktion). [3] and [5] developed a transformation of the variational inequalities 4 and 6 into a set of nonlinear equations, that can be solved numerically. The detailed derivation of this transformation is described in [5] or [6]. The complete resulting problem is given by equation 7:

$$\begin{aligned}
&\boldsymbol{M}\ddot{\boldsymbol{q}} - \boldsymbol{h} - (\boldsymbol{W}_N + \boldsymbol{H}_R;\quad \boldsymbol{W}_T)\begin{pmatrix} \boldsymbol{\lambda}_N \\ \boldsymbol{\lambda}_T \end{pmatrix} = \boldsymbol{0} \\
&\ddot{g}_{Ni} = \boldsymbol{w}_{Ni}^T\ddot{\boldsymbol{q}} + \overline{w}_{Ni};\quad \ddot{\boldsymbol{g}}_{Ti} = \boldsymbol{W}_{Ti}^T\ddot{\boldsymbol{q}} + \overline{\boldsymbol{w}}_{Ti} \\
&\lambda_{N,i} = (\lambda_{N,i} - r\ddot{g}_{N,i})^+ = \tau_{N,i}^+ \\
&\boldsymbol{\lambda}_{T,i} = \begin{cases} \boldsymbol{\tau}_{T,i} & \text{if} \quad |\boldsymbol{\tau}_{T,i}| < \mu_i \tau_{N,i}^+ \\ \mu_i \tau_{N,i}^+ \frac{\boldsymbol{\tau}_{T,i}}{|\boldsymbol{\tau}_{T,i}|} & \text{if} \quad |\boldsymbol{\tau}_{T,i}| \geq \mu_i \tau_{N,i}^+ \end{cases} \qquad \text{with: } \boldsymbol{\tau}_{T,i} = (\boldsymbol{\lambda}_{T,i} - r\ddot{\boldsymbol{g}}_{T,i})
\end{aligned} \tag{7}$$

The matrix $\boldsymbol{H}_R$ contains the tangential friction forces of the sliding contacts (see eq. 2). $r > 0$ is an arbitrary factor, that is needed for the numerical evaluation, to weight the accelerations and the forces. The operator "+" denotes the positive vector part: $x_i^+ = 0$ if $x_i \leq 0$ and $x_i^+ = x_i$ if $x_i > 0$. The solution of the equations 7 is based on a special iterative algorithm. With the resulting accelerations $\ddot{\boldsymbol{q}}$ a numerical integration can be performed. If the state of the system changes, for example if a transition from sliding to sticktion occurs, the resulting constraint forces and the relative accelerations indicate a consistent transition. The secondary conditions and the term $\boldsymbol{W\lambda}$ have to be adapted.

Multiple impacts with friction

If two bodies approach with a negative relative velocity ($\dot{g}_{Ni} < 0$), an impact occurs, causing an impulse transfer in the normal direction, and due to friction in the contact also a transfer in tangential direction. As this impact interacts with other active unilateral constraints in the system, a formulation capable of managing multiple impacts with friction is needed. The present model is based on an impact theory by [4], originally developed for two dimensional models, resulting in a linear complementary problem. This chapter describes its application for three dimensional problems, with two degrees of freedom in the tangential plane of the contact.

Time integration of the equations of motion over the impact interval results in an idealized, time discrete impulse equation. Due to Poisson's impact law, we divide the impact into two phases: the phase of compression and the phase of expansion. The indices A, C and E denote the beginning of the compression, the end of the compression and the end of the expansion.

$$\lim_{t_1 \to t_2} \int_{t_1}^{t_2} [M\ddot{q} - h - W\lambda]\, dt = 0 \quad \Rightarrow \quad \begin{aligned} M(\dot{q}_C - \dot{q}_A) - W\Lambda_C &= 0 \\ M(\dot{q}_E - \dot{q}_C) - W\Lambda_E &= 0 \end{aligned} \tag{8}$$

$\dot{q}$ denotes the generalized velocities, the impulse $\Lambda = \lim_{t_1 \to t_2} \int_{t_1}^{t_2} \lambda \, dt$ results from the integrated constraint force λ. The impulses Λ_C are transfered in normal and tangential direction during the compression phase, Λ_E during the expansion. The end of the compression phase is defined by $\dot{g}_{NCi} = 0$, with a normal impulse $\Lambda_{NCi} \geq 0$. To enable separation at the end of compression –a possible event in multiple contact systems– we allow $\dot{g}_{NCi} \geq 0$, with the condition of no transferred impulse ($\Lambda_{NCi} = 0$). Written as a complementary condition, the equation results in:

$$\Lambda_{NCi} \geq 0\,, \quad \dot{g}_{NCi} \geq 0\,, \quad \dot{g}_{NCi} \cdot \Lambda_{NCi} = 0\,. \tag{9}$$

During the expansion, the stored impulses in normal direction Λ_{NCi} are partially recovered: $\Lambda_{NEi} = \varepsilon_i \Lambda_{NCi}$, with a normal velocity $\dot{g}_{NEi} \geq 0$, ε denotes the coefficient of restitution. Due to multiple contact effects, we demand $\Lambda_{NEi} \geq \varepsilon_i \Lambda_{NCi}$, to avoid a negative normal velocity at the end of the compression. In this case $\dot{g}_{NEi} = 0$ is required:

$$\dot{g}_{NEi} \geq 0\,, \quad (\Lambda_{NEi} - \varepsilon_i \Lambda_{NCi}) \geq 0\,, \quad \dot{g}_{NEi} \cdot (\Lambda_{NEi} - \varepsilon_i \Lambda_{NCi}) = 0 \tag{10}$$

In tangential direction the Coulomb friction law is considered, stored tangential impulses are neglected. We use the same relation as in equation 5, but formulated with impulses instead of forces. This friction law has to be applied to the compression and the expansion phase:

$$\begin{aligned} &\text{sticktion:} \quad |\Lambda_{T,i}| < \mu_i \Lambda_{N,i} \quad \Rightarrow \quad |\dot{g}_{T,i}| = 0 \\ &\text{sliding:} \quad\;\; |\Lambda_{T,i}| = \mu_i \Lambda_{N,i} \quad \Rightarrow \quad \dot{g}_{T,i} = -\rho \Lambda_{T,i}\,;\ \rho \geq 0 \end{aligned} \tag{11}$$

This formulation for the tangential friction effects is an approximation of the Coulomb friction law, as transitions from sticktion to sliding or sliding to sticktion during one impact phase are neglected. The state of the contact configuration is kept constant during an impact phase [2, 4]. In addition, the application of this impact law to spatial contacts neglects a variation of the sliding direction of the relative tangential velocity during an impact phase. Nevertheless, the friction phenomena in multiple impacts can be handled by the presented equations, but they should be regarded as an independent tangential impact law, that often coincides with the Coulomb friction law and considers the main physical effects. The numerical handling of these equations is identical with the methods used for the continuous contacts (see equation 7).

3 Experimental setup and measurements

For the verification of the developed model of the transportation process an experimental vibratory feeder was built, allowing different measurements concerning the impact model and the average transportation rates. Figure 5 shows the principle of the used device. The track, fixed on leaf springs is exited with an electro magnetic shaker with a frequency about 50 Hz. The eigenfrequency of the system is at 52 Hz. The resulting vibration amplitude reaches a maximum value of about 2 mm. The track has an inclination angle $\alpha = 3^0$, the angle between the track and the direction of the vibration is $\beta = 15^0$.

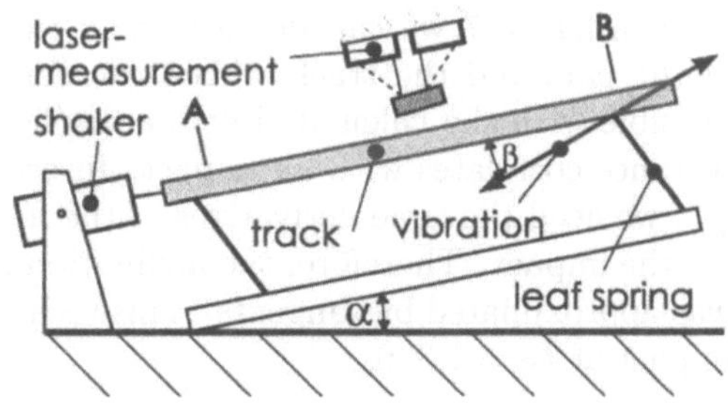

Figure 5: Principle of the Experimental Vibratory Feeder

Verification of the impact model

For the accurate contactless measurement of the motion of the transported part six laser distance sensors were applied. For the vibration measurement of the track an eddy current sensor is used. The arrangement of these sensors and the geometry of the part and the track is depicted in figure 6. The laser sensors measure the distance along the emitted beam (continuous line, figure 6) at the points indicated with "⊗" on the surfaces of the part. Due to the measurement range of 10 mm of the laser sensors, we can observe the parts movement in a small range along the transportation direction. The solution of a set of nonlinear equations enables the reconstruction of the six coordinates of position and orientation of the transported part. Together with the position measurement of the vibrating track, all the normal distances of the potential contact points between the track and the part can be computed. Disappearing distances indicate an impact between two touching bodies. Figure 7 shows the results of an exemplary measurement inside the maximum range of the sensors. The direct measurement of the distances along the laser beam (1) is little meaningful, but the calculation of the coordinates (2,3) and the distances (4) of the potential contact points show very good the unsteady changes of the motion due to the impacts between the part and the vibrating track.

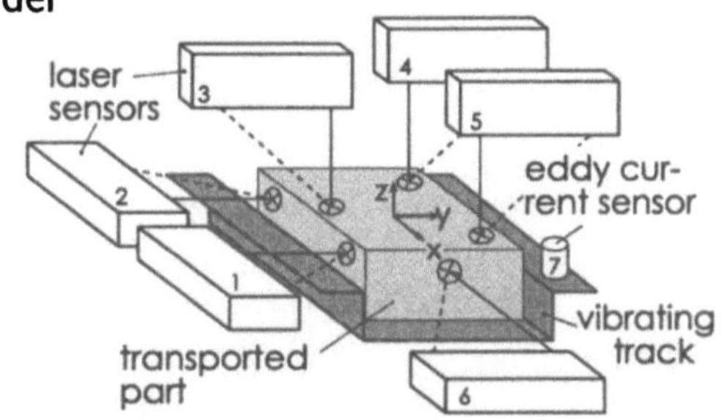

Figure 6: The Track and the Transported Part with the Arrangement of six Laser Sensors and an Eddy Current Sensor

The equations 8 to 11 establish an impact law, formulated with velocities and impulses. It maps the generalized velocities before the impact $\dot{\boldsymbol{q}}_A$ to the velocities after the impact $\dot{\boldsymbol{q}}_E$. The coordinates $\boldsymbol{q}$ remain constant. Consequently, the measurement of $\dot{\boldsymbol{q}}_A$ and $\dot{\boldsymbol{q}}_E$ enables a verification of the theory. In addition, we have to know the position $\boldsymbol{q}$ during the impact and the location of the impact on the surfaces of the part and the track. From the measured data in figure 7 we can deduce easily the vector $\boldsymbol{q}$, indicating the position of the part and the track. Also, the detection of the contacting surface is possible with the calculated distances (figure 7,(4)), because a disappearing distance correlates with an impact. In order to obtain the velocities $\dot{\boldsymbol{q}}_A$ and $\dot{\boldsymbol{q}}_E$, we need the time derivatives of the measured motion $\boldsymbol{q}(t)$ before and after the impact. Therefore, the inclination of $\boldsymbol{q}(t)$ before and after the impact was approximated by tangents. This technique is indicated in figure 7 for the impact at $t_S = 1.336$.

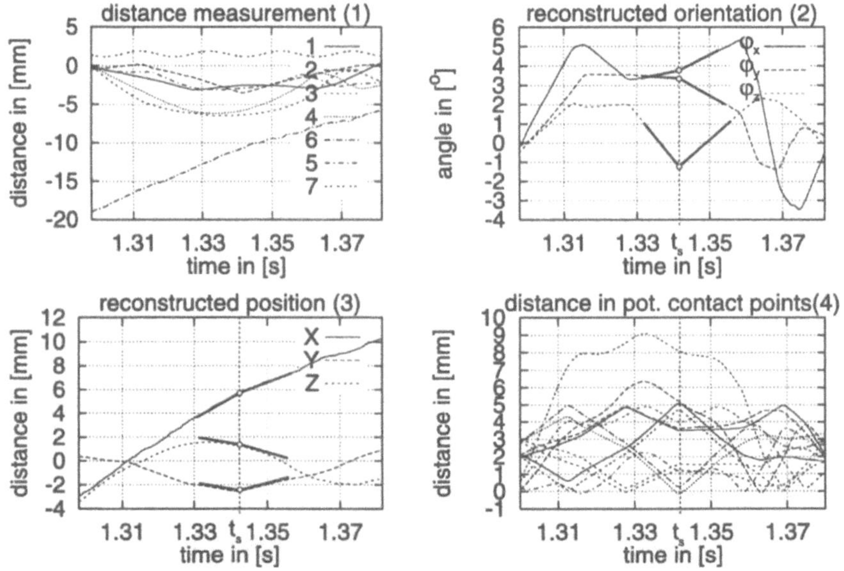

Figure 7: Exemplary Motion Measurement of the Transported Part

The theoretical model of the impact is based on the coefficient of restitution ε and the coefficient of friction μ. These values μ and ε are unknown for the applied materials (part: PVC, track: aluminium). Using an optimization we can identify these magnitudes in order to fit every measured impact optimally (least square method). Therefore, we define the criterion f, to evaluate the deviation between the measured and the calculated results $\dot{\boldsymbol{q}}_{E,meas}$ and $\dot{\boldsymbol{q}}_{E,calc}$.

$$f = \sum_{i=1}^{6} \left[\frac{|\dot{q}_{E,calc,i} - \dot{q}_{E,meas,i}|}{(|\dot{q}_{A,meas,i}| + |\dot{q}_{E,meas,i}|)/2} \right]^2 \tag{12}$$

With the impact model we map the values of $\dot{\boldsymbol{q}}_{A,meas}$ on $\dot{\boldsymbol{q}}_{E,calc}$. The function f is the sum of a quadratic normalized error of the components $|\dot{q}_{E,calc,i} - \dot{q}_{E,meas,i}|$. The target of the optimization is the detection of ε- and μ-values minimizing f.

We performed the described method for 35 different measured single impacts and obtained 35 different couples for ε and μ. The values of these ε are in a range form 0.7 to 1.0, the values of μ are between 0.07 to 0.48. Consequently, the second step is the derivation of average values for the two coefficients, fitting all the measured impacts with a minimal error. Therefore we define the criterion $F = \sum_{k=1}^{35} f_k^2$, summarizing the quadratic errors f_k^2 (equation 12) of the 35 impacts. An optimal μ-ε-couple minimizes the criterion F. These average values are $\bar{\mu} = 0.24$ and $\bar{\varepsilon} = 0.85$.

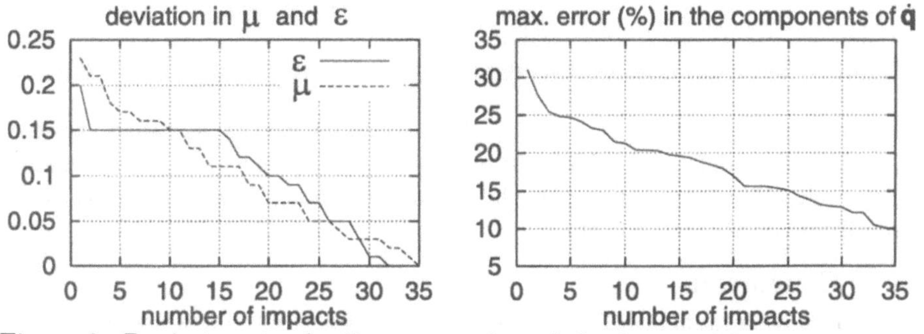

Figure 8: Deviations in the Reconstruction of the Impact Measurement by the Impact Model

The discrepancy of these values to the individual ones and the deviation of the measured and the calculated values of $\dot{q}_E$ are two criteria for the evaluation of the impact model. Figure 8 shows the corresponding results. On the left the deviation of the individual μ and ε values form the average $\bar{\mu}$ and $\bar{\varepsilon}$ is depicted. The diagram shows the minimal deviation value versus the number of impacts with this value. On the right, figure 8 shows the maximum discrepancy between measurement and theory of the six velocity coordinates $\dot{q}_E$ of every single impact, the calculation of the impacts was based on $\bar{\mu}$ and $\bar{\varepsilon}$. The main source of error in the measurements is the accuracy of the distance measurement (sampling rate: 1 kHz) and the time derivation of the position signal. Each of these effects implicates an error of 3 – 4% in the velocity information, used for the reconstruction of the impacts. Consequently, an error of about 8% is caused by the measurement, the rest (figure 8, right) is originated by the impact model.

Verification of the transportation rate

The measurement of the average transportation velocity is a possibility to verify the whole mechanical model describing the feeding process. Therefore, we used two light barriers (figure 5, "A" and "B") to detect the duration of the transportation process over a distance of 350 mm. These measurements allow a derivation of an average transportation rate. In order to guarantee stable values, sufficient measurements were performed. Figure 9 shows these average values versus the vibration amplitude of the track. The crosses "+" at every measured amplitude represent the results of multiple measurements. Negative transportation rates correlate with a transportation process from B to A, a positive sign indicates a direction from A to B. The according simulation results are also depicted in figure 9. They were obtained using the

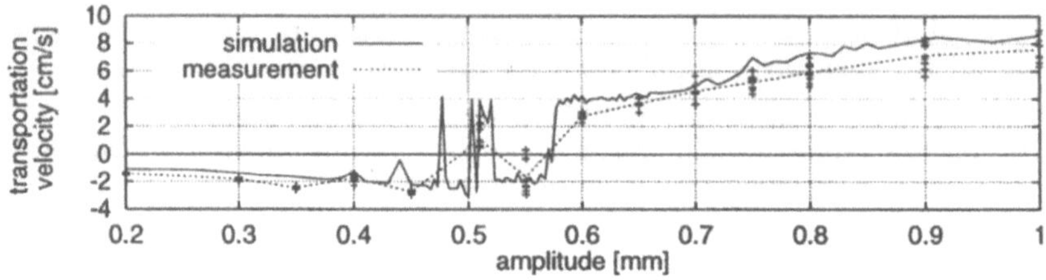

Figure 9: Simulation and Measurement of the Average Transportation Rate

coefficients $\varepsilon = 0.84$ and $\mu = 0.23$, which correlate with the identified values of the impact model. The calculated transportation rates fit the measurements very well. They even show the break in the transportation process at an amplitude of about 0.55 mm.

4 Conclusions

The presented model of the transportation process in a vibratory feeder considers the spatial geometry of the surfaces and the main physical effects of the unilateral contacts. The formulation is capable to deal with multiple parts and multiple dependent contacts. The verification of the transportation rates with the experimentally identified impact coefficients ε and μ demonstrates the quality of the proposed contact and impact model.

Acknowledgement

The research work presented in this paper is partially supported by a contract with the DFG (Deutsche Forschungsgemeinschaft, SFB 336).

References

1 Boothroyd G, Poli C R, Murch L E, Handbook for Feeding and Orienting Techniques for Small Parts, Department of Mechanical Engineering, University of Massachusetts, Amherst, MA, 1976.

2 Glocker Ch, Dynamik von Starrkörpersystemen mit Reibung und Stößen. Fortschr.-Ber. VDI, Reihe 18, Nr. 182, Düsseldorf: VDI-Verlag 1995, Dissertation, TU-München.

3 Klarbring A, Mathematical Programming and Augmented Lagrangian Methods for Frictional Contact Problems, in: Proceedings Contact Mechanics International Symposium, A. Curnier, ed., PPUR, Lausanne, Switzerland, October 7-9, 1992, S. 409-422.

4 Pfeiffer F, Glocker Ch, Multibody Dynamics with Unilateral Contacts. John Wiley & Sons Inc., 1996.

5 Wösle M, Pfeiffer F, Dynamics of Multibody Systems Containing Dependent Unilateral Constraints with Friction, in Journal of Vibration and Control 2: 161-192, 1996.

6 Wolfsteiner P, Pfeiffer F, Dynamics of a Vibratory Feeder, In: Procedings of the 1996 ASME 16th Biennal Conference on Vibration and Noise, Sacramento, California, 1997.

The effectiveness of the impact damper with granular material

I Yokomichi Y Araki** M Aisaka† and H Kusano**

The multiple -particle impact damper consists of a bed of granular materials moving in a container fixed to the resonantly vibrating system. This type of impact damper substitutes for the solid impact damper that causes excessive noise in its operation, and deteriorates the damping performance under gravity. The problem is to determine the effectiveness of the impact damper with granular materials as a damper mass, for reducing the vibration of a single degree-of-freedom system to a prescribed value when the excitation is simple harmonic. In the analysis, the particle bed is assumed to be a mass which moves uni-directionally in a frictionless container and collides plastically with the ends of container. The solutions for some possible types of the steady state impact motions, and frequency response are obtained. This study deals with the two different cases, i.e., the dampers are applied to a vertical and a horizontal vibrating systems. Experimental models were tested to verify the analysis, and further to modify them for the case of the horizontal system.

1 Introduction

The impact damper is a simple yet effective vibration absorber which consists of a solid particle constrained to move in a container attached to the primary vibrating body. Many authors have analyzed its damping characteristics by theory, experiment, and numerical simulation [1-3]. Partially filling the container of an impact damper with small mass elements(e.g., sand or steel shot) results in the multiple-particle impact damper. Although this type of damper has been successfully used to minimize the Space Shuttle Main Engine vibration[4], virtually no analytical study of this subject is available.

It is the intention of this study to develop a lumped-mass approach capable of predicting the motion of the impact damper with granular material, taking account of the pure plastic impact of granular material, and present an exact solution for the periodic motion of the system. The investigation is applied for the two cases. In the first case, the impact damper is applied to the vertical system under gravity, and the second to the horizontal system where the particle bed tends to segregate due to gravity in a container. We conduct theoretical analyses for the periodic motions of the damper system with the colliding motion of particle bed against the container ends (bottom and top, or left and right). It is shown that the theory predicts the experiments, and the impact damper with properly selected mass ratios and with clearance sizes effectively suppresses the resonant peaks within the range of frequency.

* Dept of Mechanical Engineering, Kitakyushu College of Technology, Kitakyushu, Japan
† Dept of Research & Experiment, Isuzu Motor Co. Kanagawa, Japan
** Dept of Control Engineering , Kyushu Institute of Technology, Kitakyushu, Japan

2 Analysis for the vertical system

2.1 Equations of motion The schematic model of a single-degree-of-freedom system equipped with the impact damper with granular materials is shown in Fig. 1. The impact damper comprises a particle bed of total mass m_d constrained to move vertically, with clearance d^*, in the container attached to a primary vibrating body of mass M. Considering some of the characteristics of a granular material in a oscillating container, there may be frictional interaction between particles and container as well as the particles themselves. A strict analytical approach prefers the description of each particle motion. We would expect, however, the granular material to move as a slug if the external friction were less than the internal friction. For this situation, it is possible to approximate the particle bed to a lumped mass with a zero coefficient of restitution, because all impact energy is dissipated by local particle motions.

The possible state of motion can then be divided into the two motion intervals. For the relative rest intervals, the particle bed moves together with the end walls, while for the separate intervals the particle bed moves freely in the clearance. The equations of motion for each interval are given as follows:

for the relative rest intervals(on bottom wall: $i = 1$, on top wall: $i = 2$)

$$(M + m_d)\ddot{x}^*_{ci} + c\dot{x}^*_{ci} + kx^*_{ci} = F\sin(\omega t - \varphi) - m_d g(primary\ \ mass) \quad (1)$$
$$x^*_1 = x^*_2 = x^*_{ci}(damper\ \ mass)(i = 1, 2)$$

and for the separate intervals(upward flight : $i = 1$, downward flight: $i = 2$)

$$M\ddot{x}^*_{1i} + c\dot{x}^*_{1i} + kx^*_{1i} = F\sin(\omega t - \varphi)(primary\ \ mass) \quad (2)$$
$$m_d\ddot{x}^*_{2i} = -m_d g(damper\ \ mass)(i = 1, 2),$$

where M, m_d, c, k are mass of primary body, that of damper, damping, stiffness, x_{ci} denotes displacement for the relative rest interval, x_{1i}, x_{2i} are those for separate intervals, t = time, ω = exciting frequency, φ = phase between exciting force and impact at bottom, F = exciting force, and g = acceleration due to gravity.

Introducing non-dimensional motion, equations (1) and (2) take the form

$$x''_{ci} + (2\varsigma_c/r_c)x'_{ci} + (1/r_c^2)x_{ci}$$
$$= (1/r_c^2)\sin(\tau - \varphi) - K_R\mu/(1+\mu) \quad (3)$$
$$x_1 = x_2 = x_{ci}(i = 1,2)$$
$$x''_{1i} + 2(\varsigma/r)x'_{1i} + (1/r^2)x_{1i}$$
$$= (1/r^2)\sin(\tau - \phi) \quad (4)$$
$$x''_{2i} = -K_R(i = 1,2)$$

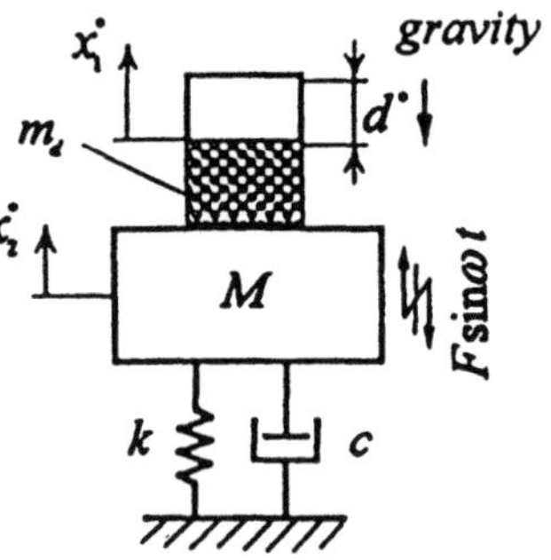

Fig. 1 Model of the impact damper with a bed of granular material

where a prime indicates differentiation with $\tau(=\omega t)$, the displacements

$x_1 = x_1^{\cdot}/\delta_{st}, x = x_2^{\cdot}/\delta_{st}, x_c = x_c^{\cdot}/\delta_{st}$ are the non-dimensional ones, and the parameters are

$$\varsigma_c = \varsigma/\sqrt{1+\mu}, r_c = r\sqrt{1+\mu}, d = d^{*}/\delta_{st}, \delta_{st} = F/k, \varsigma = c/c_c, r = \omega/\omega_n$$

$$\omega_n = \sqrt{k/M}, \mu = m_d/M, K_R = g/(\delta_{st}\omega^2).$$

Integrating Eqs. (3) and (4) yields the following general solutions:

$$x_{ci} = \exp(-\theta_c\tau)(A_{ci}\sin\alpha_c\tau + B_{ci}\cos\alpha_c\tau) + A_{pc}\sin(\tau-\varphi-\beta_c) - \gamma \ (i=1,2). \quad (5)$$

$$x_{1i} = \exp(-\theta\tau)(A_i\sin\alpha\tau + B_i\cos\alpha\tau) + A_p\sin(\tau-\varphi-\beta) \ (i=1,2) \quad (6)$$

$$x_{21} = -K_R(\tau-\tau_0)^2/2 + D_1(\tau-\tau_0) + E_1$$

$$x_{22} = -K_R(\tau-\tau_2)^2/2 + D_2(\tau-\tau_2) + E_2$$

where

$$\theta = \varsigma/r, \alpha = \sqrt{1-\varsigma^2}/r, \theta_c = \varsigma_c/r_c, \alpha_c = \sqrt{1-\varsigma_c^2}/r_c, \gamma = r^2\mu K_R$$

$$A_p = 1/\sqrt{(1-r^2)+2\varsigma r}, \beta = \tan^{-1}\{2\varsigma r/(1-r^2)\}$$

$$A_{pc} = 1/\sqrt{(1-r_c^2)+2\varsigma_c r_c}, \beta_c = \tan^{-1}\{2\varsigma_c r_c/(1-r_c^2)\}$$

and , τ_0 and τ_2 are, respectively, the times of separation of damper mass from bottom and top wall. The unknown constants $A_{ci}, B_{ci}, A_i, B_i, D_i, and\ E_i\ (i=1,2)$ in the solutions are determined from the initial conditions of each motion interval.

2.2 Steady state periodic motion As might be expected, there exist different types of motion in the impact vibration with particle bed, depending upon the parameters of system and excitation. Among the major impact vibrations that exist in the vertical damper system are the following:

[1]Type *I* motion. The damper mass leaves the bottom end at time τ_0 when the acceleration of the primary body reaches $-g$. The mass then hits top end at time τ_1, and keep contact with it until the acceleration reaches g at time τ_2

[2]Type *II* motion. Motions without sustained contact on top end; the damper mass rebounds just after impact at top end (Fig.2).

[3]Type *III* motion. Motions where the damper mass hits bottom end alone when the clearance is large enough damper mass not to hit top end.

It remains now to find the steady-state solutions to the motions of the main system subjected to the impact damper. The method for calculating the response is based on adjoining the successive motion intervals over one cycle of excitation by the conditions of periodicity. To illustrate the procedure , consider the periodic motion of Type *II* shown in Fig.2. Due to the nature of the periodic motion, the

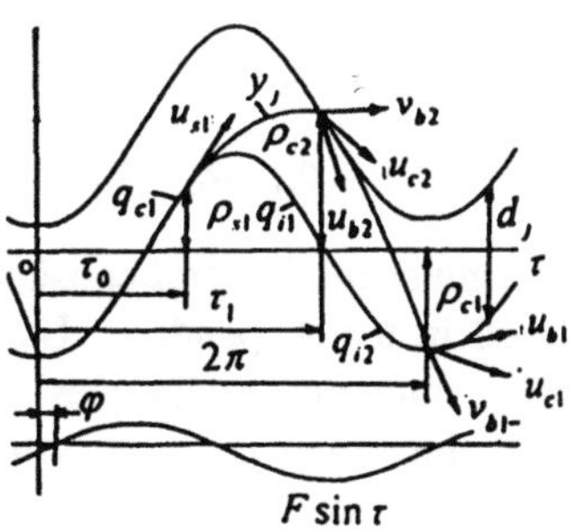

Fig. 2 Type *II* periodic motion

solutions in Eqs. (5) and (6) must satisfy the following boundary conditions:

At $\tau = 0$ (just after impact on bottom end):

$$x_{c1}(0) = \rho,\ x'_{c1}(0) = u_{c1},\ u_{c1} = (\mu v_{b1} + u_{b1})/(1+\mu) \tag{7}$$

At $\tau = \tau_0$ (separation from bottom end):

$$x_{c1}(\tau_0) = x_{11}(\tau_0) = x_{21}(\tau_0) = \rho_s,\ x'_{c1}(\tau_0) = x'_{11}(\tau_0) = x'_{21}(\tau_0) = u_{sd},\ x''_{c1} = -K_R \tag{8}$$

At $\tau = \tau_1$ (impact at top end):

$$x_{11}(\tau_1) = x_{12}(\tau_1) = \rho_{c2},\ x_{21}(\tau_1) = x_{22}(\tau_1) = \rho_{c2} + d,\ x'_{11}(\tau_1) = u_{b2},\ x'_{21}(\tau_1) = v_{b2},$$
$$x'_{12}(\tau_1) = x'_{22}(\tau_1) = u_{c2},\ u_{c2} = (\mu v_{b2} + u_{b2})/(1+\mu) \tag{9}$$

At $\tau = 2\pi$ (just before impact on bottom end):

$$x_{12}(2\pi) = x_{22}(2\pi) = \rho_{c1},\ x'_{12}(2\pi) = u_{b1},\ x'_{22}(2\pi) = v_{b1} \tag{10}$$

Substituting Eqs. (5) and (6) into the relations (7)-(10) leads to a set of three coupled nonlinear algebraic equations concerning $\tau_0, \tau_1, and\ \varphi$ in the form

$$\alpha_c A_{c1} - \theta_c B_{c1} - R_1 A_2 + R_2 B_2 - \{\mu A_p \cos(\tau_1 - \varphi - \beta) + A_p \cos(\varphi + \beta)\}/(1+\mu)$$
$$+ A_{pc} \cos(\varphi + \beta_c) + K_R \mu (2\pi - \tau_1)/(1+\mu) = 0 \tag{11}$$

$$R_3 A_1 - R_4 B_1 + a_{45} A_2 + a_{46} B_2 + (\tau_1 - \tau_0) A_p \cos(\tau_0 - \varphi - \beta) + A_p \sin(\tau_0 - \varphi - \beta)$$
$$- A_p \sin(\tau_1 - \varphi - \beta) - K_R (\tau_1 - \tau_0)^2 / 2 - d = 0 \tag{12}$$

$$R_5 A_2 - B_{c1} - R_6 B_2 + (2\pi - \tau_1) A_p \cos(\tau_1 - \varphi - \beta) + A_p \sin(\tau_1 - \varphi - \beta)$$
$$+ A_{pc} \sin(\varphi + \beta_c) - K_R (2\pi - \tau_1)^2 / 2 + d + \gamma = 0 \tag{13}$$

together with the linear algebraic equation

$$\begin{bmatrix} 0 & a_{12} & 0 & 0 & a_{15} & a_{16} \\ a_{21} & a_{22} & a_{23} & a_{24} & 0 & 0 \\ a_{31} & a_{32} & a_{33} & a_{34} & 0 & 0 \\ 0 & 0 & a_{43} & a_{44} & a_{45} & a_{46} \\ a_{51} & a_{52} & 0 & 0 & 0 & 0 \\ 0 & 0 & a_{63} & a_{64} & a_{65} & a_{66} \end{bmatrix} \begin{bmatrix} A_{c1} \\ B_{c1} \\ A_1 \\ B_1 \\ A_2 \\ B_2 \end{bmatrix} = \begin{bmatrix} b_1 \\ b_2 \\ b_3 \\ b_4 \\ b_5 \\ b_6 \end{bmatrix} \tag{14}$$

that determines the integral constants in Eqs. (5) and (6). The remaining terms are as defined in the Appendix. Using the values of the specified parameters and employing numerical techniques for the solution of nonlinear algebraic Equations (11)-(14), the values of $\tau_0, \tau_1, and\ \varphi$ can be determined. Then, the displacement of the main mass M can be found by evaluating $x_1(\tau), x_2(\tau), and\ x_c(\tau)$, respectively, for each motion interval over one cycle of the excitation.

2.3 Theoretical results It is now possible to construct the resonance curves of the system, by calculating the displacement of the main vibrating body for frequency ration r. Magnification factor X_m / δ_{st} is used to represent the amplitude. The computation has been carried out using the values: $\varsigma = 0.0065$, $\mu = 0.06$.

An example of the resonance curve is shown in Fig.3 for a mass ratio $\mu = 6\%$ and clearance $d = 20$, together with two kinds of amplitude curves for the sake of comparison, one for the absence of the impact damper (i.e., without damper), the other for the main mass equal to $M + m_d$ (i.e., m_d -stuck). It can be seen that periodic motions prevail over a wide range of resonant frequencies, as opposed to the case: $e \neq 0$ [5][6], where stable motion of the vertical impact damper is restricted to a narrow band-width below resonance due to gravity. The type of the periodic motion can be distinguished by plots of τ_0, τ_1 *and* τ_2 on top of the graph. These plots show that Type *I* motion with a comparatively large amplitude exists near m_d -stuck resonance region, while type *II* motion reduces the primary resonant amplitude appreciably, and type *III* exists outside of the primary resonance regions, approaching m_d -stuck curve. Furthermore, to illustrate the effectiveness of the clearance for small mass ratio on the order of 5 percent, some results were obtained as shown in Fig.4. It is observed that the best amplitude reduction is achieved with a suitable choice of the clearance from $d = 30$ *to* 40, where type *II* motion alone exists within the resonance region, thus reducing the amplitude of type *III* motion with $d = \infty$ to about one fourth.

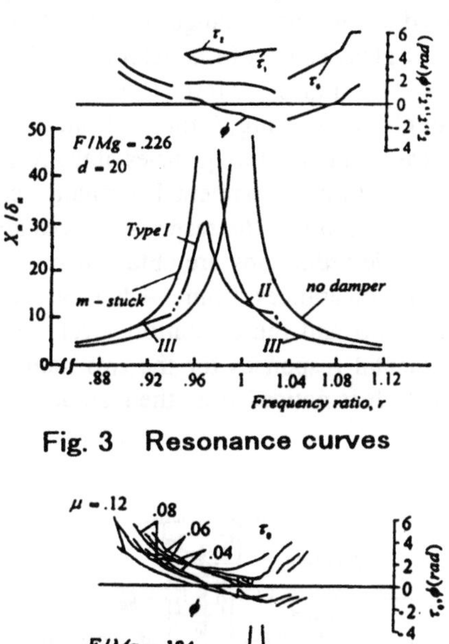

Fig. 3 Resonance curves

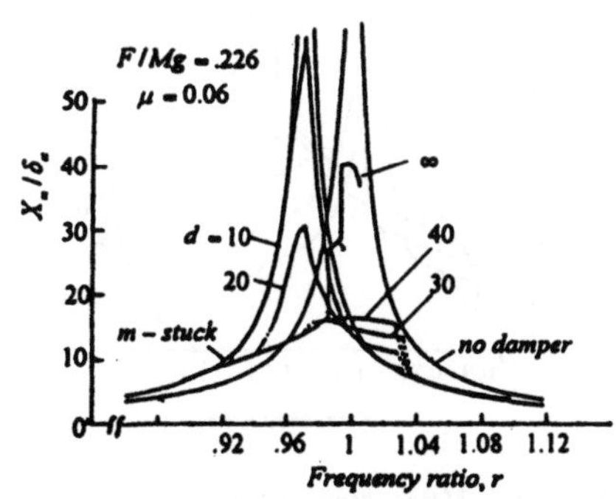

Fig. 4 Variation of response Amplitude with clearance

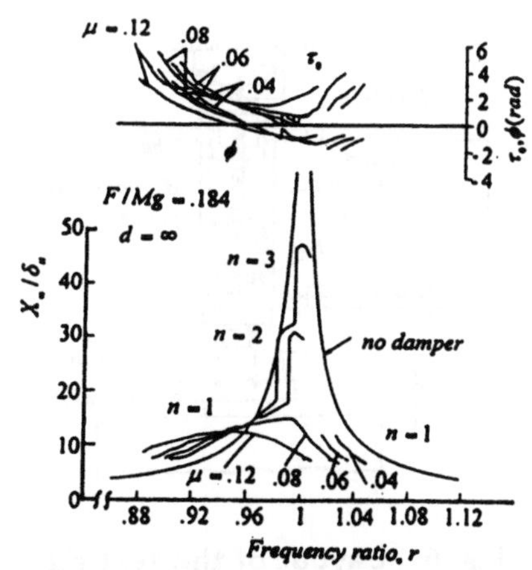

Fig. 5 Variation of response amplitude with mass ratio (*Type III*)

Fig.5 demonstrates the effect of mass ratio on the amplitude reduction for the case when the container clearance is large enough, that is, Type *III* motion with $n=1,2,$ and 3 prevails in the resonance region, where integer n indicates the multiple of the period of excitation to that of the damper motion.

The solutions obtained by Eqs.(11)-(14) are assumed to be stable, although the stability analysis of the solutions was not made due to the complexity of the system. However, the stability of the motion can be solved qualitatively by considering the kinematic picture of the perturbed motion where a damper mass moves together with vibrating body after impact [7]. The kinematic non-viable solutions were also found in calculation, but the viable solutions alone are presented.

2.4 Experiment and results The schematic view of the test rig is shown in Fig.6. The primary vibrating body consists of an acryl pipe of 40mm diameter with retaining flange ends, and it is supported by leaf springs. The lower flange constitutes the bottom end, while the top end is composed of a movable plate installed at the upper flange. Using a feed screw, the clearance can be adjusted. A bed of glass beads of diameter about 2mm is put in the container as an impacting body. The sinusoidal force F is fed to mass M by a sweep electro-dynamic exciter through a coil spring; F ranges from 0.5 to 2.0 N. Experiments were conducted for two cases of natural frequencies of the resonant structure: $\omega_n/2\pi = 7.4$ *and* 26.8 *Hz*.

In Fig.7 the experimental resonance curves are illustrated with d as a parameter. It can be seen that the response curves are affected by the clearance ratio, and high amplitude reductions are obtained under a reasonable clearance. In this case, the optimal clearances from $d \approx 40$ *to* 60 are obtained, although they depend on the mass ratio and the excitation level. The comparison between predicted and experimental responses for the two natural frequencies will be as shown in Fig.8a and 8b. It is noticed from the response curves shown in Fig. 8a, b that the increase in natural frequency results in good quantitative agreement between predictions and experiments.

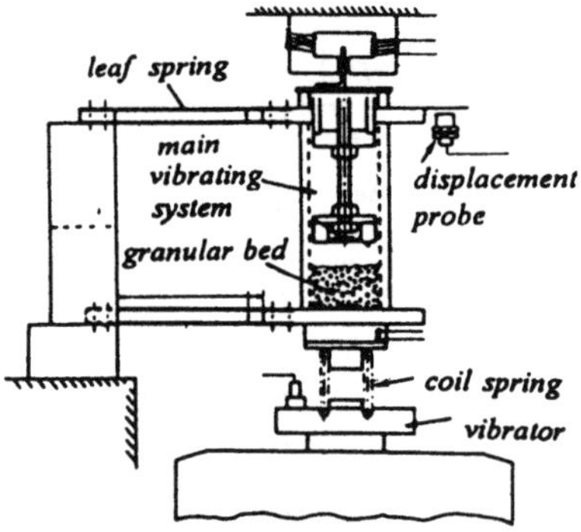

Fig. 6 Layout of the test rig

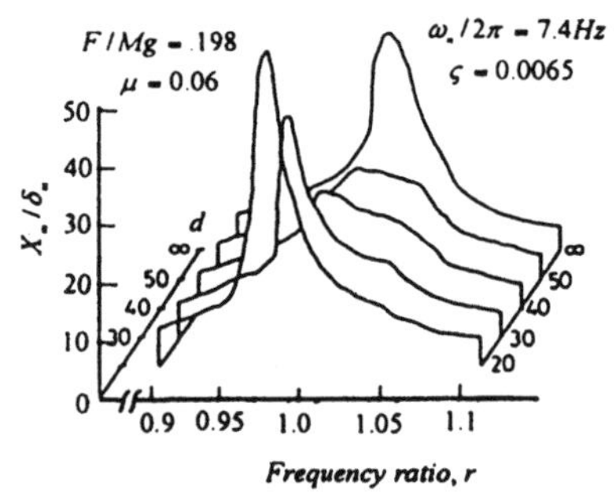

Fig. 7 Variation of response Amplitude with clearance (*Experimental*)

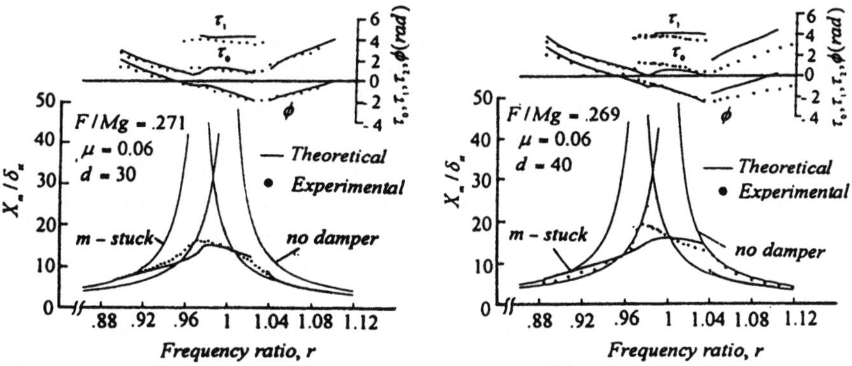

(a) $\omega_n/2\pi = 26.8Hz$ (b) $\omega_n/2\pi = 7.4Hz$

Fig. 8 Experimental and predicted resonance curves for different natural frequencies

The discrepancy noticeable at low natural frequency arises from the different situation dealt with both experimentally and theoretically. While the predicted curves has been computed with the damper mass assumed to be a single mass, the particle bed tends to disperse especially for the large clearance at low frequency, so that it will cause reductions in both effective clearance and impact force available for suppressing the motion of the main body.

3 Frequency response in horizontal system

3.1 Periodic motions The equations of motion for the impact damper applied to the horizontal vibrating system are obtained easily by letting the term of gravity $K_g = 0$ in Eqs.(3) and (4). The periodic solutions are to be calculated using the conditions of periodicity for the following two types of steady state motions.

Type *HI* motion causes two symmetric impacts per cycle of excitation and relative rests of the particle bed on the left and right ends of the container (Fig.9a). In Type *HII* motion, the particle bed moves apart at the instant just after collision, if the velocity of the main body decreases after impact (Fig.9b). By the similar procedure employed in the analysis for the vertical system, the above mentioned two types of periodic solutions for the horizontal system are obtained.

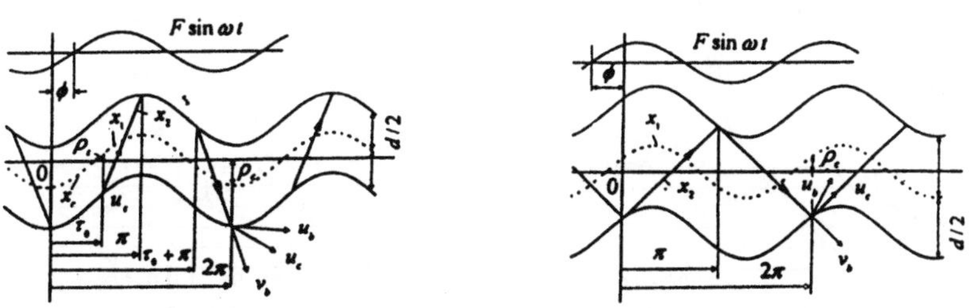

(a) *Type HI motion* (b) *Type HII motion*

Fig. 9 Symmetric two types of periodic motions for the horizontal system

3.2 Results of experiment and correction of parameter in prediction The comparison of the response between the experimental results and those by prediction with the specified mass ratio μ and clearance d is shown in Fig. 10. It is apparent that there are some differences between them for the large values of d and for the lower acceleration level. There was observed in experiments a considerable reduction in both impacting mass and the clearance, since the particle bed, in the horizontal system, tends to expand in the direction of stroke and stagnate, due to gravity, on the lower side wall of the container. On the basis of the experimental observation, to obtain better predictions, the damper parameters should be modified.

Let the expressions to modify mass ratio μ and clearance d be as follows:

$$\mu_c = \mu(1-\alpha d) \tag{15}$$

$$d_c = d(K+\beta_1 K_f - \beta_2\mu - \beta_3 d) \tag{16}$$

where $K_f = \delta_a\omega_n^2/g$. The modifying factors in relations (15) and (16) were found from the experimental results under conditions: $\omega_n/2\pi(=f_n) = 7Hz, K_f = 0.2, 0.3; \mu = 6, 8, 10\%;$ *and* $d = 20, 30, 40$ to be $\alpha = 0.004, K = 1.13, \beta_1 = 1, \beta_2 = 3.25,$ *and* $\beta_3 = 0.013.$

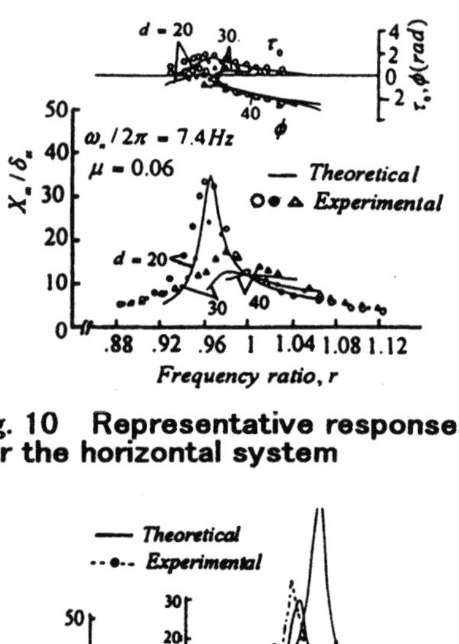

Fig. 10 Representative responses for the horizontal system

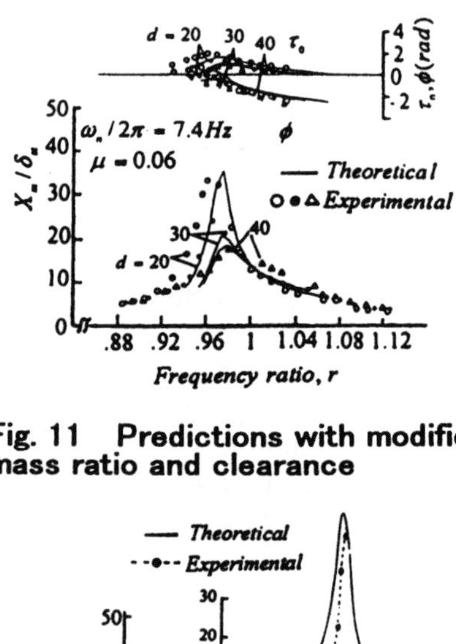

Fig. 11 Predictions with modified mass ratio and clearance

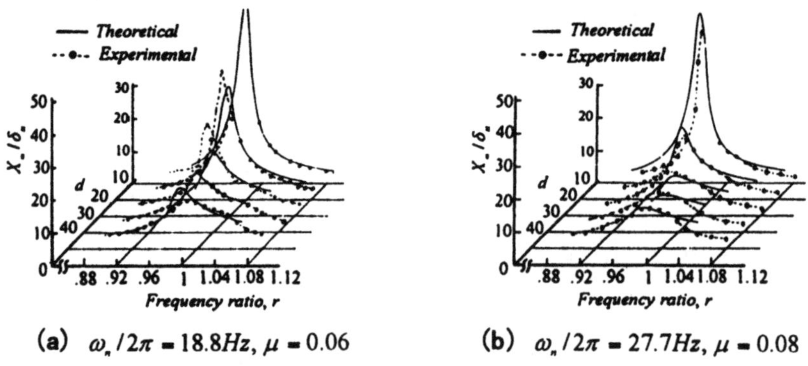

(a) $\omega_n/2\pi = 18.8Hz, \mu = 0.06$ (b) $\omega_n/2\pi = 27.7Hz, \mu = 0.08$

Fig. 12 Damping performance for the horizontal system

Applying the results of the present analysis to the prediction yields the comparison shown in Fig.11 for the frequency response.

Some examples of the resonance curves of the horizontal impact damper with granular beds are shown in Fig.12a, and 12b for two natural frequencies $f_n = 18\ and\ 27.7Hz$. It can be seen that, also for the horizontal system, the periodic motions can be set up in the resonance range, and good damping effect are achieved at an optimum setting of clearance $d_{\varphi} \approx 40$. If the clearance ratio is too small, there will occur Type *HI* motion, thereby deteriorating the damping effect considerably. The predictions with the modified damper parameters show good agreement with experiments, and the validity of the formulas has been corroborated.

4 Summary and conclusions

Theoretical and experimental analyses have been made to clarify the effectiveness of the impact damper using granular materials as a damper mass. This type of damper improves the operating conditions of the conventional solid impact damper which produces loud noise and large acceleration level at collision, and reduces frequency ranges of periodic impact motion for the vertical system. The results are summarized as follows:

In the responses of the impact damper with granular materials, it is shown that the periodic motions can be captured over a wide frequency range. And good damping effects are obtained with suitable choice of mass ratio and container clearance.

In the case when the particle bed damper is applied to the vertical system, there exist the following thee types of motions: Type *I* motion with the higher amplitude, involving relative rests on the top and bottom ends, exists below the primary resonance. Type *II* motion with relatively low amplitude appears within the resonance. Away from the resonance type *III* motion predominates with collisions only on the bottom end under high level of excitation.

In the horizontal system, for the better prediction, the mass ratio and clearance have been modified to include the effect of stagnation and expansion of the granular bed in the container. The two symmetric motions, Type *HI* and Type *HII* were analyzed, the latter of which was found to suppress the resonant amplitude effectively within resonance.

It has also been shown that the damping performance of the impact damper with granular materials depends mainly upon the mass ratio, clearance, and the excitation level. As a result, the optimal clearance ratio has been found which flattens the response curves within an overall resonance range.

References

1 Grubin C, On the theory of the acceleration damper,Trans.ASME,Journal of Applied Mechanics 23(3):373-378,1956.

2 Masri S F and Caughey T K, On the stability of the impact damper. Trans. ASME, Ser.E,Journal of Applied Mechanics 33(3):586-592,1966.

3 Masri .S F, Analytical and experimental studies of multiple-unit impact damper, Journal of the Acoustic Society of America 45(5):1111-1117,1969.
4 Panossian H V, Structural damping enhancement via non-obstructive particle damping technique, Trans. ASME, Journal of Vibration and Acoustics 114(1): 101-105,1992.
5 Kuphal K, Uber die Beeinflussung von Schwingungen durch einen Stosskorper II, Z.angew. Math., Mech.45(6):419-431,1965.
6 Sadek M M, and Mills B, Effect of gravity on the performance of an impact damper:Part 1. Steady-state motion, J.mech. Engng Sci.12(4):268-277,1970.
7 Kobrinsky A E, Dynamics of Mechanisms with Elastic Connections and Impact System, Iliffe books, London, 1969.

APPENDIX

The coefficients R_i's, a_{ij}'s and b_j's for equations (11)-(14) are shown below.

$$R_1 = (\theta a_{15} - \alpha a_{16} - \mu a_{65})/(1+\mu),\ R_2 = (\mu a_{66} - \alpha a_{15} - \theta {}_{16}a)/(1+\mu)$$
$$R_3 = (\tau_1 - \tau_0)a_{33} - a_{23},\ R_4 = (\tau_1 - \tau_0)a_{34} + a_{24}$$
$$R_5 = -(2\pi - \tau_1)a_{65} + a_{43},\ R_6 = (2\pi - \tau_1)a_{66} + a_{46}$$
$$a_{12} = 1, a_{15} = -\exp(-2\pi\theta)\sin 2\pi\alpha,\ a_{16} = -\exp(-2\pi\theta)\cos 2\pi\alpha$$
$$a_{21} = \exp(-\theta_c\tau_0)\sin\alpha_c\tau_0,\ a_{22} = \exp(-\theta_c\tau_0)\cos\alpha_c\tau_0$$
$$a_{23} = -\exp(-\theta\tau_0)\sin\alpha\tau_0,\ a_{24} = -\exp(-\theta\tau_0)\cos\alpha\tau_0$$
$$a_{31} = \alpha_c a_{22} - \theta_c a_{21},\ a_{32} = -\alpha_c a_{21} - \theta_c a_{22},\ a_{33} = \alpha a_{24} - \theta a_{23}$$
$$a_{34} = -\alpha a_{23} - \theta a_{24},\ a_{43} = \exp(-\theta\tau_1)\sin\alpha\tau_1,\ a_{44} = \exp(-\theta\tau_1)\cos\alpha\tau_1$$
$$a_{45} = -a_{43}, a_{46} = -a_{44}, a_{51} = (\theta_c^2 - \alpha_c^2)a_{21} - 2\theta_c\alpha_c a_{22}, a_{52} = (\theta_c^2 - \alpha_c^2)a_{22} + 2\theta_c\alpha_c a_{21}$$
$$a_{63} = (\alpha a_{44} - \mu a_{33} - \theta a_{43})/(1+\mu), a_{64} = -(\mu a_{34} + \alpha a_{43} + \theta a_{44})/(1+\mu)$$
$$a_{65} = \theta a_{43} - \alpha a_{44},\ a_{66} = \alpha a_{43} + \theta a_{44}$$
$$b_1 = A_{pc}\sin(\phi + \beta_c) - A_p\sin(\phi + \beta) + \gamma$$
$$b_2 = A_p\sin(\tau_0 - \phi - \beta) - A_{pc}\sin(\tau_0 - \phi - \beta_c) + \gamma$$
$$b_3 = A_p\cos(\tau_0 - \phi - \beta) - A_{pc}\cos(\tau_0 - \phi - \beta_c), b_4 = 0$$
$$b_5 = A_{pc}\sin(\tau_0 - \phi - \beta_c) - K_R$$
$$b_6 = A_p\cos(\tau_1 - \phi - \beta) - [\mu A_p\cos(\tau_0 - \phi - \beta) + A_p\cos(\tau_1 - \phi - \beta)$$
$$-\mu K_R(\tau_1 - \tau_0)]/(1+\mu)$$

Dynamics of gear-pair systems with backlash

S. Theodossiades and S Natsiavas**

Dynamics of gear-pair systems involving backlash and time-periodic stiffness is investigated. First, the equation of motion is presented in a strongly nonlinear form, with periodic stiffness and external forcing terms. Then, several types of periodic steady-state response are identified and determined by applying a methodology, that is suitable for piecewise linear systems with time-periodic coefficients and excitation. This methodology is complemented by appropriate analytical procedures, revealing the stability properties of the located periodic solutions. In the second part of the work, numerical results are presented. First, typical series of response diagrams are shown, illustrating the effect of the damping and the forcing parameters on the periodic response. These response diagrams are accompanied by results obtained by direct integration of the equation of motion. These results demonstrate that for some parameter combinations the dynamical system examined can exhibit more complicated and irregular response, including crises and intermittent chaos.

1 Introduction

Gear mechanisms have found extensive application in modern power transmission systems (eg., (1]). In many cases, the special geometrical characteristics of the gear teeth affect the dynamical and vibration behavior of geared systems in a significant way. As a consequence, research in the area of the dynamics of mechanical systems involving gear mechanisms has been intensive, especially over the last forty years. These studies focused on involved models of geared systems. However, in the presence of gear backlash, which is either introduced intentionally at the design stage or caused by manufacturing errors or wear, the equations of motion of such systems become strongly nonlinear. Another important complication arises from the variable number of gear pairs that are in contact, causing a variation of the equivalent gear meshing stiffness. These two factors introduce serious difficulties in the analysis.

Recently, the interest towards developing a better understanding of gear vibration has been renewed. In particular, the main effort of some new studies was to throw more light into fundamental issues related to interaction of nonlinear effects such as gear backlash and support nonlinearities with time-periodic variations of the gear meshing stiffness [2]. Under some assumptions, the important dynamics associated with this interaction is captured by models expressed by piecewise linear equations of motion with time periodic coefficients and external forcing. The determination of the vibration response of these models was performed either by direct integration or by applying harmonic balance methodologies. The main objective of the present study is to determine periodic motions of gear-pair systems by applying an analytical

---

**Department of Mechanical Engineering, Aristotle University, 54006 Thessalonik4, Greece*

methodology. This method was presented in [3] and determines approximate periodic solutions and stability properties for piecewise linear systems with time-periodic coefficients.

The equation of motion is first presented in the following section, together with several types of periodic motions. These motions are first defined and then determined by applying analytical methodologies which are appropriate for piecewise linear systems with time-periodic coefficients [3]. In the third section, these methodologies are completed by stability analysis techniques. Next, numerical results are presented. Initially, the effect of the damping and the external forcing parameters on the periodic motions is illustrated by a series of response diagrams. Then, direct integration is performed in parameter ranges with unstable periodic solutions, revealing the presence of other type of motions. In the final section, the summary of the work is presented.

2 Mechanical model – periodic motions

The equation of motion of the gear-pair system investigated in the present study appears in the form

$$\ddot{u} + 2\zeta\dot{u} + w(t)g(u) = f(t), \tag{1}$$

where the function $w(t)$ is periodic with fundamental period $T = 2\pi/\Omega$ and represents the gear stiffness, while the backlash is modelled by the function

$$g(u) = \begin{cases} u-1, & u \geq 1 \\ 0, & |u| < 1 \\ u+1, & u \leq -1 \end{cases}$$

This model takes into account the transmission error, representing geometrical errors of the teeth profile and spacing. As a result, the forcing term $f(t)$ includes low frequency harmonic components due mainly to external torque fluctuations as well as high frequency components due to the geometric irregularities of the mating gear teeth.

There appear two serious difficulties in trying to solve equation (1). The first is caused by the presence of backlash, which renders this equation strongly nonlinear. The second complication arises from the presence of the time-periodic stiffness terms. As a consequence, there exists no systematic analytical methodology leading to exact solutions of equation (1). Also, in any attempt to capture approximate solutions, the frequency ranges of interest should be specified from the outset, in order to preclude the appearance of small divisor terms in the analysis. Therefore, among all the possible cases, the present study focuses on revealing the important dynamics of the gear-pair system in one of them, presenting large practical interest. Namely, the following special form of the original equation of motion is considered

$$\ddot{u} + 2\varepsilon\mu\dot{u} + (1 + 2\varepsilon\cos\Omega t)g(u) = f_0 + \varepsilon f_1\cos(\Omega t + \theta), \tag{2}$$

where ε is a small positive number, while the forcing frequency is chosen as

$$\Omega = 1 + \varepsilon\sigma. \tag{3}$$

For this system, approximate periodic solutions can be obtained by applying a method presented in [3]. This method combines characteristics of the approaches employed for piecewise linear systems with time-invariant coefficients [4-6] with classical perturbation techniques applied to dynamical systems with time-varying coefficients [7]. Next, several possible types of periodic motions of the equation of motion (2) are first defined. These solutions are then determined by applying standard perturbation techniques and imposing appropriate sets of periodicity and matching conditions.

a. Type I periodic motions.

For some combinations of the system parameters, the oscillations remain always within the $u \geq 1$ range. This is classified as a type I motion and involves no impacts. In this case, the dynamics is governed by the linear equation of motion

$$\ddot{u} + 2\varepsilon\mu\dot{u} + (1 + 2\varepsilon\cos\Omega t)(u-1) = f_0 + \varepsilon f_1 \cos(\Omega t + \theta) \qquad (4)$$

with periodic coefficients and external excitation. Then, employing standard perturbation approaches, it turns out that the approximate solution of equation (4) within the frequency ranges specified by (3) takes the form

$$\begin{aligned} u(t) = 1 + f_0 + a\cos(\Omega t - \gamma) - \varepsilon a\left[\cos\gamma - \tfrac{1}{3}\cos(2\Omega t - \gamma)\right] \\ - \varepsilon^2\left[\tfrac{4}{9}a\sigma\cos(2\Omega t - \gamma) - \tfrac{1}{24}a\cos(3\Omega t - \gamma) + \tfrac{4}{9}f_0\cos 2\Omega t \right. \\ \left. - \tfrac{2}{9}f_1\cos(2\Omega t + \theta)\right] + O(\varepsilon^3) \end{aligned} \qquad (5)$$

The amplitude and phase of the oscillatory terms are found to satisfy a set of autonomous equations with form

$$\dot{a} = g_1(a,\gamma), \qquad a\dot{\gamma} = g_2(a,\gamma) \qquad (6)$$

Clearly, constant solutions of the last set of slow-flow equations correspond to periodic motions of the mechanical oscillator, with fundamental frequency Ω.

b. Type II periodic motions.

When the periodic response assumes maximum displacement values larger than 1 and minimum values below 1 but above -1, it is characterised as a type II motion and involves one-sided impacts of the mechanical system. Taking into account the information depicted in Figure 1, the equation of motion within the time interval $0 \leq t_1 \leq t_{1c}$ is first expressed in the form:

$$\ddot{u}_1 + 2\varepsilon\mu\dot{u}_1 + \left[1 + 2\varepsilon\cos(\Omega t_1 + \varphi)\right](u_1 - 1) = f_0 + \varepsilon f_1\cos(\Omega t_1 + \theta + \varphi)$$

The constant phase φ is introduced so that the initial displacement is one. Here, an approximate solution is obtained in the regular perturbation series form

$$u_1(t_1;\varepsilon) = u_{01}(t_1) + \varepsilon u_{11}(t_1) + \varepsilon^2 u_{21}(t_1) + O(\varepsilon^3) \qquad (7)$$

Since the time interval examined is small, there is not enough time for secular terms of this series expansion to grow and cause problems. On the other hand, small divisor terms are properly eliminated, by making use of the frequency condition (3). Therefore, applying standard methodologies, the solution of equation examined can eventually be put in the form

$$u_1 = A_1 a_1(t_1;\varphi) + B_1 b_1(t_1;\varphi) + c_1(t_1;\varphi) \qquad (8)$$

where $a_1(t_1;\varphi), b_1(t_1;\varphi)$ and $c_1(t_1;\varphi)$ are known functions of the system parameters and the phase φ.

Next, the response within $0 \leq t_2 \leq t_{2c}$ is determined by considering the corresponding form of the equation of motion, which becomes

$$\ddot{u}_2 + 2\zeta\dot{u}_2 = f_0 + \hat{f}_1\cos(\Omega t_2 + \theta + \varphi) \qquad (9)$$

with $\hat{f}_1 = \varepsilon f_1$. The last equation accepts the following exact solution form

$$u_2(t_2) = A_2 e^{-2\zeta t_2} + B_2 + C_2 t_2 + D_2 \cos(\Omega t_2 + \theta + \varphi - \psi) \qquad (10)$$

with

$$C_2 = f_0/2\zeta, \quad D_2 = \hat{f}_1\Big/\left(\Omega\sqrt{\Omega^2 + 4\zeta^2}\right), \quad \sin\psi = 2\zeta/(-\Omega). \qquad (11)$$

Based on the above formulation, the determination of type II periodic motions has been reduced to the evaluation of the constants A_1, B_1, A_2, B_2, φ and the crossing time t_{1c}. These constants can be determined by applying the following set of periodicity and matching conditions:

$$u_1(0) = u_1(t_{1c}) = u_2(0) = u_2(t_{2c}) = 1$$

$$\dot{u}_1(0) = \dot{u}_2(t_{2c}) \equiv \dot{u}_{10}, \quad \dot{u}_1(t_{1c}) = \dot{u}_2(0) \equiv \dot{u}_{20}$$

where $t_{2c} = nT - t_{1c}$ and n is an integer. The resulting set of six transcendental equations can be reduced to two equations for φ and t_{1c} only. Numerical solution of these equations determines φ and t_{1c}. The corresponding values of A_1, B_1, A_2 and B_2 are determined by back substitution.

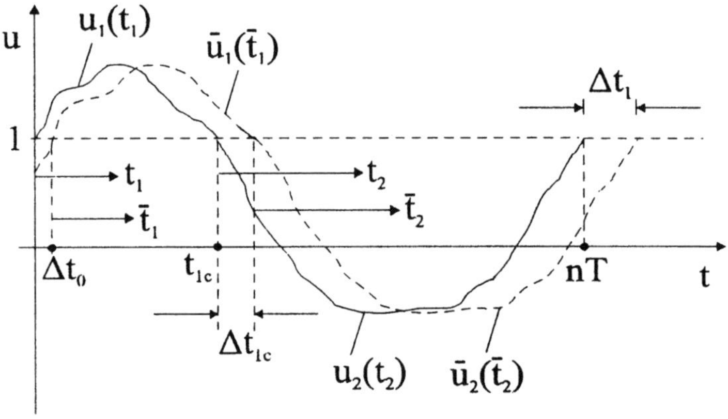

Figure 1. Type II periodic steady-state motion and a neighbouring motion

c. Type III periodic motions.

When the maximum displacement in a periodic motion of equation (1) is larger than 1 and its minimum value is less than -1 (Figure 2), it involves double impacts and is characterised as a type III motion. Application of a methodology similar to that presented for the type II motions, leads to determination of this motion as a solution of a set of twelve transcendental equations. This set results by imposing the following periodicity and matching conditions:

$$u_1(0) = u_1(t_{1c}) = u_2(0) = u_4(t_{4c}) = 1$$

$$u_2(t_{2c}) = u_3(0) = u_3(t_{3c}) = u_4(0) = -1$$

$$\dot{u}_1(0) = \dot{u}_4(t_{4c}), \quad \dot{u}_1(t_{1c}) = \dot{u}_2(0), \quad \dot{u}_2(t_{2c}) = \dot{u}_3(0), \quad \dot{u}_3(t_{3c}) = \dot{u}_4(0)$$

and includes as unknowns the eight coefficients of the homogeneous solution parts in the four solution segments, the phase φ and the three crossing times

t_{1c}, t_{2c} and t_{3c}. In this case, the numerical effort is reduced to the solution of four equations, involving only t_{1c}, t_{2c}, t_{3c} and φ as unknowns.

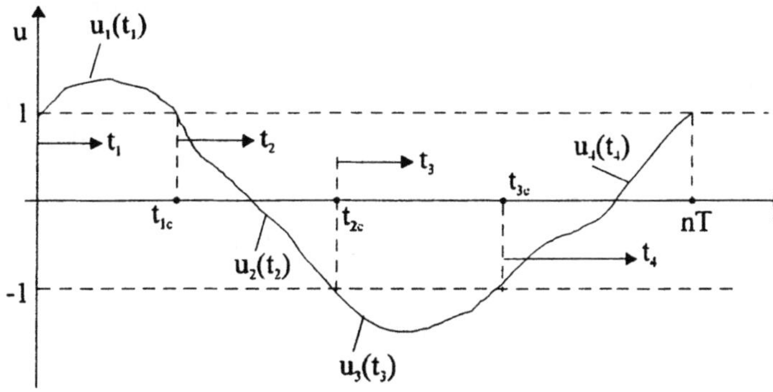

Figure 2. Type III periodic steady-state motion

d. Type IV periodic motions.
For the cases where the constant component of the forcing term is negligible, the dynamical system examined may exhibit periodic motions with displacement amplitude confined between -1 and 1. In such cases, the equation of motion takes the form

$$\ddot{u} + 2\zeta\dot{u} = \hat{f}_1 \cos(\Omega t + \theta)$$

and accepts an exact solution which has identical form with (10), but with $A_2 = C_2 = 0$. This means that

$$u(t) = B_2 + D_2 \cos(\Omega t + \theta - \psi)$$

where the amplitude D_2 and the phase ψ are given by (11), while the constant B_2 is determined by the initial conditions.

3 Stability analysis

First, it can readily be shown that the **type IV** periodic solutions are always stable, provided that the damping parameter is positive. In addition, the stability properties of **type I** motions can be determined by applying the classical method of linearization [7]. In the present case, it can be shown that the constant solutions of equations (6) are stable, provided that the damping parameter is positive.

The stability analysis for the remaining types of periodic motions requires the application of a different methodology. The basic ideas of this methodology have been presented in detail in [3]. Next, the stability analysis of **type II** motions is presented briefly. In Figure 1, the solution segments $u_1(t_1)$ and $u_2(t_2)$ represent the two pieces of a type II periodic motion, while the pieces $\bar{u}_1(\bar{t}_1)$ and $\bar{u}_2(\bar{t}_2)$ represent a motion starting from neighbouring initial conditions. First, applying the initial conditions of the periodic solution

$$u_1(0) = 1, \qquad \dot{u}_1(0) = \dot{u}_{10}$$

and the following conditions for the perturbed solution

$$\bar{u}_1(0) = 1,\ \dot{\bar{u}}_1(0) = \dot{u}_{10} + \Delta v_0,\ \bar{u}_1(\bar{t}_{1c}) = 1,\ \dot{\bar{u}}_1(\bar{t}_{1c}) = \dot{u}_{20} + \Delta v_{1c}$$

Taylor expanding the last conditions and keeping only up to first order terms leads to a system of two equations, which can be cast in the matrix form

$$\underline{\varepsilon}_{1c} = Q_1\, \underline{\varepsilon}_0 \qquad (12)$$

In the last equation, the vectors $\underline{\varepsilon}_0 = \left(\Delta\varphi_0 \quad \Delta v_0\right)^T$, $\underline{\varepsilon}_{1c} = \left(\Delta\varphi_{1c} \quad \Delta v_{1c}\right)^T$ represent the difference of the perturbed solution from the periodic solution at $t = 0$ and $t = t_{1c}$, respectively, with $\Delta\varphi_{1c} = \Omega\,\Delta t_{1c}$.

Likewise, the difference between the periodic and the perturbed solution in the second interval of the motion is obtained in the similar form

$$\underline{\varepsilon}_1 = Q_2\, \underline{\varepsilon}_{1c} \qquad (13)$$

where the vector $\underline{\varepsilon}_1 = \left(\Delta\varphi_{2c} \quad \Delta v_{2c}\right)^T$ includes the difference at times $t = t_{2c}$ and $\Delta\varphi_{2c} = \Omega\,\Delta t_{2c}$. Therefore, equations (12) and (13) yield

$$\underline{\varepsilon}_1 = Q\, \underline{\varepsilon}_0 \qquad (14)$$

with $Q \equiv Q_2\, Q_1$. Equation (14) determines the error at the end of the first response period as a function of the original error in the initial conditions. Then, the error after k response periods will be

$$\underline{\varepsilon}_k = Q^k\, \underline{\varepsilon}_0 \,.$$

and will disappear gradually provided that all the eigenvalues of matrix Q have modulus less than one. In such cases, the periodic solution examined is stable. If at least one eigenvalue of matrix Q has modulus larger than one, the periodic solution is unstable. Finally, when the eigenvalue of Q with maximum modulus equals one, the solution exhibits a bifurcation [7].

The stability analysis of **type V** periodic motions requires only slight modifications. In addition, the stability analysis of **type III** periodic motions is also very similar. The main difference is due to the fact that these motions consist of four discrete segments. As a consequence, application of the same methodology over a response period leads to a relation identical to (14), where the new matrix Q is now a product of four matrices, instead.

4 Numerical results

First, Figure 3 presents results obtained by applying the procedure developed for harmonic steady-state response. More specifically, it shows response diagrams for a system with parameters: ε=0.03, μ=0.8, f_1=2.5 and five values of

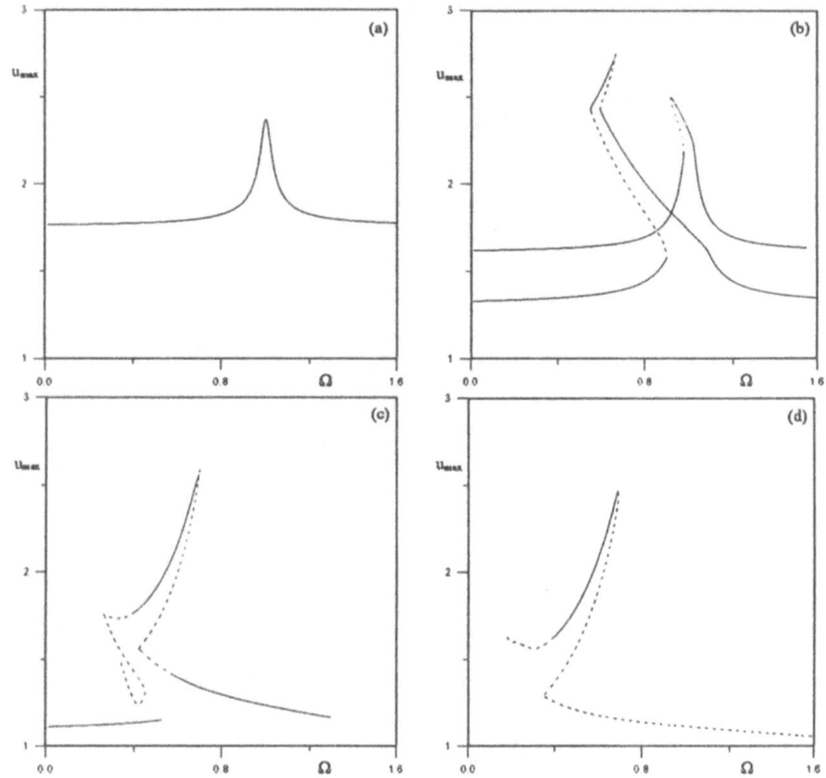

Figure 3. Frequency-response diagram of periodic motions for (a) $f_0=0.75$, (b) $f_0=0.6$ and 0.3, (c) $f_0=0.075$, (d) $f_0=0.03$

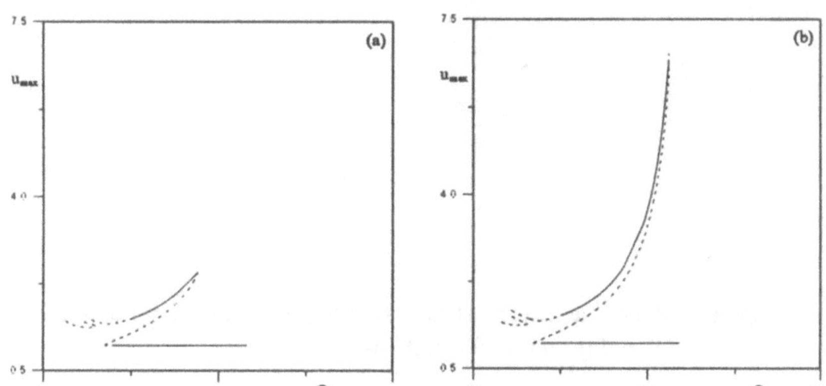

Figure 4. Effect of damping parameter (a) $\mu=0.8$, (b) $\mu=0.1$

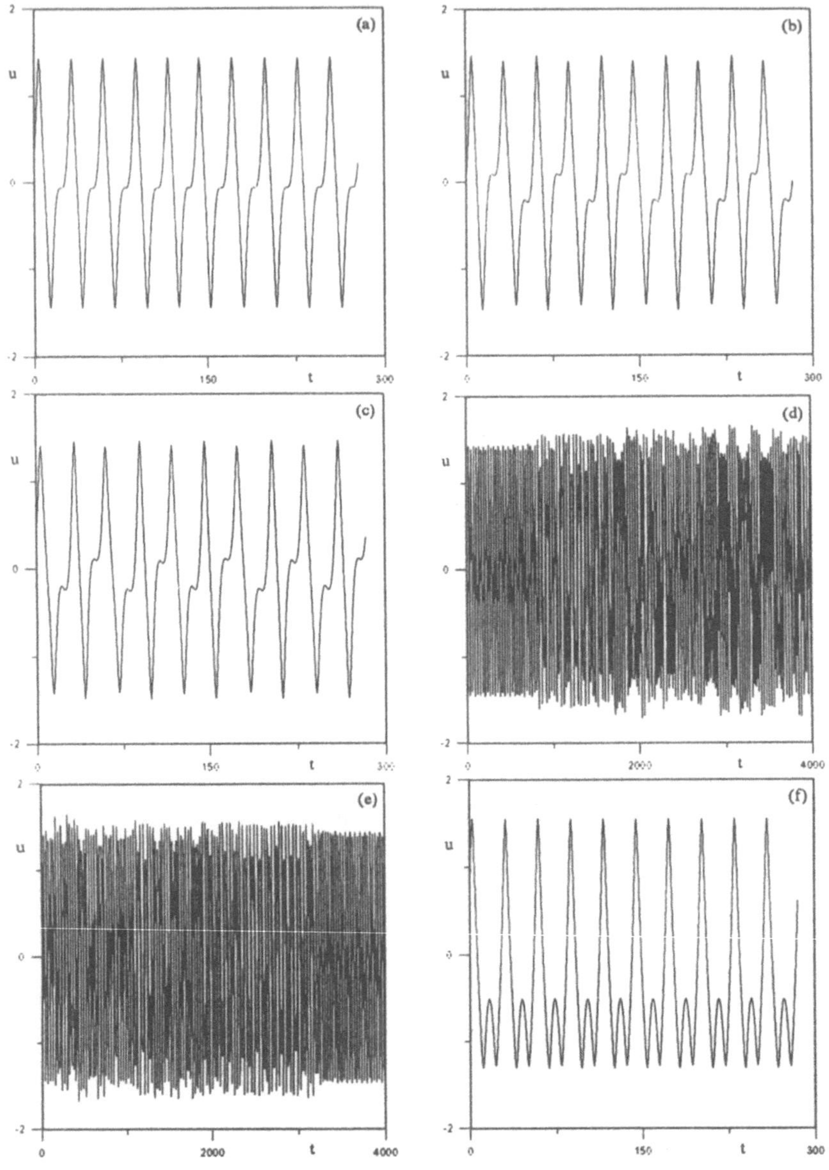

Figure 5. Response history at (a) σ=-25.8, (b) σ=-25.935, (c) σ=-25.95, (d) and (e) σ=-25.9562, (f) σ=-25.99

f_0. The solid/broken curves represent stable/unstable periodic solution branches, respectively. For relatively large values of f_0, only type I motions (involving no impacts) are possible and the resonance occurs at Ω=1, since the system is linear

(Figure 3a). As the value of f_0 decreases, type I motions withdraw gradually from the picture, while type II and III motions appear, involving one- and two-sided impacts (Figures 3b-3d). Among them, the type II motions present softening behavior, while the type III motions exhibit hardening behavior, as for piecewise linear oscillators with constant coefficients [6].

Likewise, Figure 4 reveals effects of the damping parameter on the system steady-state response. Figure 4a is obtained for the same parameters as those of the previous figure, but with $f_0=0$, while for Figure 4b the damping parameter value is reduced to $\mu=0.1$. Clearly, as the value of the damping level decreases, the solution amplitude increases, as expected.

Besides the existence of harmonic and subharmonic motions, the oscillator examined may exhibit more complicated and rich responses. This was realised by performing numerical integration of the original equation of motion. For instance, Figure 5 shows a typical sequence of response histories, obtained for the system of Figure 4b. First, Figure 5a shows an n=1 (harmonic) motion, existing at $\sigma=-25.8$. By gradually decreasing σ, this motion gives way to an n=2 motion (Figure 5b), which in turn gives its place to an n=4 motion (Figure 5c). At about $\sigma=-25.9562$, the system undergoes a boundary crisis and exhibits apparently chaotic response [8]. In fact, in some frequency ranges the system n=4 behavior is intermittently interrupted by intervals of chaotic response (Figures 5d and 5e). Eventually, the system settles to a new n=1 motion (Figure 5f).

5 Summary and conclusions

A new methodology was.applied for determining periodic steady-state response of gear-pair systems with backlash, modelled as piecewise linear dynamical systems with weakly periodic coefficients. This method shares and combines characteristics of classical perturbation approaches applied to oscillators with time varying coefficients as well as of exact methods employed for piecewise linear systems with constant coefficients. Through the application of a proper analytical scheme, the task of locating periodic steady-state motions of such systems was reduced to the solution of relatively small sets of algebraic equations. In addition, appropriate methodologies were applied for investigating the stability properties for each type of the located periodic solutions.

In the second part of the study, numerical results were presented. First, frequency-response diagrams were shown, illustrating effects of the important technical parameters on the system dynamics. In particular, as the constant external forcing parameter decreases, the original motions start developing one- and two-sided impacts. In addition, as the damping parameter decreases, the solution amplitude increases. Moreover, for relatively small levels of damping, results obtained by direct integration of the equation of motion demonstrated a rich dynamic behavior, involving period doubling cascades, leading to the existence of boundary crises and transient chaos.

References

1. Chen, C.-S., Natsiavas, S. and Nelson, H.D., Stability analysis and complex dynamics of a gear-pair system supported by a squeeze film damper. *ASME J. of Vibration and Acoustics* **119**, 85-88 (1997).

2. Kahraman, A. and Blankenship, G.W., Experiments on nonlinear dynamic behavior of an oscillator with clearance and periodically time-varying parameters. *J. of Applied Mechanics* **64**, 217-226 (1997).
3. Natsiavas, S., Theodossiades, S. and Goudas, I., Dynamic analysis of piecewise linear oscillators with time periodic coefficients. *Submitted for publication.*
4. Masri, S.F. and Caughey, T.K., On the stability of the impact damper. *J. of Applied Mechanics* **33**, 586-592 (1966).
5. Shaw, S.W. and Holmes, P.J., A periodically forced piecewise linear oscillator. *J. of Sound and Vibration* **90**, 129-155 (1983).
6. Natsiavas, S. and Gonzalez, H., Vibration of harmonically excited oscillators with asymmetric constraints. *J. of Applied Mechanics* **59**, S284-S290 (1992).
7. Wiggins, S., *Introduction to Applied Nonlinear Dynamical Systems and Chaos.* New York: Springer-Verlag (1990).
8. Bajaj, A.K., Examples of boundary crisis phenomenon in Structural Dynamics. *International Series of Numerical Mathematics* **97**, 27-36 (1991).

Nonlinear stability analysis of a single mass rotor contacting a rigid backup bearing

H Ecker [1]

This paper investigates the steady-state response of a rigid, single mass rotor with imbalance eccentricity supported by an active magnetic bearing with nonlinear characteristics. The rotor may have intermittent contact with an axially collocated, fixed, rigid and circular backup bearing. A radial offset position of the backup bearing center with respect to the magnetic bearing center is assumed. Parameter studies are carried out, especially for the excitation frequency and the friction conditions at the contact point. For frequencies ranging from the onset of contact up to the critical speed various kinds of periodic, non-periodic and quasi-periodic solutions can be observed. Within the parameter range investigated, a two-periodic orbit with one contact was found to be the dominant stable orbit for low excitation frequencies.

1 Introduction

Backup bearings (also called auxiliary bearings) are necessary to prevent the non-rotating parts of an Active Magnetic Bearing (AMB) from solid contact with the rotor. They are in regular use during the transient behavior of a rotor when a magnetic bearing starts or stops operating and a transfer of support from the backup bearing to the AMB (or vice versa) takes place. This can happen either under regular operating conditions or in case of an AMB failure. Numerical and experimental investigations have been carried out on the transient behavior of a rotor dropping into its backup bearings, see for example [5] and others.

However, a rotor might also have intermittent or permanent contact with the backup bearing during steady-state operation due to abnormal operating conditions. This can occur, for instance, due to a bearing overload resulting in an off-center orbit of the rotor which cannot be compensated by the magnetic bearing controller, or a high rotor unbalance leading to a steady-state orbit exceeding the backup bearing clearance. Misalignement of the effective centers of an AMB and the backup bearing in combination with a small clearance between rotor and backup bearing may also cause contact and rub between a rotor and a backup bearing.

Investigations of the steady-state response of a rotor experiencing stator contact have been carried out by a number of authors in the past. Recent

[1]Dr. Horst Ecker, Vienna University of Technology, Institute for Machine Dynamics and Measurement, Wiedner Hauptstr. 8-10/E303, A-1040, Vienna, Austria.

numerical works have been presented by [6] and [9], employing a modified Jeffcott-rotor and detailed models for the nonlinear contact forces. Only very few results have been published on a rotor with operating AMB and backup bearing contact. In [2] and [3] the same mathematical model as in the present investigation was used by the author, but different sets of parameters were investigated. The lack of studies on abnormal operating conditions justifies basic studies on simple models in order to better understand the complex nonlinear behavior of a rotor contacting the backup bearing boundaries.

2 Mathematical model

2.1 Rotor and Active Magnetic Bearing

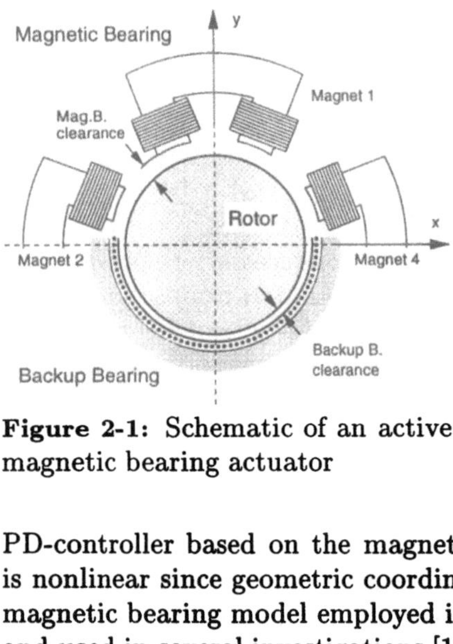

Figure 2-1: Schematic of an active magnetic bearing actuator

Figure 2-1 shows a schematic diagram of a planar, rigid rotor with magnetic bearing actuator and backup bearing and represents the type of model used in this investigation. Only the upper part of the magnetic bearing actuator and the lower part of the backup bearing is shown in Fig. 2-1. Sensors, controller and amplifiers are not shown as well. The actuator consists of two pairs of opposed electromagnets arranged symmetrically about x-axis and y-axis. Each pair of magnets is independently controlled by an idealized PD-controller based on the magnetic flux in the airgap. However, control is nonlinear since geometric coordinate coupling is taken into account. The magnetic bearing model employed in this investigation was developed in [8] and used in several investigations [1], [7]. For details on the analysis it is referred to previously published results [8]. Although the model has been kept simple for the purpose of this basic investigation, applications like flywheels in energy storage systems can be described by this model.

The dimensionless rotor positions X and Y are derived from the physical positions x_{rot} and y_{rot} of the rotor center of rotation by non-dimensionalizing with the magnetic bearing clearance c_{amb} as $X = x_{rot}/c_{amb}$, $Y = y_{rot}/c_{amb}$. Similarily, the dimensionless imbalance eccentricity $E = e_{rot}/c_{amb}$ is obtained. Differentiation with respect to dimensionless time τ is indicated as $()'$. The period $T = 2\pi/\Omega$ is defined by the dimensionless imbalance excitation frequency Ω , which is non-dimensionalized with ω_n being the natural frequency of the linear system without coordinate coupling. Parameters K and G are dimensionless coefficients for proportional and derivative feedback, respectively. Finally, the symbol A denotes the coordinate coupling parameter.

The resulting dimensionless equations of motion of the rotor subject to forces from the magnetic bearing actuator and rotor imbalance eccentricity can be written in the form

$$
\begin{aligned}
X'' &= \frac{A}{2K} X \left(T^2 + (KTY)^2 + 2KGTYY' + G^2 Y'^2 \right) - \\
&\quad \frac{1}{K} \left(KT^2 X + GTX' \right) + 4\pi^2 E \cos 2\pi\tau, \qquad (2.1)
\end{aligned}
$$

$$
\begin{aligned}
Y'' &= \frac{A}{2K} Y \left(T^2 + (KTX)^2 + 2KGTXX' + G^2 X'^2 \right) - \\
&\quad \frac{1}{K} \left(KT^2 Y + GTY' \right) + 4\pi^2 E \sin 2\pi\tau. \qquad (2.2)
\end{aligned}
$$

Since a rigid, single mass rotor is assumed, shaft elasticity is not taken into account. Note also that the static weight of the rotor is not considered in this set of equations for reasons of simplicity. Similar equations including weight have been derived in [1] and will be used in subsequent investigations. In the present case the dynamical properties of the AMB are identical with respect to both principle axes (x- and y-axis). Due to the coordinate coupling nonlinearity the bearing model is not isotropic but symmetric with respect to the principle axes, for details see [8]. Although this anisotropy of the AMB-model is weak for moderate eccentricities, it becomes important in the investigation of steady state orbits, when the backup bearing center has little or no offset. See [2] for a more detailed investigation.

2.2 Backup bearing

A rather simple model for the backup bearing is used in this investigation. Since the rotor model is a 2D-model that does not take into account the axial dimension of the rotor, the backup bearing is collocated with the magnetic bearing. It is assumed that the backup bearing is a sleeve bearing with a fixed and rigid ring. This assumption is valid only if the mass and stiffness ratios between the backup bearing housing and the AMB-suspended rotor are sufficiently high. There is no major difficulty with introducing bearing housing mass and stiffness to the present model, as done in [6] and others. As this investigation is mainly focused on the effect of impact friction and bearing offset, the model is kept as simple as possible.

Figure 2-2 shows the geometric situation of the rotor within the backup bearing. In this figure the rotor is represented by its axis and only the clearance C_{bb} of the backup bearing is shown. The center of the circular shaped backup bearing is located off-center with respect to the origin of the coordinate system of the magnetic bearing by X_{bb}, Y_{bb} in x- and y-direction, respectively. In a real application this radial offset is a very common situation for several reasons, as for example manufacturing imperfections or rotor deflection in combination with dislocated backup and magnetic bearings.

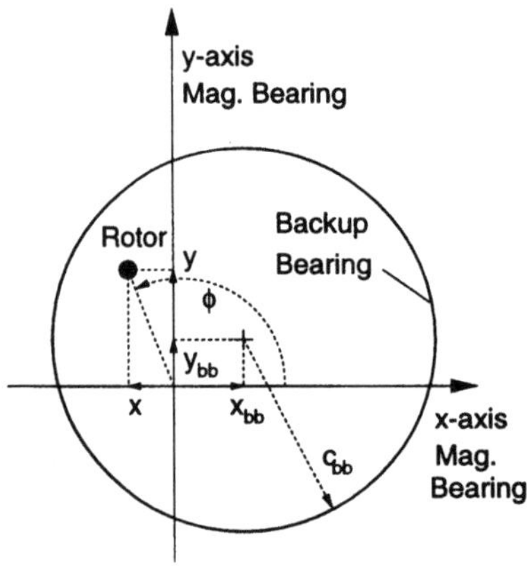

Figure 2-2: Geometric situation showing the rotor axis within the backup bearing clearance and the radial offset of the backup bearing center

Another reason may be the fact that the center of bearing-stiffness of an AMB usually does not coincide with the geometric center of the bearing. This center of bearing-stiffness is defined as a rotor axis position where all AMB-generated forces acting on the rotor are in equilibrium. The inner surface of the backup bearing, which gets in contact with the rotor is assumed as a surface with dry friction. The friction parameter will be varied from low values up to moderate values. Therefore, a ball bearing is schematically shown in the lower part of Fig. 2-1. However, the complex dynamic behavior of the inner race of a ball bearing that is contacted by a spinning rotor is beyond the scope of this investigation.

2.3 Model of contact

The contact between the rotor and the backup bearing is modeled as an impact of infinitesimally short period of time, occuring between two rigid surfaces with Coulomb friction coefficient μ. Different kinds of impact are possible when friction is taken into account. "Sticking impact" occurs when the contact points on both bodies do not move relative to each other within a tangential plane to the contact surfaces. The dimensionless contact momenta per unit mass S_r, S_t, represented in a local (r-t)-coordinate system at the contact point (CP), can be calculated from

$$S_r = (1+\varepsilon)\ {}_b v_{CP-r} \tag{2.3}$$

$$S_t = (1+\varepsilon)\ {}_b v_{CP-t} \qquad \text{if} \qquad S_t \le \mu S_r \tag{2.4}$$

with ${}_b v_{CP-r}$, ${}_b v_{CP-t}$ being the radial and tangential velocity of the contact point on the rotor before the impact, respectively. Energy dissipation during the impact is accounted for by introducing a coefficient of restitution ε. But if the additional condition of Eq.(2.4) does not hold, a sticking contact is not possible. This will occur for low values of the friction coefficient and/or for a large tangential velocity ${}_bv_{CP-t}$ at the contact point, compared to the radial velocity ${}_bv_{CP-r}$. In this case a "sliding contact" occurs and Eq.(2.4) has to be replaced by

$$S_t = \mu S_r \qquad \text{if} \qquad {}_bv_{CP-t} > \mu\ {}_bv_{CP-t}. \tag{2.5}$$

Since, in general, the radius of the rotor at the bearing station R_{bb} is much larger than the clearance of the backup bearing C_{bb}, $R_{bb} \gg C_{bb}$, and since the rotational frequency of the rotor is constant, only sliding contact is considered in this investigation. Consequently, the velocity of the rotor after the contact ${}_av_{r,t}$ is calculated from

$$ {}_av_r = -\varepsilon\ {}_bv_r, \tag{2.6}$$

$$ {}_av_t = {}_bv_t - \mu(1+\varepsilon)\ {}_bv_r. \tag{2.7}$$

As this investigation is focused on steady state conditions and a sufficiently high rotary inertia of the rotor is assumed, a change in the moment of momentum of the rotor due to an impact is neglected.

3 Numerical solution method

The nonlinear equations of motion of the rotor, Eqs.(2.1) and (2.2), were solved by numerical simulation with a state-of-the-art Runge-Kutta algorithm. A state vector was picked from a sufficiently large steady-state orbit without rotor-bearing contact and chosen as initial condition for a single simulation run. In the presented parameter studies the excitation frequency Ω was increased step-by-step. The state vector at the end of the previous simulation run was used as an initial vector for a subsequent run. Usually, the simulation was stopped after 300 revolutions (periods) of the rotor.

Numerically, the most crucial point in this investigation is the precise calculation of the time instant at which a contact occurs. Since the domains of attraction for certain periodic solutions are rather narrow, results can change qualitatively if numerical errors introduced at the instant of contact are not sufficiently small. Therefore, a root finding procedure was started each time when a contact event was detected after an integration step. After calculating the exact contact time, Eqs.(2.6) and (2.7) were evaluated and integration was restarted with new initial conditions after the impact.

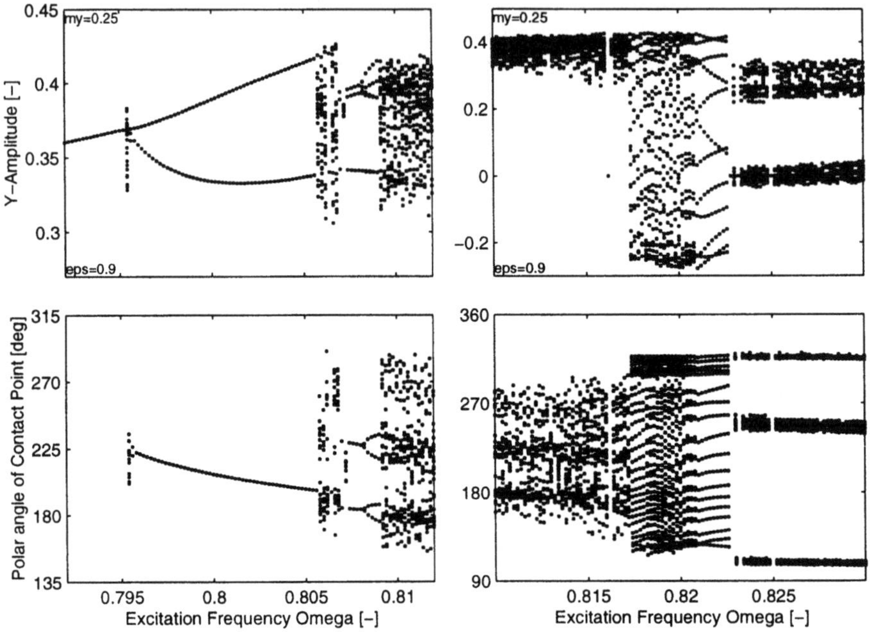

Figure 4-1: Bifurcation diagram for $\mu = 0.25$ and $X_{bb} = Y_{bb} = 0.09$

4 Results

The following results are based on a set of default system parameters as given in Table 4-1. Changes in the listed parameter values are explained in the figure captions.

Figures 4-1 show bifurcation diagrams for the Y-amplitude of a rotor in a backup bearing with offset values as indicated. Note the different frequency ranges and the different scales on the vertical axes of the left and the right figures. The diagrams at the bottom of Fig. 4-1 show the corresponding angular position of the rotor ϕ at the contact. In the case of a periodic

Table 4-1: Default values for non-dimensional system parameters

Stiffness parameter	K	3	Coordinate coupling	A	0.15
Damping param.	G	0.2	Imbalance eccentricity	E	0.2
Clearance AMB	C_{amb}	1	Backup bearing offset	X_{bb}	0.09
Clearance BB	C_{bb}	0.5	Backup bearing offset	Y_{bb}	0.09
Friction coefficient	μ	0.25	Excitation phase angle		
Restitution coeff.	ε	0.9	to sample Y-amplitude	$2\pi\tau$	$\pi/2$

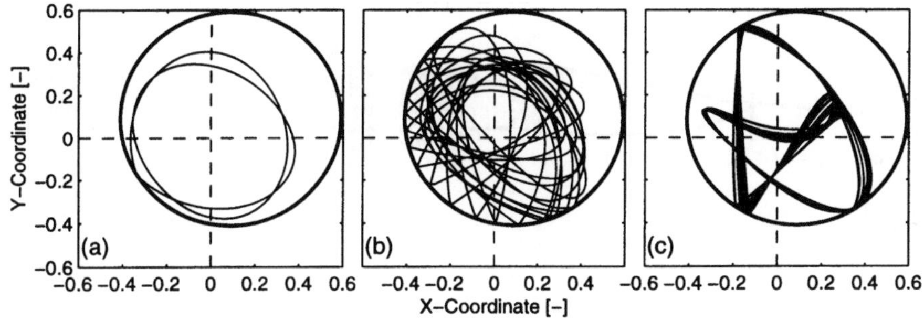

Figure 4-2: Orbits of the rotor axis within backup bearing clearance corresponding to Fig. 4-1: (a) $\Omega = 0.80$, (b) $\Omega = 0.822$, (c) $\Omega = 0.83$

solution it is shown primarily how many contacts occur during a single period of a periodic orbit.

The one-periodic solution for excitation frequencies $\Omega < 0.795$ corresponds to an orbit of the rotor without backup bearing contact. After a bifurcation at $\Omega \simeq 0.795$ a stable two-periodic orbit develops. However, above $\Omega \simeq 0.805$ this periodic solution was not attracted anymore within 300 periods. A small window with multiperiodic solutions is observed in the frequency range of $0.807 < \Omega < 0.809$. A rather large window with a number of different multiperiodic solutions occurs in the vicinity of $\Omega = 0.82$. Above $\Omega \simeq 0.823$ quasi-periodic solutions were found.

Figure 4-2 shows various orbits for selected frequencies, corresponding to the bifurcation map in Fig. 4-1. Note the difference between the two-periodic orbit (a) with one contact and the complicated twelve-periodic orbit with eighteen contacts at $\Omega = 0.822$. Orbit (c) at $\Omega = 0.83$ is a quasi-periodic orbit with three contacts.

This investigation is mainly focused on the influence of the friction coefficient μ and the backup bearing offset X_{bb}, Y_{bb}. The restitution coefficient and the parameters of the magnetic bearing model are kept constant at their values given in Table 4-1. References [2] and [3] present some results for a variation of these parameters.

In the course of this investigation the friction parameter was varied from $\mu = 0$ to $\mu = 0.45$. Some typical results are shown in Fig. 4-3 for two different values of μ at a fixed restitution coefficient of $\varepsilon = 0.9$.

For certain excitation frequencies the simulation had to be terminated before completing 300 periods because the limitations of the model were exceeded. These frequency ranges are indicated by dashed vertical lines in Fig. 4-3. The termination was caused by a sequence of impacts that occured with rapidly decreasing time intervals between the contact events. In most cases, especially for high values of the friction coefficient μ, the anti-clockwise direction

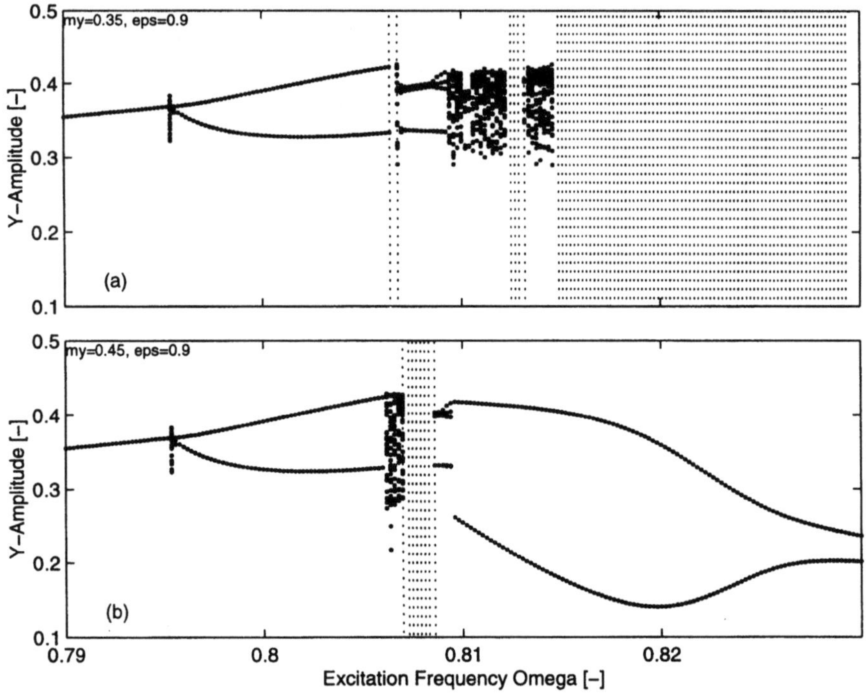

Figure 4-3: Bifurcation diagrams for different sets of parameters: (a) $\mu = 0.35$ and (b) $\mu = 0.45$

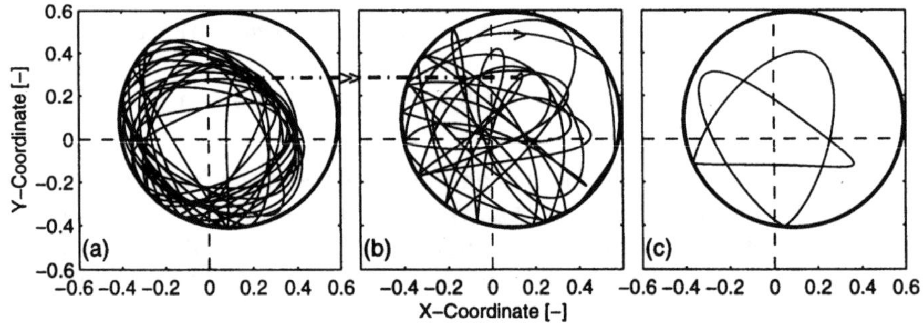

Figure 4-4: Orbits of the rotor axis within backup bearing clearance corresponding to Fig. 4-3: (a) $\mu = 0.35, \Omega = 0.815$: non-periodic orbit developing backward whirl (b) ending up in permanent contact with backup bearing. (c) $\mu = 0.45, \Omega = 0.815$: two-periodic orbit with two contacts.

of the rotor orbit reversed and the rotor started a backward whirl motion with rapidly increasing velocity. Since the assumption of a sliding contact might be violated for a continued backward whirl the simulation was stopped.

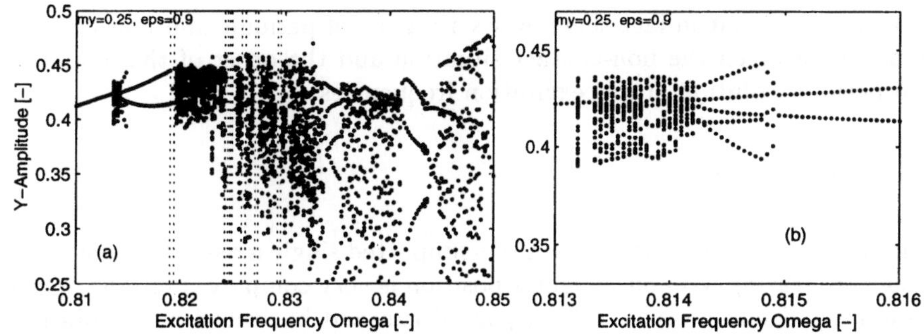

Figure 4-5: (a) Bifurcation diagram for default parameters and backup bearing offset $X_{bb} = Y_{bb} = 0.05$ (b) details of onset of contact

For a friction coefficient of $\mu = 0.35$ Fig.4-4a shows the anti-clockwise motion of the rotor center before the onset of a backward whirl. Figure 4-4b shows the continuation of the motion and how a backward whirl is developed until the simulation was terminated.

However, in some cases the rotor maintained a forward whirl but started having (at least partially) permanent contact with the bearing, and therefore the limitations of the model in use were exceeded. This contact situation is known as "partial rub", see [6].

From Fig. 4-1 and 4-3a one might conclude that with increasing values of μ the regime of either backward whirl solutions or permanent rub increases to lower frequencies. However, the result of Fig. 4-3b for $\mu = 0.45$ shows, that this is not true in general. For this set of parameters only a very small frequency range of non-periodic solutions is observed. Two different types of two-periodic orbits are dominating, exhibiting a rather different result compared with the results as previously shown. Figure 4-4c shows the two-periodic orbit with two contacts that is observed at $\Omega = 0.815$ for a friction coefficient of $\mu = 0.45$.

In Figs. 4-1 and 4-3, an extremely small window of different solutions was found at the very beginning of the two-periodic solution regime between $\Omega \simeq 0.795$ and $\Omega \simeq 0.806$. To show more details of this small region of a grazing bifurcation the backup bearing offset was reduced and set to $X_{bb} = Y_{bb} = 0.05$. The resulting bifurcation diagram is shown in Fig. 4-5. Note that the onset of contact has shifted to a higher excitation frequency since the reduced offset permits larger orbits without contact.

Figure 4-5a exhibits basically the same characteristic patterns that have been observed already for a larger backup bearing offset. However, the window in question at the onset of contact has become wider, compared to previous results. Figure 4-5b is a detailed view of this frequency range that reveals once more multi-periodic and non-periodic solutions. From this result it can be concluded, that in fact there exists a regime of periodic and non-periodic solutions between the non-contact situation and the onset of the dominant two-periodic solution at low excitation frequencies.

5 Conclusions

The steady-state behavior of an AMB-supported single mass rotor contacting a backup bearing with offset center position is very complex. Periodic, quasi-periodic and non-periodic orbits, partial rub and backward whirl solutions were found in the frequency range $\Omega \leq 0.85$. Some of the results are very sensitive to initial conditions and accurate numerical methods have to be used. In a future work path-following methods will be applied to get a better understanding of the presented bifurcation diagrams and basins of attractions should be calculated to determine which solution is the most likely one in the case of multiple stable solutions.

References

1 Ecker H, Knight J D, Wu L, Nonlinear dynamic simulation of an active magnetic bearing with non-symmetric coordinate coupling forces, ASME IGTI Congress, (Orlando, FL, June), 1997.

2 Ecker H, Steady state orbits of an AMB-supported rigid rotor contacting the backup bearings, MAG'97, Industrial Conference on Magnetic Bearings, (Alexandria, VA, Aug.), 129-138, 1997.

3 Ecker H, Nonlinear stability analysis of a single mass rotor contacting a rigid auxiliary bearing, IFToMM - Proc. Fifth Int. Conf. on Rotor Dynamics, (Darmstadt, Germany, Sept.), 790-801, 1998.

4 Free A M, Flowers G T, Trent V S, Dynamic modelling and response characteristics of a magnetic bearing rotor system with auxiliary bearings, Proc. 15th Biennial ASME Vibr. Conf. (Boston, Sept.), 1995.

5 Fumagalli M, Schweitzer G, Motion of a rotor in retainer bearings, Proc. of 5th International Symposium on Magnetic Bearings (Kanazawa, Japan, August), 509-514, 1995.

6 Isaksson J L, On the dynamics of a rotor interacting with non-rotating parts, Ph.D. thesis, Linköping University, Sweden, 1994.

7 Knight J D, Ecker H, Simulation of nonlinear dynamics in magnetic bearings with coordinate coupling, Proc. 1996 Summer Computer Simulation Conf., (Portland,OR, July), 466-471, 1996.

8 Knight J D, Walsh T F, Virgin L N, Dynamic analysis of a magnetic bearing system with flux control, 2nd Int. Symposium on Magnetic Suspension Technology (Seattle, WA, July), 359-366, 1993.

9 Wang X, Noah S, Dynamic response of a rotor landed on auxiliary bearings, The 1995 ASME Design Technology Conf. (Boston, MA, Sept.), 1995.

Influence of contact and impacts on the dynamics of an elastic rotor with an elastic retainer bearing

Georg Wegener, Richard Markert**

This study deals with the nonlinear vibrations of a Jeffcott-rotor with an elastic retainer bearing with particular emphasis on the influence of impacts between rotor and retainer bearing. Impacts occur during the non-stationary transitions between states of motion with and without contact and also if the retainer bearing is misaligned towards the rotor axis. Experimental results are compared to computer simulations. For the computer simulations impacts were taken into account by contact models which approximate the force-deformation-relation in the contact point by means of simple linear or nonlinear springs and dampers. Impacts turn out to be particularly important if the retainer bearing is heavy in comparison to the rotor or if the retainer bearing is strongly misaligned.

1 Introduction

Retainer bearings are additional bearings, which have no supporting effect during normal operating conditions with small rotor deflections. For large rotor deflections however, the rotor gets in contact with the retainer bearing and the retainer bearing gives additional support [1]. Rigid retainer bearings are often used as security devices for the case of a failure of the rotor system. An example for this application are rotors with active magnetic bearings [2]. The present study however concerns with elastically mounted retainer bearings serving as devices for limiting the deflections of elastic rotors.

The dynamics of rotors in retainer bearings is in many aspects very similar to other problems of rotor-to-stator-contact and thus various results from publications in these fields (e.g. [3] and [4]) hold for rotors with retainer bearings as well. However, the order of magnitude of some parameters differs considerably. The simplest model for the dynamics during contact is to consider contact as a change of the rotor's dynamic parameters. Yet in this study a separate set of equations of motion is used for the retainer bearing, which is coupled to the equations of motion of the rotor in case of contact.

Although the coupled system consisting of the rotor and the retainer bearing is nonlinear in case of contact, for constant rotational speeds purely harmonic

* *Darmstadt University of Technology , Mechanik II, Hochschulstr. 1, 64289 Darmstadt, Germany. E-mail: markert@mechanik.tu-darmstadt.de*

motions exist. They can be determined analytically as was shown in [5], [6] and [7].

The present study deals with the effects of impacts in rotor–retainer bearing systems. It extends the results the authors presented in [8]. The theoretical results are qualitatively compared to experimental results. Particular emphasis is put on the sensitivity of the theoretical results to variations of the contact model. Although highly irregular and chaos-like motions caused by rotor-stator impacts have been reported by different authors ([9], [10], and others), they turn out to be of minor significance for rotors with retainer bearings.

2 Modeling the Rotor and the Retainer Bearing

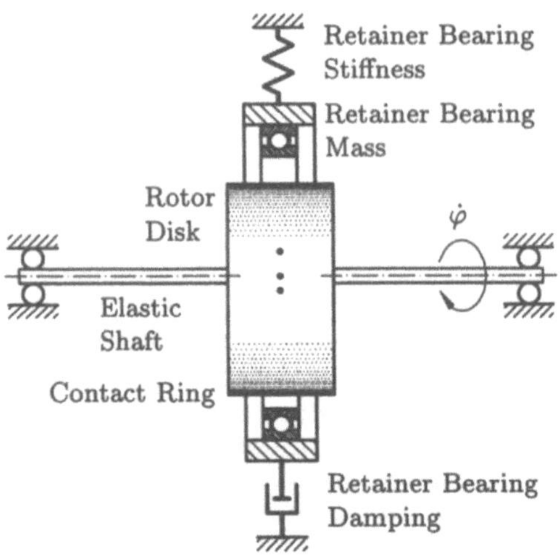

Figure 1: Jeffcott rotor with retainer bearing

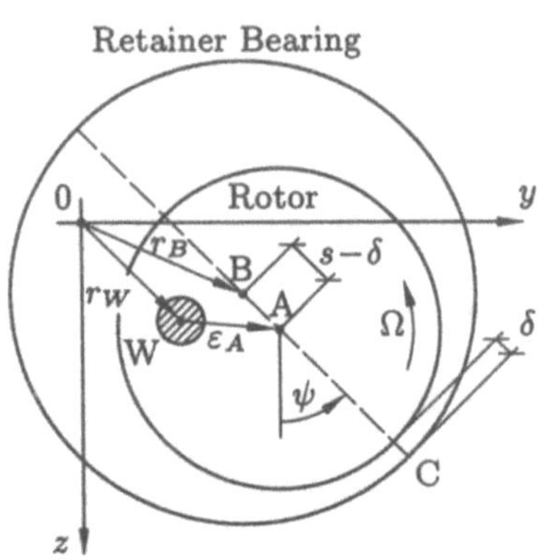

Figure 2: Kinematics of rotor and retainer bearing

A simple model for a rotor in a retainer bearing also used in [11] is shown in figure 1: The Jeffcott rotor consists of an elastic and massless shaft (stiffness k_W), which carries an unbalanced disk (mass m_W, mass eccentricity ε_M, damping factor b_W). The contact ring fixed to the disk has the geometric eccentricity ε_A. The angular velocity of the rotor is given by $\dot{\varphi}(t) = \Omega(t)$ where $\varphi(t)$ is the angle of rotation. The displacement of the center A of the contact ring fixed to the rotor are connected to the displacements of the shaft center W and the rotor mass center C_M by the geometric relations

$$r_M(t) = r_W(t) + \varepsilon_M\, e^{i\varphi(t)} \qquad \text{and} \qquad r_A(t) = r_W(t) + \varepsilon_A\, e^{i\varphi(t)} \qquad (1)$$

using complex coordinates $r = z + iy$.

The retainer bearing (mass m_B) is supported by linear springs and dampers (stiffness k_B, damping factor b_B), both acting between non-rotating parts of the retainer bearing and the pedestal, but it has the misalignment r_{BStat}. The average clearance between the rotor and the retainer bearing is s and the contact force F_C is defined by one of the contact models described in section 5. Friction in the contact point is neglected in this study and therefore the contact force has no tangential component.

The nonlinear vibrations of the rotor and the retainer bearing respectively are described by complex equations of motion,

$$m_W\ddot{r}_W + b_W\dot{r}_W + k_W r_W = -F_C - m_W\varepsilon_M[e^{i\,\varphi}]^{\cdot\cdot}, \tag{2}$$
$$m_B\ddot{r}_B + b_B\dot{r}_B + k_B(r_B - r_{BStat}) = F_C\,. \tag{3}$$

Whether the rotor and the retainer bearing are in contact is determined by the smallest distance δ between the contact ring fixed to the rotor and the retainer bearing (figure 2). The actual distance δ between the contact ring and the retainer bearing is given by the equation

$$\delta\, e^{i\,\psi} = r_B - r_A + s\, e^{i\,\psi}\,. \tag{4}$$

The angle ψ defines the normal direction at the contact point in case of contact ($\delta \leq 0$) or the direction of the shortest distance in case of no contact ($\delta > 0$). If the rotor touches the retainer bearing ($\delta \leq 0$), the equations of motion for both the rotor and the retainer bearing are coupled by the contact force F_C. If the distance δ is positive the contact force F_C becomes zero and the rotor and the retainer bearing move independently.

If the contact surface is rigid and frictionless the geometric condition for the distance δ and the contact force F_C are simply

$$\begin{aligned} &\delta = 0 \quad \text{and} \quad F_C = |F_C|e^{i\psi} \quad \text{for contact and} \\ &\delta > 0 \quad \text{and} \quad F_C = 0 \qquad\quad\;\; \text{for no contact.} \end{aligned} \tag{5}$$

This description of the contact does not contain any statement about the transient behavior at the beginning of the contact when the rotor hits the retainer bearing. The difference in the normal velocities of the rotor and the retainer bearing results in impacting between the components. In section 5 some hypotheses for describing the impacts are discussed.

3 Quasi-Stationary Vibrational Behavior

Although the coupled system is nonlinear, purely harmonic solutions exist for the unbalance excited vibrations of a rotor with a retainer bearing which is centric with the shaft. Figure 3 shows the unbalance response for steady state operation (constant rotational speed Ω) calculated analytically with a technique explained in [6] and [7]. The behavior is typical for a nonlinear system with a stiffening spring characteristic. The stiffening here results from the additional stiffness acting from the retainer bearing, when the rotor deflection exceeds the clearance. The behavior shows the typical bent resonance curve, multiple solutions and jump phenomenon. The critical speed $\omega_0 = \sqrt{k_W/m_W}$

of the rotor without retainer bearing is passed with low amplitudes. The jumps in these stationary solutions indicate where to expect transient motions during slow, quasi stationary acceleration or deceleration.

In the case of a misaligned retainer bearing ($r_{BStat} \neq 0$) however, the stationary rotor vibrations are no longer harmonic if the rotor is in contact with the retainer bearing. Subharmonic vibrations with $\Omega/2$ or $\Omega/3$ can be observed [12], and simulation results as well as experiments suggest that even chaotic behavior can occur [10].

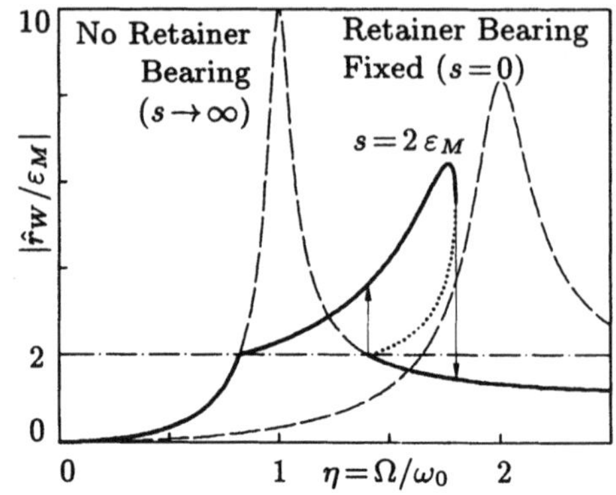

Figure 3: Unbalance response (rotor deflection) of a Jeffcott rotor with a retainer bearing

4 Experimental Results

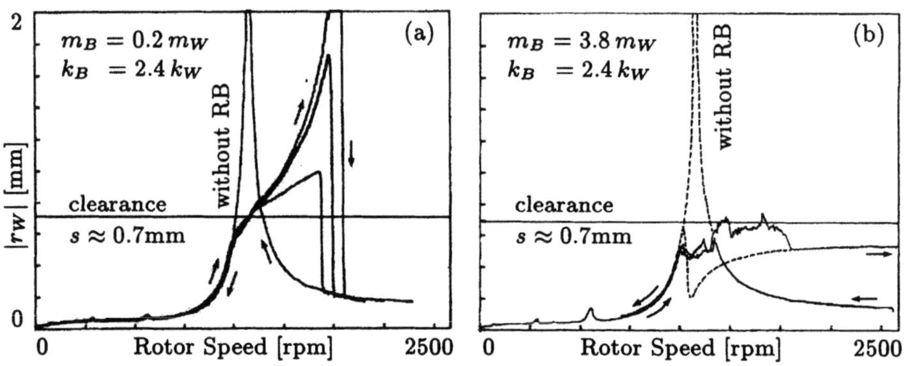

Figure 4: Experimental results: (a): smooth (lightweight retainer bearing with varied damping), (b): impacts (heavy retainer bearing).

Experimental investigations were carried out with a test rotor similar to a Jeffcott rotor [7]. The test rig allows to vary the parameters of the rotor and the retainer bearing easily. The results shown in figure 4a confirm the theoretical results for stationary operation for a lightweight retainer bearing with different damping coefficients, yet in other cases the behavior may be totally different as a consequence of even small inaccuracies.

An example for such a behavior is given in figure 4b, which is for a rotor with a considerably heavier retainer bearing. The behavior differs in two ways from that of the lightweight retainer bearing. In agreement with theoretical results for heavy retainer bearings, the rotor does not separate from the retainer bearing automatically to the self-centered state in the supercritical range. Instead, the rotor and the retainer bearing move with continuous contact around the common mass center like a Hula hoop.

The second effect is the chaos-like behavior within the speed range around the critical speed. The rotor does not move synchronously with the rotational speed as in the low or high speed ranges. Instead, the rotor and the retainer bearing bounce against each other and no periodicity can be detected. Such a behavior cannot be predicted by theoretical steady-state investigations and thus illustrates that contact models for transient behavior and impacts are necessary.

5 Contact Models

For modeling the impacts additional hypotheses are necessary describing either the behavior of the rotor and the retainer bearing during contact or defining the state of motion immediately after impact-like contact events. In the following, different approaches are given and the corresponding impulsive forces and hysteresis cycles are discussed.

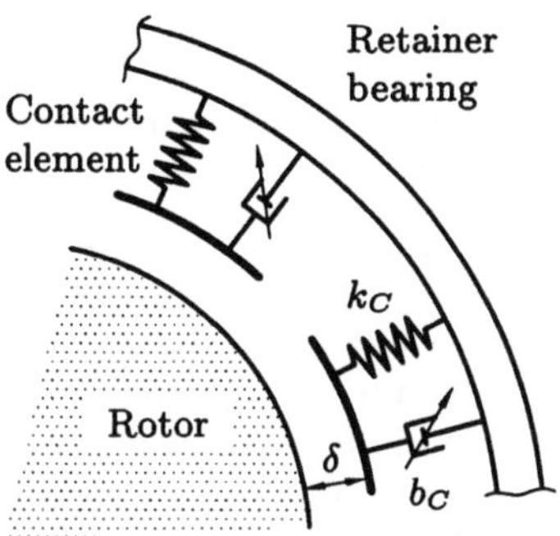

Figure 5: Equivalent contact model

The classic approach to describing impact events implies rigid contact surfaces and neglects the impact duration. This results in instantaneous changes in the velocities of both the rotor and the retainer bearing whereas their positions remain unchanged during the short contact duration. If friction is neglected, only the velocity components in normal direction $\dot{r}_{WN}$ and $\dot{r}_{BN}$ of rotor and retainer bearing respectively change abruptly, whereas the tangential components remain unchanged. The velocities $\dot{\bar{r}}$ immediately after the impact can be calculated from the corresponding values $\dot{r}$ just before the impact applying the conservation of momentum and NEWTON's impact law (coefficient of restitution e)

$$e = -\frac{\dot{\bar{r}}_{WN} - \dot{\bar{r}}_{BN}}{\dot{r}_{WN} - \dot{r}_{BN}} \tag{6}$$

to the normal components of the respective velocities.

Other approaches to the modeling of contact, which have some numerical advantages, take into account the elasticity of the contact region and therefore allow the distance δ between the rotor and the retainer bearing to become negative which means penetration in a mathematical sense. The local elasticity and dissipation are substituted by simple isotropic force-deformation-laws as shown in figure 5. The resulting contact force F_C is a function of the distance δ and the normal component $\dot{\delta}$ of the relative velocity. Neglecting friction the complex contact force can be given in the general form

$$F_C = F_{CN}(\delta, \dot{\delta})\, e^{i\,\psi}, \tag{7}$$

where F_{CN} is the real-valued magnitude of the contact force and ψ is the angle of the contact direction.

Various contact models have been applied to the rotor-retainer bearing system. All have in common, that the contact force is zero whenever the distance δ is positive. The respective formulae for the contact force are used only when δ is negative, i.e. in the case of penetration. The contact forces corresponding to the different contact models are compared in figure 6 for a single impact event.

The easiest way to include stiffness and damping is a purely linear contact model defined by

$$F_{CN} = k_C(-\delta) + b_C(-\dot{\delta}), \tag{8}$$

with a hysteresis loop shown in figure 6a. Several authors have successfully applied this simple model with even zero damping for numerical simulations. Modified models, so-called "pseudo-linear" models, can be derived from the linear model by introducing additional switching functions $\sigma(\delta, \dot{\delta})$,

$$F_{CN} = [k_C(-\delta) + b_C(-\dot{\delta})]\,\sigma(\delta, \dot{\delta}). \tag{9}$$

The switching function can be chosen such that discontinuities of the contact force or tensile contact forces are avoided. An example is shown in figure 6b.

The contact force relations can also be nonlinear in the stiffness or the damping. Examples for nonlinear stiffness relations are a hyperbola-shaped contact force characteristic

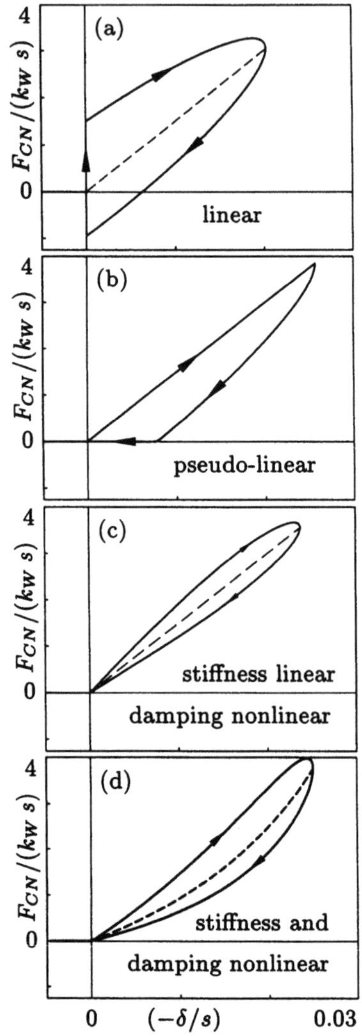

Figure 6: Hysteresis Loops (contact force vs. penetration) for different contact models

$$F_{CN,\text{elastic}} = k_n\langle-\delta\rangle \frac{d_{\max}}{d_{\max} - (-\delta)} \tag{10}$$

and the power-law

$$F_{CN,\text{elastic}} = k_n\langle-\delta\rangle^n, \tag{11}$$

which agrees with the HERTZian theory for the special case $n = 1.5$.

Examples for nonlinear damping of the contact are

$$F_{CN} = F_{CN,\text{elastic}} - b_C \frac{\langle-\delta\rangle}{d_{\text{ref}}} \dot{\delta} \tag{12}$$

and the product formulation suggested in [13]

$$F_{CN} = F_{CN,\text{elastic}} \cdot \left[e^{\beta(-\dot{\delta})} \right] . \tag{13}$$

Figures 6c and 6d show the hysteresis loops for two nonlinear contact models: figure 6c a combination of a linear spring with the nonlinear damping according to (13), figure 6d a combination of the nonlinear stiffness from (10) with the damping from (13). A more detailed comparison of hysteresis loops and time-histories of the corresponding contact forces for various contact models is given in the recent publication [8].

A good practical measure for the comparison of impact models is NEWTON's coefficient of restitution e defined in (6). Although usually used for instantanuous rigid impacts it can as well be calculated from the time-histories obtained from contact models with deformation. The examples shown in figure 7 are the purely linear, the pseudo-linear and the nonlinear contact forces from figures 6a, 6b and 6d respectively. The time histories were obtained by numerical simulation with a standardized initial velocity. As in the hysteresis loops, the effects of the contact model are superimposed by the effects of the rotor stiffness and the retainer bearing stiffness, whereas the damping of rotor and retainer bearing was neglected for this investigation.

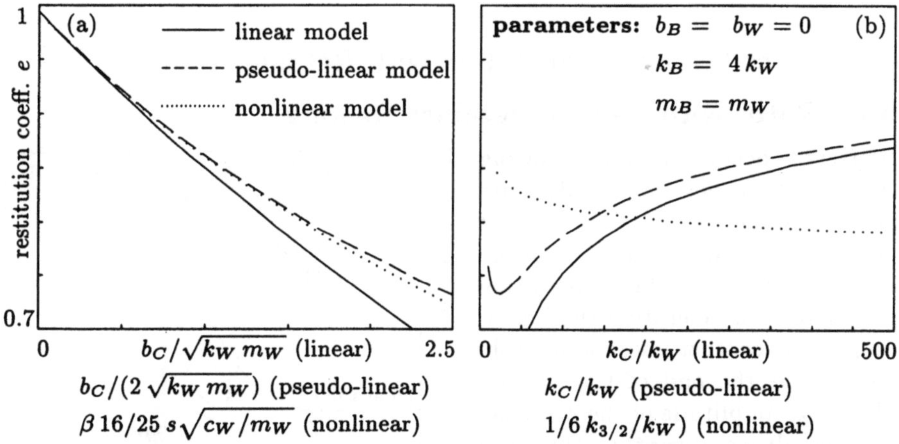

Figure 7: Coefficient of restitution e for different contact damping models. (a): e versus contact damping, (b): e versus contact stiffness. initial conditions: $r_{W0} = s$, $\dot{r}_{W0} = \frac{1}{2} s\,\omega_0$, $r_{L0} = 0$, $\dot{r}_{L0} = 0$

Figure 7a shows, that in all models increasing damping results in a decreasing coefficient of restitution e. A more surprising detail is, that the restitution coefficients for zero damping are a little below unity for all models. This results from the finite impact duration, since the momentum decreases as the deflections of the springs supporting the rotor and the retainer bearing increase.

In general, for both the linear and the pseudo-linear model an increasing contact stiffness leads to an increasing restitution coefficient. This results from the fact, that a stiffer spring causes a shorter contact duration and a smaller penetration which leads to a reduced dissipation. If in the pseudo-linear model the damping acts only during the restitution phase or only during the compression phase, the damping values required for the same dissipation are approximately twice as big as for the purely linear model. Another property ot the pseudo-linear model is that the contact force is set to zero when the sum of elastic contact force and contact damping force would be zero. This effect, which causes an increase of the restitution coefficient, turns out to be stronger when the stiffness is relatively small. This causes the decrease of the restitution coefficient in the area with $k_C < 30\,k_W$.

In case of the nonlinear contact model the influence of the stiffness on the restitution coefficient is small. A more detailed investigation shows, that the slight slope of the curve for the restitution coefficient e is caused only by the influence of the stiffnesses of the rotor and the retainer bearing and not by the contact stiffness. The reason for this is due to the fact that the damping influence is in the nonlinear model proportional with the elastic portion of the contact force and thus compensates the fact that the impact duration decreasees with increasing stiffness.

6 Impact Dynamics of Rotors with Retainer Bearing

6.1 Rotor with Centric Retainer Bearing

Impacts occur mainly during the transitions between steady-state motions with and without contact between rotor and retainer bearing. As can be seen from figure 8, during a slow rotor acceleration the beginning of the contact phase is smooth, whereas at the end of the contact phase the amplitude of the rotor deflections drops abruptly. During a slow deceleration the rotor deflections grow rapidly at the beginning of the contact phase whereas the separation at the end of the contact phase is smooth. The most intense impacts occur during an acceleration immediately after the separation of rotor and retainer bearing. A detailed study of this phase of motion in [8] has shown no significant differences for different contact models and damping values.

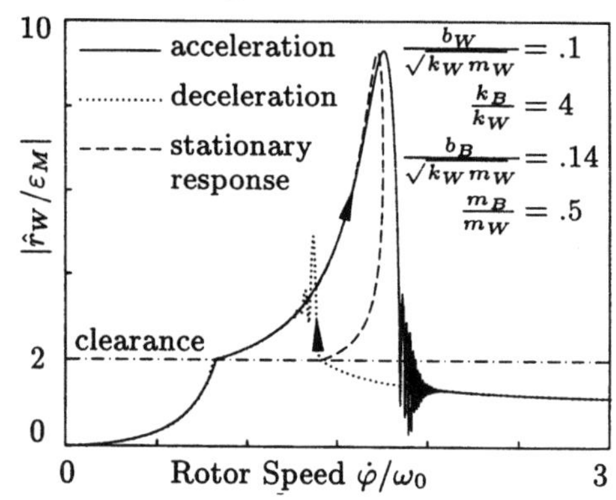

Figure 8: Non-stationary deflection of a Jeffcott rotor with a centric retainer bearing ($r_{BStat} = 0$)

6.2 Rotor with Non-Centric Retainer Bearing

If the retainer bearing is not centric with the rotor axis ($r_{BStat} \neq 0$) impacts occur much more frequently, sometimes even during steady-state motions at constant rotational speeds. Various authors have studied effects of rotor-stator-rub situations similar to this case [12], [9] and [10]. Phenomena like rotor-sub-synchronous vibrations and partly even chaos-like motions have been reported for such a misalignment.

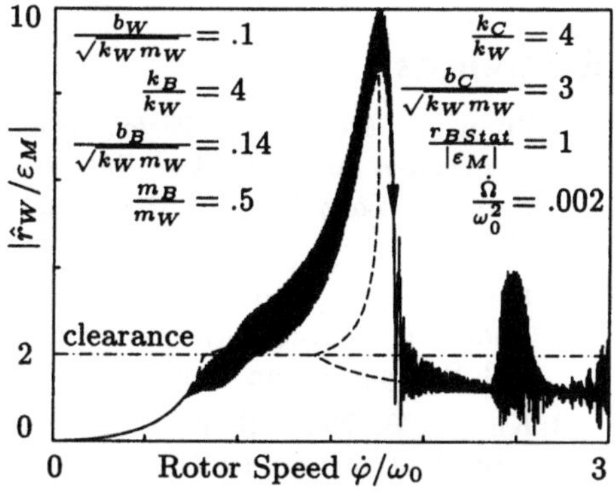

Figure 9: Non-stationary deflection of a Jeffcott rotor with non-centric retainer bearing

The rotor deflection shown in figure 9 for the slow acceleration of a rotor with a strongly misaligned retainer bearing ($r_{Bstat} = |\varepsilon_M|$) show a second maximum in a range of higher rotational speeds. This maximum is the resonance of the rotor-subsynchronous vibrations of order two.

Whereas the global behavior of the system like the resonance diagram remains practically unaffected by variations of the contact model and parameters, more detailed investigations show noticable differences. An example is the behavior at constant rotational speed of a rotor-retainer bearing system with and of one without contact damping illustrated in figures 10a and 10b. The rotor orbits were both determined by numerical integration and show a time span after the oscillations caused by the initial conditions have decayed. The orbit of the system with contact damping shows quasi-periodic behavior with the fundamental frequency of half the rotational frequency, whereas the system without contact damping shows no easily recognizable structure. In general the tendency is, that motions tend to be the more regular the more dissipation is acting during impacts.

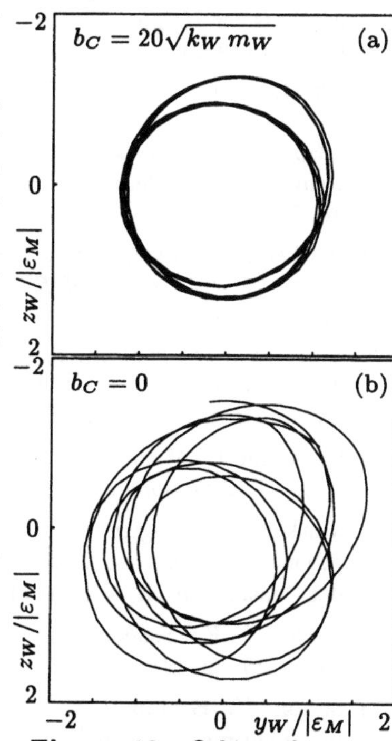

Figure 10: Orbits of a rotor with non-centric retainer bearing (rotor speed $\Omega = 2.5\,\omega_0$)

Acknowledgement

The research project is funded by the German Federal Research Society DFG.

References

1 Zippe, G., Meinke, P.: *Verfahren und Vorrichtung zum Durchlaufen kritischer Drehzahlen langgestreckter Rotoren. (Method and Device for Passing Critical Speeds of Long Rotors.)* Patent specification (FR Germany) DE 2632586 C2 (1983).

2 Fumagalli, M., Varadi, P., Schweitzer, G.: *Impact Dynamics of High Speed Rotors in Retainer Bearings and Measurement Concepts.* Proc. of the Fourth Intl. Symposium on Magnetic Bearings, Zurich (1994) pp. 239-244.

3 Kim, Y. B., Noah, S. T.: *Steady-State Analysis of a Nonlinear Rotor-Housing System.* J. of Eng. for Gas Turbines and Power *113,* 4 (1991) pp. 550-556.

4 Ehrich, F. F., O' Connor, J. J.: *Stator Whirl With Rotors in Bearing Clearance.* Journal of Engineering for Industry *89* B (3) (1967) pp. 381-390.

5 Black, H. F.: *Interaction of a Whirling Rotor With a Vibrating Stator Across a Clearance Annulus.* J. of Mech. Engineering Science *10,* 1 (1968) pp. 1-12.

6 Edbauer, R., Meinke, P., Müller, P. C., Wauer, J.: *Passive Durchlaufhilfen beim Durchfahren biegekritischer Drehzahlen elastischer Rotoren. (Passive Methods for Helping Elastic Rotors to Pass the Critical Speeds.)* VDI-Berichte Nr. 456 (1982) pp. 157-166.

7 Abraham, D., Markert, R., Witfeld, H.: *Experimentelle Untersuchungen zu aktiven und passiven Resonanzdurchlaufhilfen für extrem elastische Rotoren. (Experimental Investigations of Active and Passive Methods for Helping Extremely Elastic Rotors to Pass the Resonances.)* VDI-Berichte Nr. 695 (1988) pp. 145-169.

8 Markert, R., Wegener, G.: *Transient Vibrations of Elastic Rotors in Retainer Bearings.* Proceedings of ISROMAC-7, Hawaii, Bird Rock Publ. House (1998), pp. 764-774.

9 Gonsalves, D., Neilson, R. D., Barr, A. D. S.: *Response of a Discontinuosly Nonlinear Rotor System.* Proc. of the Intl. Conference on Rotating Machine Dynamics, Springer Publishers, London (1992) pp. 182-189.

10 Goldman, P., Muszynska, A.: *Chaotic Behavior of Rotor/Stator Systems With Rubs.* Transactions of the ASME *116* (1994) pp. 692-700.

11 Markert, R., Wegener, G.: *Dynamik von elastischen Rotoren in Fanglagern. (Dynamics of Elastic Rotors in Retainer Bearings.)* Schwingungen in rotierenden Maschinen III, Vieweg (1995) pp. 20-30.

12 Ehrich, F. F.: *Subharmonic Vibration of Rotors in Bearing Clearance.* ASME Paper No. 66-MD-1 (1966).

13 Simon, U., Brommundt, E.: *Periodische Bewegungen einer Pendelzentrifuge in einem mehreckigen Fanglager. (Periodic Motion of a Pendulum Centrifuge in a Polygonal Retainer Bearing.)* Schwingungen in rotierenden Maschinen IV, Vieweg (1997) pp. 181-188.

Wavelets transform in analysis and control of a piezoelectric vibroconverter with impacts

V.F. Poterasu[*]

Wavelets provide a new tool for the analysis of vibrations especially for localized phenomena as resonance or impact, i.e. discontinuous actions. The study is exemplified on a piezoelectric vibroconverter with an impact rheological model. It consists of a piezoelectric body and another rigid body prestressed to each other by means of a connecting element. During the vibrations of the piezoelectric vibroconverter between its rigid hand and the rigid body the impact interaction takes place. The impact stationary regime and the shock phenomena are analyzed by means of Haar wavelet method. We suppose that the vibrations can be controlled by a feedback system. By means of Haar wavelets we arrive to a signal flow graph of a position control system.

1 Introduction

Vibrodrives are the mechanisms converting high-frequency vibrations of an input line into the directed stepwise or continuous motion of an output link by means of essentially nonlinear mechanical interaction forces.

Vibrodrives possess several specific features making difficult the immediate applications of the existing theory of vibroconveyers and vibroimpact systems. The characteristics of vibrodrives are small vibration amplitudes (several microns) and high frequencies therefore during the interaction between input and output links a contact surface elasticity is to be taken into account. Microimpacts can not be assumed instantaneous, as an impact time takes a significant part of a vibration period. Consequently, a stereomechanical model characterized by a velocity restitution ratio in general is insufficient for presenting the contact phenomena in vibrodrives. Moreover, a resonant modal vibration of the input links usually is employed in vibrodrives, therefore the distributed parameter or finite element models are necessary.

Together with the stereomechanical impact models, the combined local and wave contact interaction models have been taken into account, the viscoelastic surface rheology approach being commonly employed.

A vibroconverter should be considered as the main part of a vibrodrive. Vibroconverters produce mechanical vibrations under the high frequency alternating voltage applied to their electric terminals. The operations of piezoelectric vibroconverters are based upon the excitation of their modal

[*] *Theoretical Mechanics Department, Technical University Iasi, Romania*

vibrations employing an inverse piezoelectric effect. Recently, mathematicians, scientists and engineers have been active in seeking new methods for representing oscillatory, discontinuous or intermittent functions or complex combinations of these: the wavelet transform, new functions with several applications from signal theory, pattern recognition, fluid turbulence, geophysics, etc.

2 Finite element model of a piezoelectric vibroconverter

In a three dimensional continuum the equilibrium is expressed by the differential equation

$$(1) \qquad B_n^T \sigma + p = \rho \ddot{u}$$

σ – the stress vector, p – the body force vector, n – the displacement vector. The electrical equilibrium

$$(2) \qquad B_\Phi^T D = 0$$

where $D^T = [D_1\ D_2\ D_3]$ is the electrical displacement and

$$B_\Phi^T = \left[\frac{\partial}{\partial x_1}\ \frac{\partial}{\partial x_2}\ \frac{\partial}{\partial x_3}\right]$$

The boundary conditions for surface traction

$$(3) \qquad \left(\bar{t} - t\right) = 0 \qquad T_{\sigma n}\sigma = \bar{t}$$

$T_{\sigma n}$ - normal transformation matrix for σ, $\bar{t}^T = \left[\bar{t}_n \quad \bar{t}_{t1} \quad \bar{t}_{t2}\right]$ the vector of applied surface traction.

The electrical boundary conditions

$$(4) \qquad \left(\bar{Q} - Q\right) = 0 \qquad \text{with} \qquad T_{dn} D = \bar{Q}$$

T_{dn} - the normal transformation matrix for D, Q – the applied surface change.

The linear coupled electromechanical constitutive equations have the form

$$(5) \qquad \begin{aligned} \sigma &= C\varepsilon - eE \\ D &= e^T\varepsilon + \mu E \end{aligned}$$

C – the elasticity matrix, e – the piezoelectric matrix, μ - the dielectric matrix.

The strain – displacement matrix

$$(6) \qquad \varepsilon = B_n u$$

and analogous for the electric field

$$(7) \qquad E = B_\Phi \Phi$$

The coupled electromechanical functional

$$(8) \qquad \begin{aligned} &\int_V \delta u^T \rho \ddot{u}\, dV + \int_V (B_u \delta u)^T C B_u u dV + \int_V (B_u \delta u)^T e B_\Phi \Phi dV + \\ &\int_V (B_\Phi \delta\Phi)^T e^T B_u u dV - \int_V (B_\Phi \delta\Phi)^T \mu B_\Phi \Phi dV - \int_V \rho u^T \bar{p} dV - \\ &\int_{\Gamma_t} \delta u^T \bar{t} d\Gamma - \int_{\Gamma_q} \delta\Phi \bar{Q} d\Gamma = 0 \end{aligned}$$

the kinematic boundary conditions:

(9) $u = \bar{u} \qquad \delta u = 0$ on surface Γ_t

the boundary conditions for the electric potential

(10) $\Phi = \bar{\Phi} \qquad \delta\Phi = 0$ on surface Γ_q

For discretization we approximate the mechanical displacements and the electric potential in a finite element by using the interpolation functions G

$$(11) \qquad u_i = \sum_{i=1}^{N} G_j u_{ij} \qquad \Phi = \sum_{i=1}^{N} G_j \Phi_j$$

where N is the number of element nodes and j is the nodal index.

For the electromechanical problem we extend the mechanical element model vector q_e by adding the electric potential Φ to each node.

$$(12) \qquad q_e^T = \left[u_1;\ \Phi_1 \ldots u_j;\ \Phi_j \ldots u_N;\ \Phi_N\right]$$

in matrix notation

$$(13) \qquad \begin{bmatrix} u \\ \Phi \end{bmatrix} = \begin{bmatrix} G_1 I & 0 & G_2 I & 0 & \ldots & G_N I & 0 \\ 0 & G_1 & 0 & G_2 & \ldots & 0 & G_N \end{bmatrix} q_e = G q_e$$

and

$$(14) \qquad \begin{bmatrix} \varepsilon \\ E \end{bmatrix} = \begin{bmatrix} B_u G_1 & 0 & B_u G_2 & 0 & \ldots & B_u G_N & 0 \\ 0 & B_\Phi G_1 & 0 & B_\Phi G_2 & \ldots & 0 & B_\Phi G_N \end{bmatrix} q_e = $$

$$= B_{u\Phi G} q_e$$

From (8), (13) and (14) we obtain the differential equation system

$$(15) \qquad M_e \ddot{q}_e + R_e \dot{q} + K_e q_e = f_e$$

where M_e is the element mass matrix

$$(16) \qquad M_e = \int_{V_e} \rho G^T G dV$$

K_e is the element stiffness / elasticity matrix

$$(17) \qquad K_e = \int_{V_e} B_{u\Phi G}^T \begin{bmatrix} C & e \\ e^T & -\mu \end{bmatrix} B_{u\Phi G} dV$$

and f_e is the load vector

$$(18) \qquad f_e = \begin{bmatrix} \int_{V_e} G^T \bar{p} dV + \int_{Q_e \Gamma} G^T \bar{t} d\Gamma \\ \int_{\Gamma_{eq}} G^T \bar{Q} d\Gamma \end{bmatrix}$$

3 Analysis of lumped – parameters by Haar wavelets

Instead of seeking to break down the signal into its harmonics, which are global functions that go on forever, the signal is broken down into a series of local basis functions called wavelets. Each wavelet is located at a different position on the time axis and it is local in the sense that it decays to zero when sufficiently far from its center. At a finest scale, wavelets may be very short indeed, at a coarse scale they may be very long. Any particular feature of a signal can be identified from the scale and position of the wavelets into which it is decomposed. Thus, we may analyze any nonlinear non-stationary signals

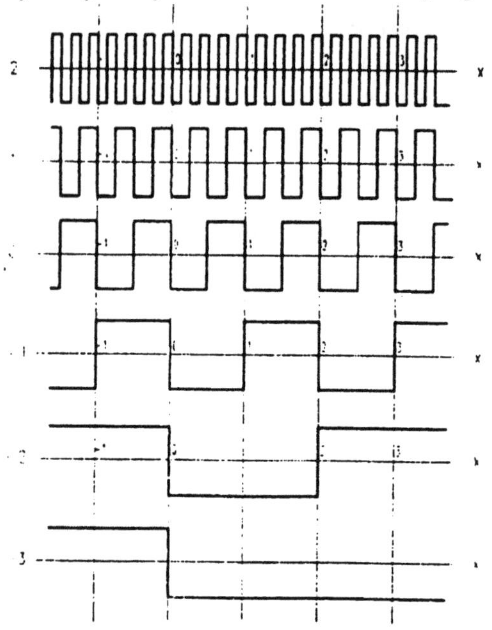

The orthogonal set of Haar wavelets are defined in the Fig. 1, a group of square waves with magnitude ±1 in some intervals and zero elsewhere. Just These zeros make the Haar transform faster than other square functions such as Walsh's. In general is, like a kind of local, filtered Fourier transform and is obtained by dilating, translating and rotating the analysing wavelet and then convolving it with the signal.
By looking at the modulus of transform as a function of position and scale one can investigate the dependence of quantities such as energy on scale and location. It is always possible to recover a smooth version of the energy spectrum from a wavelet transform by integrating it over all locations. The fundamental difference between Fourier transform and the wavelet transform is that Fourier transform breaks up a signal into contributions from different sine waves of various phases and frequencies, while the wavelet transform breaks up the signal into contributions from wave pulses (mother wavelets) which are dilated, translated and in two or more dimensions, if the wavelet is anisotropic, or rotated.
Plots of the phase as a function of position and scale can be used to detect singularities.
Several books as C.K. Chui [1], Y.T. Chan [2], D.E. Newland [3], M.V. Wickerhonser [4] study various aspects of wavelets.

$$(19) \qquad h_n(t) = h_1\left(2^j t - \frac{k}{2^j}\right)$$

$$n = 2^j + k \qquad j \ge 0 \qquad 0 < k \le 2^j$$

h_0 is also included to make this set complete.
The Haar wavelets are orthogonal to each other

$$(20) \qquad \int_0^1 h_j(t)h_l(t)dt = 2^{-j}\delta_{il} = \begin{cases} 2^{-j} & i = l = 2^j + k \\ 0 & i \ne l \end{cases}$$

and they form a very transform basis.
Any function y(t) which is square integrable in the interval [0,1] namely

$\int_0^1 y^2(t)dt$ is finite, can be expanded into Haar wavelets [5], [6].

$$(21) \qquad y(t) = c_0 h_0(t) + c_1 h_1(t) + \ldots = c^T H(t)$$

$$(22) \qquad c_i = 2^j \int_0^1 y(t) h_j(t) dt$$

The Haar coefficient c_i can be obtained by applying eq. (21) directly, however it is more convenient to evaluate it by matrix inversion

$$(23) \qquad c^T = y(t) H^{-1}(t)$$

called the forward transform. Since H and H^{-1} contain many zeros, the Haar transform becomes much faster than the Fourier transform and it is even faster than the Walsh transform.

In order to solve the differential dynamic equation models we need to perform integration. If we denote

$$(24) \qquad \int_0^1 H(t)dt = PH(t)$$

for an m-th order system with $m = 2^j$, j a positive integer, the P_m is given as

$$(25) \qquad P_m = \frac{1}{2m} \begin{bmatrix} 2mP_m/2 & -H_m/2 \\ H_m^{-1}/2 & 0 \end{bmatrix}$$

For the state space equation of motion with feedback law we assume that n(t) is square integrable in the interval $0 \le t \le 1$. In the Haar series expression we have

$$(26) \qquad u(t)=RM(t)$$

where R is a pxm matrix, the system having n states x(t), p inputs u(t) and q outputs y(t).

To avoid dealing with impulse functions, we expand the variable x(t) instead of x(t) itself into a Haar series

$$(27) \qquad \dot{X}(t) = \overline{B}H(t)$$

Integrating yields

$$(28) \qquad X(t) = \overline{B}PH(t) + X_0$$

and from previous relations

$$(29) \qquad \overline{B}H(t) = A\overline{B}PH(t) + AX_0 + BPH(t)$$

$$(30) \qquad \overline{B} - A\overline{B}P = \{[AX_0 \;\; 0 \;\; 0 \ldots 0] + BR\} HU$$

The number of zero columns involved in eq. (30) is m-1, or

$$(31) \qquad \overline{B} - A\overline{B}P = [AX_0 \;\; 0 \;\; 0 \ldots 0] + BR = R_1$$

$$(32) \qquad [f_0 \;\; f_1 \ldots f_{m-1}] - A[f_0 \;\; f_1 \ldots f_{m-1}]P = [g_0 \;\; g_1 \ldots g_{m-1}]$$

where f_i and g_i are the (i+1)-th column of B and R_1 respectively.

$$\begin{bmatrix} M \\ m \end{bmatrix} = \begin{Bmatrix} \ddot{U} \\ \ddot{X}_2 \end{Bmatrix} + \begin{bmatrix} C \\ c \end{bmatrix} \begin{Bmatrix} \dot{U} \\ \dot{X}_2 \end{Bmatrix} + \begin{bmatrix} K \\ P \end{bmatrix} \begin{Bmatrix} U \\ X_2 \end{Bmatrix} =$$

$$(34) \qquad = \begin{Bmatrix} F_0 \\ \vdots \\ 0 \\ -F_0 \end{Bmatrix} + \begin{Bmatrix} P \\ \vdots \\ -P \\ 0 \end{Bmatrix} \sin \omega t + \begin{Bmatrix} 0 \\ \vdots \\ -F_N \\ F_N + F_T \end{Bmatrix}$$

where F_T is the friction force and F_N is the normal interaction force in the contact pair 4

$$(35) \qquad \begin{Bmatrix} (U_n - X_2)k + (\dot{U}_n - \dot{X}_2)c & \text{if } U_n \geq X_2, F_N \geq 0 \\ 0 & \text{otherwise} \end{Bmatrix}$$

U_n being the last element of the displacement vector V.

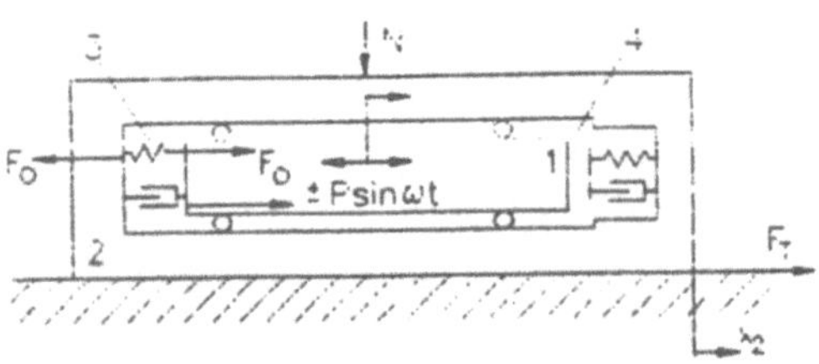

Fig.3

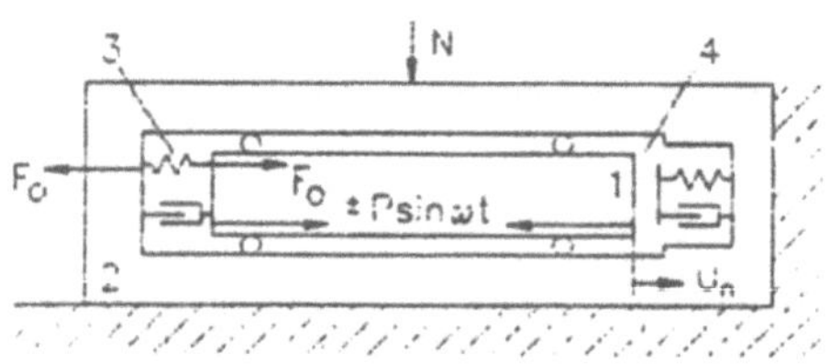

Fig.4

In the Fig. 3 is presented a simplified model, the two bodies 1 and 2 being connected by the spring 3 as a whole and the external excitation is applied by means of two harmonic forces. The impact takes place between the rigid bodies having the masses m_i and m_2, their displacement being respectively X_i and X_2.

The second simplified model represented in Fig. 4 considers the vibroconverter as a resonant vibro-impact system vibro-impact system.

In order to integrate the differential motion equation (34) we use the Haar wavelets with 6 levels during the interval (5T – 10T) where T is the period excitation and the constant resonant frequency of the free vibration,

$\omega = (k_i/m_1 + h_i/m_2)^{i/2,}\ c_i$ and dimensionless relations

$$\dot{X} = \frac{X}{S\omega}, \quad \overline{N} = \frac{N}{(m_1 + m_2)g}, \quad \overline{m} = \frac{m}{m_1 + m_2}, \quad \overline{k} = \frac{k}{k_i}, \quad \overline{c} = \frac{c}{k_i}\omega$$

where S denotes the mean peak-to-peak amplitude of vibration.

$S = \max(X_2 - X_1) - \min(X_2 - X_1)$

The results are presented in Fig. 5 for $\overline{\mathrm{N}}\mathrm{f} = 0{,}025 - 0{,}04$. The transient vibrations of the displacements for the local contact rheological model are illustrated in Fig. 6 vibroconverter prestressed against a rigid constraint.

If we have the control force exercised by the force in piezoelectric vibroconverter the eq. (34) can be transformed in the form

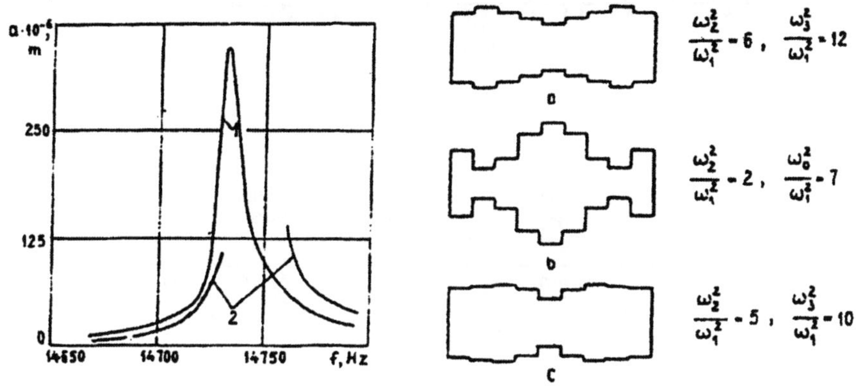

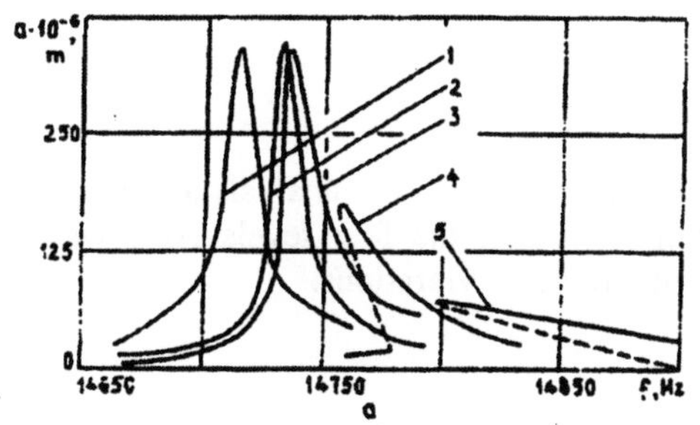

(36) $\qquad X(t) = Ax(t) + Bu(t)$

and the feedback law

(37) $\qquad y(t) = Cx(t) + Du(t)$

The signal flow graph for an illustrative example of dynamic equation is presented in Fig. 7.

Fig.5

$$X(t) = \begin{bmatrix} 0 & 18 & 0 & 0 \\ 0 & 0 & 25 & 250 \\ 0 & 0 & 0 & 250 \\ -k & -95{,}25 & 0 & -250 \end{bmatrix} x(t) + \begin{bmatrix} 0 \\ 0 \\ 0 \\ k \end{bmatrix} u(t)$$

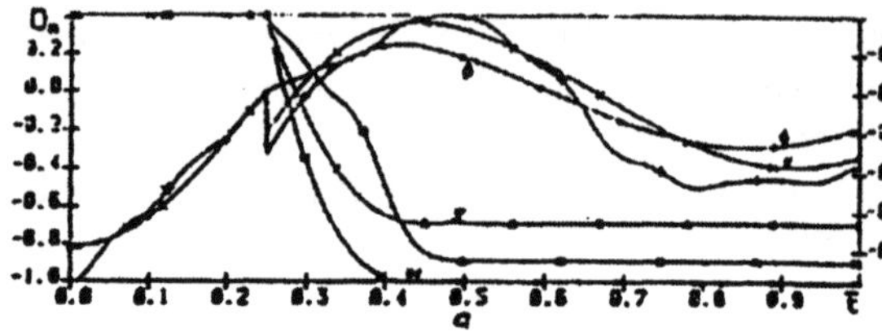

Fig. 6

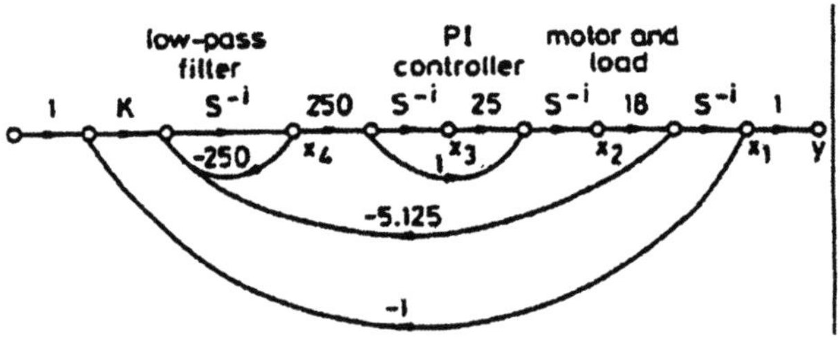

Fig.7

4 Conclusion

By means of FEM and wavelets transform we study the analysis and control of piezoelectric vibroconverter with impact. Using the four levels Haar wavelets function we may localize resonance and transient vibrations.

References

1. Chui C.K., Wavelets: A Mathematical Tool for Signal Analysis, SIAM, Philadelphia, 1997
2. Chan Y.T., Wavelet Basic, Kluwer Ac. Publ., 1995
3. Newland D.E., An Introduction to Random Vibrations, Spectral and Wavelet Analysis, Longman Pub. Group, 1993
4. Wickerhauser M.V., Adapted Wavelet Analysis from Theory to Software, A.K. Peters. Limited, 1994
5. Chen C.F., Hsiao C.H., Haar Wavelet Method for Solving Lumped and Distributed Parameter Systems, IEE Proc. Control Theory, 144, 1. 84-94, 1997
6. Strang G., Nguyen T., Wavelets and Filter Banks, Wellesley – Cambridge Press, Cambridge, MA, 1996
7. Babitsky V.I., Theory of Vibro-Impact Systems, Nauka, Moscow, 1978, (in Russian).

8. Barauskas R., etc., Design of Piezoelectric Bimorph elements, Vibrotehnika, 2(46), 118-127, 1983, (in Russian).
9. Barauskas R., Dynamic analysis and Synthesis of Elastic Structures with Unilateral Constraints, Kaunas University of Technology, 1992
10. Poterasu V.F., Optimal Control of Smart Composite Plates for Distributed Sensors Using FEM and Wavelets, Proc. Euromech 373 Colloquium, Modelling and Control of Adaptive Mechanical Structures, 11-13 March 1998, Magdeburg, 207-217.

Stability improvement of the vibro-impact discrete systems

J Awrejcewicz* and K Tomczak*

In this work a control of the one- and two-degree-of-freedom vibro-impact systems with a delay loop is presented and analysed. The aim of the delay loop application is to make the return to the vibro-impact periodic motion after the occurrence of disturbances quicker than in the case without the loop. The proposed analytical approach yields the required delay loop coefficients. The original vibro-impact map is described which allows to solve the problem analytically for near resonance case in the one-degree-of-freedom system. The numerical calculations have supported our theoretical investigations. In addition, an efficient delay loop control applied to two-degree-of-freedom system is proposed.

1 Introduction

Nowadays, in the field of the vibro-impact phenomena two main directions of investigations are dominating. The first one concerns the mathematical description of a vibro-impact motion including the restitution coefficient [1]. The experimental investigation shows that it depends on many parameters such as material of impacting bodies, their shapes and velocities [2, 3] and therefore it is difficult to define it exactly.

The second direction concerns the control of vibro-impact systems. Many references are devoted to a field of control of the nonlinear systems including the control of chaotic orbits [4, 5]. Among them we mention only the control methods with a delay loop [6], the adaptive control [7], the learning control systems [8] and others [9].

Generally, the aim of those approaches is to control rather complicated systems where a mathematical model is not known and their dynamics is tracked numerically. In contrary to those methods, in this work we propose an analytical approach to determine suitable delay loop coefficients to realise the required vibro-impact periodic dynamics for the resonance case. The obtained analytical formulas allow for a proper choice of the delay loop coefficients in order to

* Technical University of Łódź, Division of Automatics and Biomechanics (K-16), 1/15 Stefanowskiego St., 90-924 Łódź, POLAND.

achieve the required vibro-impact periodic motion quicker than in the case without a loop. When the vibro-impact periodic motion is achieved, the delay loop is automatically switched off.

2 Control of the vibro-impact systems using a feedback delay loop

We consider the following two vibro-impact systems presented in Fig. 1. In those systems m, m_1 and m_2 denote the masses, c_1 and c_2 are the damping coefficients, k_1, k_2 are the stiffness coefficients, f_0 is the friction coefficient. P_0 and w_0 denote the amplitude of force and the kinematic excitation, respectively, and φ and φ_1 are the phases. Finally, the coefficient s denotes the clearance.

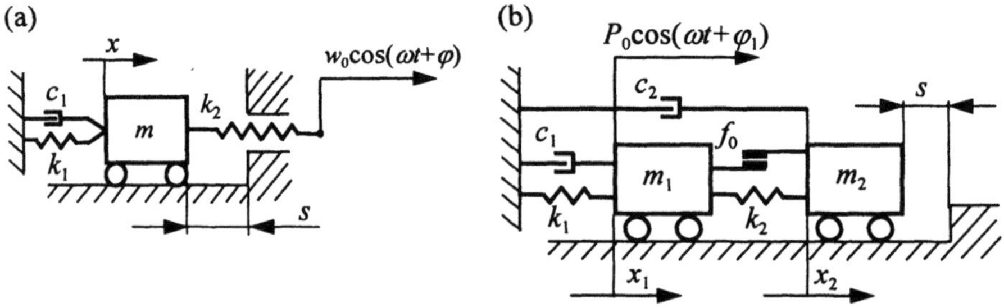

Fig. 1. Schematic view of the one-degree-of-freedom (a) and two-degree-of-freedom (b) vibro-impact system

The analysed one-degree-of-freedom system with the feedback delay loop (Fig. 2) is governed by the following equations

$$\ddot{x} + c\dot{x} + \alpha^2 x = P_0 \cos\omega t + S[x(t) - x(t-T)] + Q[\dot{x}(t) - \dot{x}(t-T)] \quad \text{for} \quad x < s, \tag{1}$$

$$x_+ = x_-, \quad \dot{x}_+ = -R\dot{x}_- \quad \text{for} \quad x \geq s,$$

where: $c = c_1/m$, $\alpha^2 = (k_1 + k_2)/m$, $P_0 = k_2 w_0/m$, $S = k_2 p/m$, $Q = k_2 q/m$, T is the period of a stabilised periodic orbit, and R denotes the restitution coefficient. The indeces "+" and "–" define positions and velocities of the body just after and before an impact, correspondingly.

It is assumed (Fig. 1) that the stabilised periodic orbit $x_0(t) = x_0(t-T)$ has the period of the excitation the same as in the system without that loop and that $x_0(t)$ is the particular solution for the system with and without the delay loop [10-12]. The delay loop starts to operate when the disturbances occur and is to act on the dynamics in order to achieve the vibro-impact periodic motion quicker than in the case without a loop.

The following system parameters were used for the simulation: $m = 1[kg]$, $c_1 = 0.2[Ns/m]$, $k_1 = 7[N/m]$, $k_2 = 1[N/m]$, $\omega/\alpha = 0.99$, $w_0 = 0.01[m]$. For the restitution coefficient $R = 0.6$ and $s = 0.0005[m]$ the vibro-impact motion with one impact during a period of excitation has occurred. The simulation model of the investigated system, as well as its control, are constructed using the MATLAB-Simulink package.

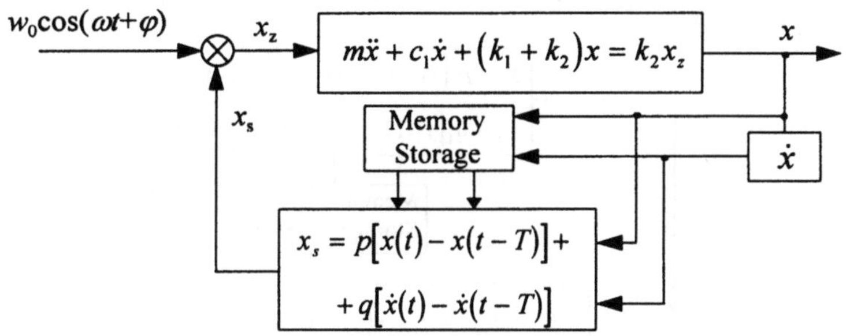

Fig. 2. Schematic view of control of the system presented in Fig. 1a

In order to accelerate a stabilization of periodic vibro-impact vibration, the following delay loop coefficients are applied: $p = 0[N/m]$ and $q = -0.08[Ns/m]$. In order to compare the simulation results, the additional parameter μ is defined which characterises the difference between two signals $x(t) - x(t-T)$. In Fig. 3 the time history of $x(t)$-$x(t$-$T)$ for the system without (Fig. 3a) and with (Fig. 3b) a feedback delay loop are presented.

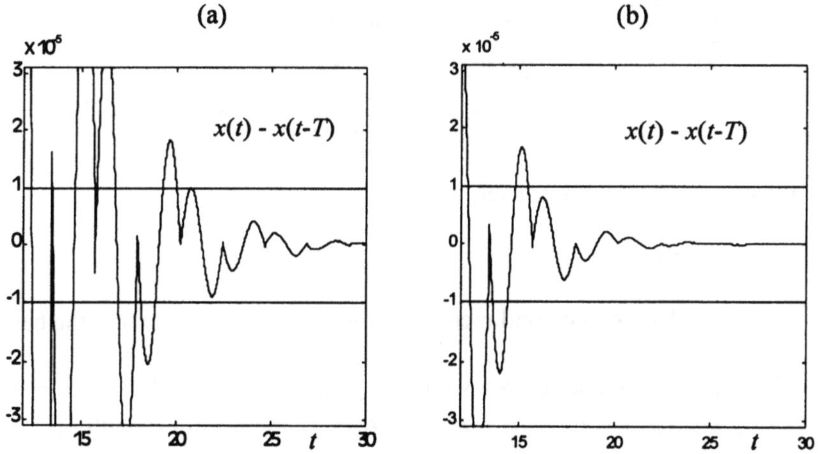

Fig. 3. Transients $x(t)$ - $x(t$-$T)$ for the system: (a) without the feedback loop ($p = 0$, $q = 0$); (b) with the feedback loop ($p = 0$, $q = -0.08$)

The results indicate that the time of stabilisation of the vibro-impact periodic motion is equal to 15.2 seconds in the case when a delay loop occurs. Without a loop the vibro-impact periodic attractor is achieved after 20 seconds. In both the cases $|\mu| = 10^{-5}$. The presented results prove efficiency of the applied feedback loop.

It is obvious that a control of the two-degree-of-freedom system can be realised in many ways. In this paper we present the possibility, where the delay loop acts on the mass m_1 (Fig. 4).

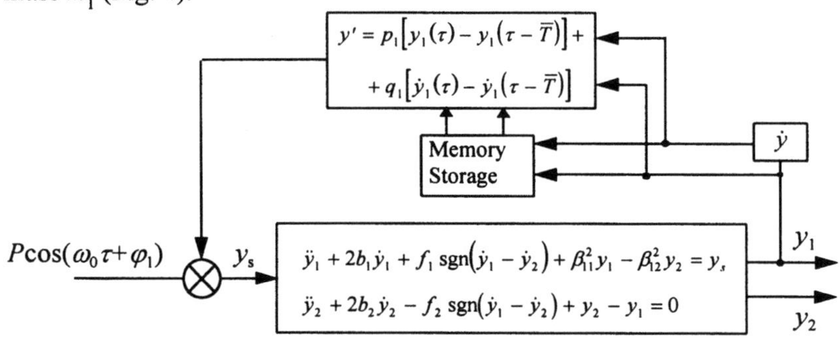

Fig. 4. Schematic view of control of the system presented in Fig. 1b

The system dynamics is governed by the following equations:

$$\ddot{y}_1 + 2b_1\dot{y}_1 + f_1\,\mathrm{sign}(\dot{y}_1 - \dot{y}_2) + \beta_{11}^2 y_1 - \beta_{12}^2 y_2 =$$
$$= P\cos(\omega_0\tau + \varphi_1) + p_1\left[y_1(\tau) - y_1(\tau - \bar{T})\right] + q_1\left[\dot{y}_1(\tau) - \dot{y}_1(\tau - \bar{T})\right],$$
$$\ddot{y}_2 + 2b_2\dot{y}_2 - f_2\,\mathrm{sign}(\dot{y}_1 - \dot{y}_2) + y_2 - y_1 = 0,$$

where: $b_1 = \dfrac{c_1}{2m_1}\sqrt{\dfrac{m_2}{k_2}}$, $b_2 = \dfrac{c_2}{2\sqrt{k_2 m_2}}$, $f_1 = \dfrac{m_2}{m_1}\dfrac{k_1}{k_2}\dfrac{f_0}{P_0}$, $f_2 = \dfrac{k_1}{k_2}\dfrac{f_0}{P_0}$,

$$\beta_{11}^2 = \frac{k_1 + k_2}{k_2}\frac{m_2}{m_1},\ \beta_{11}^2 = \frac{k_1 + k_2}{k_2}\frac{m_2}{m_1},\ \beta_{12}^2 = \frac{m_2}{m_1},\ P = \frac{m_2}{m_1}\frac{k_1}{k_2},\ \omega_0 = \omega\sqrt{\frac{m_2}{k_2}}.$$

When the mass m_2 meets the barrier, we have:

$$\begin{cases} y_{2+} = y_{2-} \\ \dot{y}_{2+} = -R\dot{y}_{2-} \end{cases} \quad \text{for } y_2 \geq \bar{s},$$

where $\bar{s} = s/x_{1st}$ denotes the nondimensional distance of the mass m_2 to the barrier.

The following system parameters were adopted for the simulation: $b_1 = 0.05$, $b_2 = 0.05$, $f_1 = 0.18$, $f_2 = 0.5$, $\beta_{12} = 1$, $\beta_{12} = 0.6$. The excitation amplitude $P = 1$ and the excitation frequency $\omega_0 = 0.885$. The restitution coefficient $R = 0.7$. The barrier is situated in the position $\bar{s} = 0.12$. The feedback loop parameters are as follows:

a) $p_1 = 0$ and $q_1 = 0$ for the system without the delay loop,

b) $p_1 = 0.05$ and $q_1 = 0$ for the system with the delay loop.
The vibro-impact periodic motion is characterised by two impacts during the period of excitation $\overline{T} = 2\pi/\omega_0$.

The simulation results are presented in Fig. 5 in the case without (a) and with (b) the delay loop, respectively. Without the loop the mass m_1 reaches a periodic orbit after 120, whereas the mass m_2 after 121.5 nondimensional units ($\mu = 10^{-3}$). With the loop the mass m_1 achieves the periodic orbit after 64, whereas the mass m_2 after the 71.5 nondimensional units.
The calculations showed that an application of the delay loop connected with only one mass (m_1) caused an acceleration in reaching the stable vibro-impact orbit of the second mass (m_2).

(a)

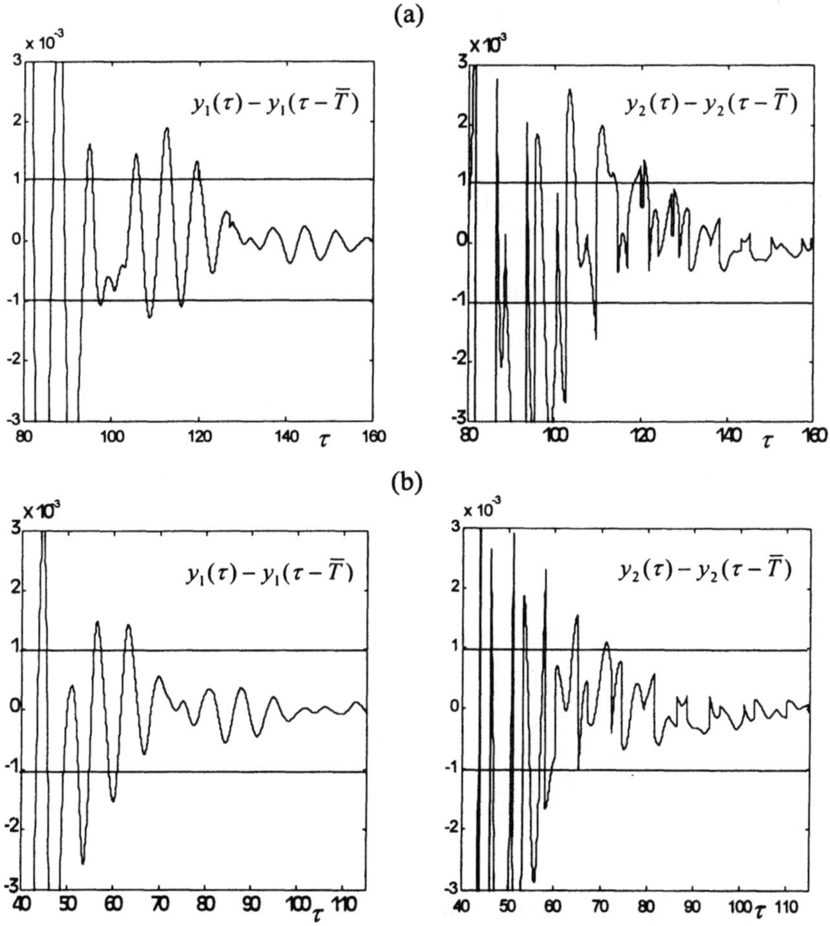

Fig. 5. The time histories $y_i(\tau) - y_i(\tau - \overline{T})$ for a vibro-impact system with a dry friction in the system without (a) and with (b) the delay loop

3 Analytical method to detect suitable feedback delay loop coefficients

Consider the periodic vibrations of the analysed one-degree-of-freedom system without the delay loop. Then, the vibro-impact dynamics (1) can be reduced to the following form:

$$\ddot{x} + c\dot{x} + \alpha^2 x = P_0 \cos\omega t \qquad \text{for} \quad x < s, \tag{2}$$

$$x_+ = x_-, \quad \dot{x}_+ = -R\dot{x}_- \qquad \text{for} \quad x \geq s.$$

The solution to the equation (2) has the form:

$$x(t) = \mathrm{e}^{-0.5ct}\left(C\cos\lambda_1 t + D\sin\lambda_1 t\right) + F_1\cos(\omega t + \theta) \quad \text{for} \quad x < s, \tag{3}$$

$$x_+ = x_-, \quad \dot{x}_+ = -R\dot{x}_- \qquad \text{for} \quad x \geq s,$$

where:

$$F_1 = \frac{P_0}{\sqrt{\left(\alpha^2 - \omega^2\right) + c^2\omega^2}}, \qquad \lambda_1 = \sqrt{\alpha^2 - \left(\frac{c}{2}\right)^2},$$

and the coefficients C, D, and θ are defined using the impact boundary conditions. The boundary conditions have the form:

$$t = 0, \quad x(t) = s, \quad \dot{x}(t) = \dot{x}_+, \text{ and } t = \frac{2\pi k}{\omega} = 2\beta, \quad x(t) = s, \quad \dot{x}(t) = \dot{x}_-. \tag{4}$$

Differentiating the equation (3) and taking into account the boundary conditions (4), we get the following four equations:

$$C + F_1\cos\theta = s, \qquad -0.5cC + \lambda_1 D - F_1\omega\sin\theta = \dot{x}_+,$$

$$e^{-\beta c}\left(C\cos 2\beta\lambda_1 + D\sin 2\beta\lambda_1\right) + F_1\cos\theta = s,$$

$$-0.5c\mathrm{e}^{-\beta c}\left(C\cos 2\beta\lambda_1 + D\sin 2\beta\lambda_1\right) + \mathrm{e}^{-\beta c}\left(\lambda_1 D\cos 2\beta\lambda_1 - \lambda_1 C\sin 2\beta\lambda_1\right) +$$
$$- F_1\omega\sin\theta = \dot{x}_-,$$

which gives

$$C = s - F_1\cos\theta, \quad D = \frac{C\left(\mathrm{e}^{\beta c} - \cos 2\beta\lambda_1\right)}{\sin 2\beta\lambda_1}, \quad \sin\theta = L\dot{x}_-, \quad \cos\theta = \frac{s}{F_1} - M\dot{x}_-,$$

$$\dot{x}_{-(1,2)} = \frac{\frac{s}{F_1}M \pm \sqrt{L^2\left(1 - \frac{s^2}{F_1^2}\right) + M^2}}{L^2 + M^2}, \tag{5}$$

and

$$L = \frac{-0.5c(R+1)\sin 2\beta\lambda_1 + \lambda_1\left[(R-1)\cos 2\beta\lambda_1 + \mathrm{e}^{\beta c} - R\mathrm{e}^{-\beta c}\right]}{\lambda_1 F_1\omega\left(2\cos 2\beta\lambda_1 - \mathrm{e}^{\beta c} - \mathrm{e}^{-\beta c}\right)}, \tag{6}$$

$$M = \frac{(R+1)\sin 2\beta\lambda_1}{\lambda_1 F_1\left(2\cos 2\beta\lambda_1 - e^{\beta c} - e^{-\beta c}\right)}.$$

In order to find the analytical solution to equation (1), the approximate analytical method has been applied assuming that:

- the difference $x(t) - x(t-T)$ is small,
- the damping c is of the same order as the formally introduced perturbation parameter ε.

From equation (1) one obtains:

$$\ddot{x} + \alpha^2 x = \varepsilon\left\{P_0 \cos\omega t + S[x(t) - x(t-T)] + Q\left[\left(1 - \frac{c}{Q}\right)\dot{x}(t) - \dot{x}(t-T)\right]\right\} \tag{7}$$

for $x < s$

$x_+ = x_-, \quad \dot{x}_+ = -R\dot{x}_-$ for $x \geq s$.

We are looking for a solution in the form: $x(t) = a\cos\psi$. We assume that the amplitude of the excitation is small and in the case, when $\omega \approx \alpha$ we use the method of equivalent linearization [3, 12]. The problem is reduced to the analysis of the following equivalent linear equation with the error of $O(\varepsilon^2)$:

$$\ddot{x} + 2h_e(a)\dot{x} + \alpha_e^2(a)x = \varepsilon P_0 \cos\omega t, \tag{8}$$

where:

$$h_e(a) = h_e = -\frac{\varepsilon}{2}\left[\frac{S}{\alpha}\sin\alpha T - c + Q(1 - \cos\alpha T)\right],$$

$$\alpha_e(a) = \alpha_e = \alpha + \frac{\varepsilon}{2}\left[\frac{S}{\alpha}(\cos\alpha T - 1) + Q\sin\alpha T\right].$$

The solution to the equivalent linear equation (8) has the form:

$$x(t) = e^{-h_e t}(C\cos\lambda t + D\sin\lambda t) + F\cos(\omega t + \theta), \quad \text{for} \quad x < s, \tag{9}$$

$x_+ = x_-, \quad \dot{x}_+ = -R\dot{x}_-$ for $x \geq s$,

where:

$$F = \frac{P_0}{\sqrt{\left(\alpha_e^2 - \omega^2\right) + 4h_e^2\omega^2}}, \qquad \lambda = \sqrt{\alpha_e^2 - h_e^2},$$

and the parameters C, D, and θ are defined using relations (5) and (6).

In order to investigate the stability, we use the following approach [1]. When the vibro-impact periodic solution (9) is disturbed because of δx_l, then it causes a change of C and D and the phase shift of δC_l, δD_l and $\delta\theta_l$, correspondingly.

The perturbated solution has the form:

$$x + \delta x_l = e^{-h_e t}\left[(C + \delta C_l)\cos\lambda t_l + (D + \delta D_l)\sin\lambda t_l\right] + F\cos(\omega t_l + \theta + \delta\theta_l). \tag{10}$$

In the above relation the time t is measured beginning from the l-th impact. The next impact occurs after the time $t_{l+1} = 2\pi l/\omega + \delta T_l$, where δT_l denotes the period change. Because t is related to the unperturbated equation, therefore $t_l = t + \delta t_l$, where $\delta t_l = 0$ for the l-th impact and $\delta t_l = \delta T_l$ for the l+1 impact.

After transformations we get:

$$\delta x_l = \mathrm{e}^{-h_e t}\left[\left(\delta C_l + \lambda D\delta t_l - h_e C\delta t_l\right)\cos\lambda t + \left(\delta D_l - \lambda C\delta t_l - h_e D\delta t_l\right)\sin\lambda t\right] + \\ - F\left(\omega\delta t_l + \delta\theta_l\right)\sin(\omega t + \theta). \tag{11}$$

Introducing the following boundary conditions for the l-th and (l+1)-th impact:

$$\begin{aligned} &l: && t = 0,\quad \delta t_l = 0,\quad \delta x_l = 0,\quad \delta\dot{x}_l = \delta\dot{x}_{l+}, \\ &l+1: && t = \frac{2\pi k}{\omega} + \delta t_l = 2\beta + \delta t_l,\quad \delta t_l = \delta T_l,\quad \delta x_l = 0,\quad \delta\dot{x}_l = \delta x_{(l+1)-}, \end{aligned} \tag{12}$$

we get the following six equations:

$$\delta C_l - F\delta\theta_l\sin\theta = 0, \qquad -h_e\delta C_l + \lambda\delta D_l - F\omega\delta\theta_l\cos\theta = \delta\dot{x}_{l+},$$

$$\delta C_{l+1} - F\delta\theta_{l+1}\sin\theta = 0, \qquad -h_e\delta C_{l+1} + \lambda\delta D_{l+1} - F\omega\delta\theta_{l+1}\cos\theta = \delta\dot{x}_{(l+1)+},$$

$$\mathrm{e}^{-h_e(2\beta+\delta T_l)}\Big[\left(\delta C_l + \lambda D\delta T_l - h_e C\delta T_l\right)\cos\lambda\left(2\beta + \delta T_l\right) + \\ + \left(\delta D_l - \lambda C\delta T_l - h_e D\delta T_l\right)\sin\lambda\left(2\beta + \delta T_l\right)\Big] - F\left(\omega\delta T_l + \delta\theta_l\right)\sin\left(\omega\delta T_l + \theta\right) = 0,$$

$$\mathrm{e}^{-h_e(2\beta+\delta T_l)}\Big[\left(-h_e\delta C_l - 2h_e\lambda D\delta T_l + C\left(h_e^2 - \lambda^2\right)\delta T_l + \lambda\delta D_l\right)\cos\lambda\left(2\beta + \delta T_l\right) + \\ + \left(-h_e\delta D_l + 2h_e\lambda C\delta T_l + D\left(h_e^2 - \lambda^2\right)\delta T_l - \lambda\delta C_l\right)\sin\lambda\left(2\beta + \delta T_l\right)\Big] + \\ - F\omega\left(\omega\delta T_l + \delta\theta_l\right)\cos\left(\omega\delta T_l + \theta\right) = \delta\dot{x}_{(l+1)-}.$$

Taking into account that $\delta\theta_{l+1} = \delta\theta_0 + \sum_{i=0}^{l}\omega\delta T_i$, we finally get the following three equations:

$$\delta C_l - F\delta\theta_l\sin\theta = 0, \\ \mathrm{e}^{-2\beta h_e}\left\{\delta C_l\cos 2\beta\lambda + \delta D_l\sin 2\beta\lambda + \frac{1}{\omega}\left(\delta\theta_{l+1} - \delta\theta_l\right)\left[\left(\lambda D - h_e C\right)\cos 2\beta\lambda + \\ - \left(\lambda C + h_e D\right)\sin 2\beta\lambda\right]\right\} - \delta C_{l+1} = 0, \tag{13}$$

$$R\mathrm{e}^{-2\beta h_e}\Big\{\left(\lambda\delta D_l - h_e\delta C_l\right)\cos 2\beta\lambda - \left(\lambda\delta C_l + h_e\delta D_l\right)\sin 2\beta\lambda + \\ + \frac{1}{\omega}\left(\delta\theta_{l+1} - \delta\theta_l\right)\Big[\left(-2h_e\lambda D + C\left(h_e^2 - \lambda^2\right)\right)\cos 2\beta\lambda + \\ + \left(2h_e\lambda C + D\left(h_e^2 - \lambda^2\right)\right)\sin 2\beta\lambda\Big]\Big\} - h_e\delta C_{l+1} + \lambda\delta D_{l+1} - (R+1)\delta\theta_{l+1}F\omega\cos\theta = 0.$$

The solutions are being sought in the form:

$$\delta C_l = a_1 \gamma^l, \qquad \delta D_l = a_2 \gamma^l, \qquad \delta \theta_l = a_3 \gamma^l, \tag{14}$$

where γ denotes the constant. Substituting (14) to (13) we get the following characteristic equation:

$$b_2 \gamma^2 + b_1 \gamma + b_0 = 0, \tag{15}$$

where:

$$b_2 = \lambda \left\{ F \sin\theta - \frac{1}{\omega} e^{-2\beta h_e} \left[(\lambda D - h_e C)\cos 2\beta\lambda - (\lambda C + h_e D)\sin 2\beta\lambda \right] \right\}, \tag{16}$$

$$b_1 = e^{-2\beta h_e} \left\{ F \sin\theta \left[(R-1)\lambda \cos 2\beta\lambda - (R+1) h_e \sin 2\beta\lambda \right] - (R+1) F\omega \cos\theta \sin 2\beta\lambda + \right.$$

$$\left. + \frac{1}{\omega} \lambda \left[(h_e C - \lambda D)\left(R e^{-2\beta h_e} - \cos 2\beta\lambda \right) - (\lambda C + h_e D)\sin 2\beta\lambda \right] \right\},$$

$$b_0 = R\lambda e^{-4\beta h_e} \left[\frac{1}{\omega} (\lambda D - h_e C) - F \sin\theta \right].$$

The problem of the investigation of this equation stability is reduced to the analysis of the roots of the second power equation (15). If these roots fulfil the following inequality:

$$|\gamma_{1,2}| < 1, \tag{17}$$

then, in accordance with expressions (14), the solutions δC_l, δD_l and $\delta \theta_l$ approach zero at $l \rightarrow +\infty$, and the solution will be called asymptotically stable. The above inequality is equivalent with the location of the roots inside the unit circle of the complex plane.

Consider the one-degree-of-freedom system presented in Section 2. The vibro-impact motion of that system is characterised by one impact during a period of excitation. Therefore, the earlier presented analytical approach can be applied to that case.

On the basis of the analysis of the stability of both the systems, the following results were obtained: $|\gamma_{1,2}| = 0.479$ for the case without the loop, and $|\gamma_{1,2}| = 0.453$ for the case with the loop. It indicates that the vibro-impact periodic orbit of the system with the delay loop is more stable, which corresponds to the numerical results.

4 Conclusions

In this paper we have presented an analytical approach to estimate the delay control coefficients for efficient stabilisation or destabilisation of the periodic orbit under consideration. Although the efficiency of the method is presented for $k = 1$ (periodic orbit with the same period as the excitation period) but our considerations are also valid for subharmonics (for arbitrarily taken $k > 1$). The validity of our analytical approach has been verified by numerical simulations.

To date, in the literature available to the authors, in order to achieve the mentioned objective, the feedback loop coefficients have been adopted in a random way, using the numerical observation. In this paper, in the case of the one-degree-of-freedom system this problem was solved analytically for near resonance case. In addition, an efficient control of the vibro-impact two-degree-of-freedom system has been proposed and illustrated.

5 Acknowledgment

This work has been financially supported by the Polish State Committe for Scientific Research Grant No 7T07A00210.

References

1 Babitskij W.I., *Theory of impact systems*, Nauka, Moscow 1978 (in Russian).
2 Gryboś R., *Theory of impact in discrete mechanical systems*, PWN, Warszawa 1969 (in Polish).
3 Awrejcewicz J., *Deterministic vibrations of the discrete systems*, WNT, Warszawa 1996 (in Polish).
4 Pyragas K., "Continuous control of chaos by self-controlling feedback", *Physics Letters A*, Vol. 170, Iss. 6, 421-428, 1992.
5 Shinbrot T., Grebogi C., Ott E. and Yorke J.A., "Using small perturbations to control chaos", *Nature*, Vol. 363 (June), 411-117, 1993.
6 Youcef-Toumi K. and Wu S.-T., "Input/output linearization using time delay control", *Journal of Dynamics Systems, Measurement and Control*, Vol. 114 (Mar.), 10-19, 1992.
7 Slotine J.-J.E. and Li W., *Applied Nonlinear Control,* Prentice-Hall, Englewood Cliffs, New Jersey 1991.
8 Arimoto S., "Learning control theory for robotic motion", *International Journal of Adaptive Control and Signal Processing*, Vol. 4, 549-564, 1990.
9 Hara S., Yamamoto Y., Omata T. and Nokano M., "Repetitive control system: a new type servo system for periodic exogenous signals", *IEEE Transactions on Automatic Control*, Vol. 33, 7, 659-668, 1988.
10 Awrejcewicz J., Tomczak K. and Lamarque C.-H., "Controlling Systems with Impacts". In: Awrejcewicz J. and Lamarque C.-H., *Proceedings of the ICNBC: the Doors to the Future*, Łódź-Dobieszków, 79-82, 1996.
11 Awrejcewicz J., Tomczak K. and Lamarque C.-H., "Multibody Vibro-Impact Dynamics", *Proceedings of the 2nd ENOC,* vol. 2, Prague, 9-12, 1996.
12 Awrejcewicz J., Andrianov I.V. and Manevitch L.I., *Asymptotic approach in nonlinear dynamics: new trends and applications*, Springer-Verlag, Berlin 1998.

Stability of periodic motions in systems with unilateral constraints

A P Ivanov *

Unilateral constraints form a vast class of restrictions which are met in mechanical engineering: it includes rigid obstacles, elastic dampers, tethers, supports, and so forth. From a physical point of view, any existing mechanical constraint is a combination of one-side ones. However, a favorite topic in text-books is the theory of (ideal) bilateral constraints, while very important in practice unilateral problems are considered in scientific journals only. The reason is a number of mathematical difficulties which are met in unilateral dynamics and which may be called discontinuities in one or another sense.

Generally speaking, there exist two basic kinds of the discontinuities in finite freedom mechanical systems. First of all, due to the imposed constraints the equations of motion can have different form in different regions of the phase space. Though the right-hand sides of the equations are discontinuous, the solution curves are continuous (non-differentiable in general). Many results on dynamics of such systems were obtained by numerous authors (see [1] for review and references). The second kind is constituted with the systems which have discontinuous solution curves. These breaks are caused by the collisions of rigid bodies. Some special tricks are needed to define usual topological concepts, such as vicinity, for disconnected sets [2].

To study periodic motions in discontinuous system, one should define a concept of stability (some definitions were given in [3–5]). As a rule, such concept agrees with stability of fixed point of the Poincaré map, associated with the phase flow. As far as such flow is differentiable, it is possible to solve the stability problem by calculating the Floquet multipliers. In this case the main difficulty is to calculate the monodromy matrix. To do it one can use formulas proposed in [6–8].

In the present paper we discuss stability problem for two cases which have not been considered in detail previously. The first of them is related to visco-elastic one-side constraints, the second – to multiple "rigid" constraints. The last case is shown to be rather singular, as in general the corresponding Poincaré map is piecewise-differentiable. Some stability conditions for such maps are obtained. They are applied to certain mechanical systems.

1 Variation of solutions in discontinuous systems

Consider an ODE system

$$\dot{\mathbf{x}} = \mathbf{F}(\mathbf{x},t), \quad \mathbf{F}(\mathbf{x}, t+\tau) \equiv \mathbf{F}(\mathbf{x},t), \quad \mathbf{q} \in R^n, \tag{1}$$

* Moscow State Academy of Device and Informatics, Stromynka 20, Moscow 107846, Russia

where $\tau > 0$ is a period.

Suppose $\mathbf{x}^*(t)$ is a periodic solution to (1), its variation $\mathbf{y} = \mathbf{x} - \mathbf{x}^*$ determines local behavior, including stability and bifurcation. In the case that the right-hand side $\mathbf{F}$ has continuous differential $\mathbf{DF}$ in a vicinity of $\mathbf{x}^*(t)$, the well-known Poincaré theorem claims that

$$\mathbf{y}(t) = \mathbf{Y}(t_0, t)\mathbf{y}_0 + o\|\mathbf{y}_0\|, \quad \mathbf{y}_0 = \mathbf{y}(t_0), \tag{2}$$

where the fundamental solution matrix $\mathbf{Y}(t_0, t)$ satisfies to

$$\dot{\mathbf{Y}} = (\mathbf{DF})\mathbf{Y}, \quad \mathbf{Y}(t_0, t_0) = \mathbf{E}_n. \tag{3}$$

Formula (2) can be extended to some discontinuous cases. The simplest of them is related to systems subject to shocks at prescribed times $t_1, t_2, \ldots$ Then the matrix $\mathbf{Y}$ is piecewise differentiable solution to (3) with $\mathbf{F} = \mathbf{F}_k$ for $t \in (t_{k-1}, t_k)$.

More important in practice is the case of discontinuity connected with switches on some surfaces, which is met in controlled systems. Let the surface of switching have form $f(\mathbf{x}) = 0$, so that we have in (1) $\mathbf{F} = \mathbf{F}^+$ for $f > 0$ and $\mathbf{F} = \mathbf{F}^-$ for $f < 0$ while for $f = 0$ the function $\mathbf{F}$ may be defined by different ways [1]. It was proved [6] that if $\mathbf{x}^*(t)$ intersects the surface $f = 0$ at $t = t'$ from negative to positive side transversally, then (2) holds for $t > t'$ where the matrix $\mathbf{Y}$ is discontinuous according to the following relation:

$$\mathbf{Y}(t_0, t' + 0) = \mathbf{B}\mathbf{Y}(t_0, t' - 0), \quad \mathbf{B} = \mathbf{E_n} + (\mathbf{F}^-, \operatorname{grad} f)^{-1}(\mathbf{F}^+ - \mathbf{F}^-)(\operatorname{grad} f)^T. \tag{4}$$

Similar result is valid in systems with unilateral constraint $f(\mathbf{x}) \geq 0$ and non-degenerate collisions upon this constraint [7, 8]. Let $\mathbf{I} = \mathbf{I}(\mathbf{x}^-)$ be the impulse, then $\mathbf{x}^+ = \mathbf{x}^- + \mathbf{I}(\mathbf{x}^-)$, where t' is the collision time and $\mathbf{x}^\pm = \mathbf{x}(t' \pm 0)$. The function $\mathbf{I}(\mathbf{x}^-)$ is differentiable provided collision is simple, i.e. two rigid bodies meet one another at a regular point of their surfaces. Then the jump of matrix $\mathbf{Y}$ is described by formula (4) where

$$\begin{aligned} &\mathbf{B} = \mathbf{E_n} + \mathbf{I_x} + (\mathbf{F}^-, \operatorname{grad} f)^{-1}(\mathbf{F}^+ - \mathbf{F}^- - \mathbf{I_x}\mathbf{F}^-)(\operatorname{grad} f)^T, \\ &\mathbf{I_x} = \partial \mathbf{F}/\partial \mathbf{x}, \quad \mathbf{F}^\pm = \mathbf{F}(\mathbf{x}^\pm, t'). \end{aligned} \tag{5}$$

There exist degenerate cases of discontinuity which can be met in multibody dynamics and which are far more complicated from a mathematical point of view [9]. Suppose for instance that the function $\mathbf{F}(\mathbf{x}, t)$ is discontinuous on two surfaces $f_1(\mathbf{x}) = 0$ and $f_2(\mathbf{x}) = 0$ while the studied trajectory $\mathbf{x}^*$ intersects both of them simultaneously. Then perturbed trajectory $\mathbf{x}(t)$ might have triple behavior: along with simaltaneous intersection of the surfaces $f_{1,2} = 0$, subsequent intersections of them (in one or another order) are possible too. Each of these three scenarios can be studied by means of eq.(4). As the calculations show, the results do not coincide in general. Thus the jump matrix $\mathbf{B}$ in (4) depends on initial condition $\mathbf{y}_0$ and variation $\mathbf{y}(t)$ is not differentiable with respect to $\mathbf{y}_0$ for $t > t'$ (in fact, it is piecewise differentiable).

The case of multiple collisions (when three or more bodies are in touch) is even more singular [10]. The reason is that in general $\mathbf{y}(t)$ has not to be small after such collision, and infinitesimal analysis is inapplicable. In special orthogonality case the impulsive reactions are mutually independent, and the above sketched method is relevant: the variation would be piecewise differentiable.

2 Stability analysis

Discontinuity of variation causes in simple cases (4), (5) no problem in stability analysis. Indeed, in accordance with (2) usual linearization technique is applicable. One should calculate matrix $\mathbf{Y}(t_0, t_0 + N\tau)$ (here $N\tau$ is period of the solution $\mathbf{x}^*(t)$) using (3)–(5) and then determine the eigenvalues of this matrix ρ_j $(j = 1, \ldots, n)$ (the Floquet multipliers). Sufficient conditions of (asymptotic) stability have form

$$|\rho_j| < 1 \quad (j = 1, \ldots, n). \tag{6}$$

If at least one of ineqs (6) makes opposite sense, the considered motion is unstable.

In the case of complex discontinuities (simultaneous intersection of several switching surfaces or multiple collision where the orthogonality conditions are satisfied), the matrix $\mathbf{Y}(t_0, t_0 + N\tau)$ in (2) depends on $\mathbf{y}_0$:

$$\mathbf{Y}(t_0, t_0 + N\tau) = \mathbf{A}_k, \text{ if } \mathbf{y}_0 \in G_k \quad (k = 1, \ldots, r) \tag{7}$$

where the unit of sets G_k is a vicinity of the origin, cf.[10]. (The directions for the calculation of matrices $\mathbf{A}_k$ can be found in the previous section.)

The connection between stability of the periodic motion and eigenvalues of matrices $\mathbf{A}_k$ is indirect: neither ineqs (6) (for all k) ensure stability, nor vice versa. The following results were proved in [10]:

Theorem 1. If $||\mathbf{A}_k|| < 1$, $(k = 1, \ldots, r)$, then the motion $\mathbf{x}^*(t)$ is asymptotically stable (here the norms of matrices are calculated relative to arbitrary norm in R^n).

Theorem 2. If at least one of the matrices $\mathbf{A}_k$ has an eigenvalue $\rho > 1$ with eigenspace belonging to the interior of the corresponding set G_k, then the motion $\mathbf{x}^*(t)$ is unstable.

Note that there exist many cases which are not covered by these two results (for instance, $\rho < -1$), thus the problem is still open.

In the remainder of the paper we shall consider some examples of mechanical systems with one-side constraints and show how the general results may be applied to the particular kinds of discintinuity and to the stability analysis in such systems.

3 Systems with elastic constraints

Consider a mechanical system with generalized coordinates $\mathbf{q} \in R^m$ subject to unilateral constraint

$$f(\mathbf{q}) \geq 0. \tag{8}$$

The inequality (8) can be treated in two ways in accordance with physical premises: as elastic or absolutely rigid. These two possibilities may be illustrated by the example of two balls on steel base: one of them is rubber and another is steel. Another example is a tethered system in space: when the tether is short, it can be modeled as non-extendable, but if it is of several kilometers long, the deformations must be taken in account. Note that the boundary between these two models is rather conventional.

By the first approach, the inequality (8) can be violated, by that a positive reaction of the constraint appears:

$$\mathbf{R} = \begin{cases} 0, & \text{if } f(\mathbf{q}) > 0, \\ \mathbf{R}(\mathbf{q}, \dot{\mathbf{q}}) & \text{if } f(\mathbf{q}) \leq 0, \end{cases} \qquad (\mathbf{R}(\mathbf{q}, \dot{\mathbf{q}}), \operatorname{grad} f(\mathbf{q})) \geq 0 \tag{9}$$

By the second approach, an attempt to violate the ineq. (8) will be accompanied with an impact. After the impact the generalized velocities $\dot{\mathbf{q}}$ will be compatible with the constraint.

We discuss at first the elastic constraints (9). The appropriate examples are: suspended bridges, oscillators with elastic stops etc. The equations of motion can be presented in Lagrangian form

$$\frac{d}{dt}\left(\frac{\partial T}{\partial \dot{\mathbf{q}}}\right) - \frac{\partial T}{\partial \mathbf{q}} = \mathbf{Q} + \mathbf{R}, \tag{10}$$

where T is kinetic energy, $\mathbf{Q} \in R^n$ – generalized force, and $\mathbf{R}$ has form (9). It is easy to convert (10) to standard form (1) where $\mathbf{x} = (\mathbf{q}, \dot{\mathbf{q}})$ and $n = 2m$.

Suppose the periodic trajectory $\mathbf{q}^*(\mathbf{t})$ belongs to the region where ineq. (8) is fulfilled strictly. Then $\mathbf{R} \equiv \mathbf{0}$, variation $\mathbf{y}$ is smooth and no special analytical tricks are needed. Similar arguments are applicable to such motions where the constraint (8) is permanently strained. Thus it is sufficient to discuss only those motions which include both cases in formula (9). In view of periodicity, such solutions involve transitions from relaxed constraint to strained one and backwards. The transition of the first kind can be accounted by means of formula (4), while the relaxation of the constraint may have dual character. In the first place, the value $f(\mathbf{x}^*(t))$ might change sign from negative to positive. In this case the variation jump should be accounted by means of (4). Secondly, the direction of reaction force might become incompatible with imposed unilateral constraint, i.e. $(\mathbf{R}(\mathbf{q}, \dot{\mathbf{q}}), \operatorname{grad} f(\mathbf{q})) < 0$. Here no discontinuity occurs and matrix $\mathbf{Y}$ is determined from (3) without jumps. Note that the second scenario is more usual in practice since for dissipative resistance the reaction force vanishes, the constraints slacken. We arrive at the following conclusion.

Theorem 3. Variation of solution to system (1) with unilateral constraint (8), (9), can be described by (2), (3) with jumps (4) at such moments when the value $f(\mathbf{x}^*(t))$ changes its sign. There is no jump when the reaction force vanishes.

Example. Consider rectilinear motion of a particle under periodic excitation in the presence of constraint (8), (9) with visco-elastic reaction force (Kelvin - Voight media). We place the origin on the surface of media and suppose that mass equals unity. The equations of motion are

$$\ddot{q} = P(t) + R(q, \dot{q}), \quad R = \begin{cases} 0, & \text{if } q > 0, \\ -c^2 q - 2k\dot{q} & \text{if } q \leq 0, \end{cases} \quad R(q, \dot{q}) \geq 0 \qquad (11)$$

where k and c are coefficients of damping and elasticity, $k < c$. We set $\mathbf{x} = (q, \dot{q})$, $\mathbf{F} = (x_2, P(t) + R(x_1, x_2))^T$ to obtain standard form (1).

Suppose $q^*(t)$ is a τ - periodic solution for (11), with $q > 0$ for $t \in (t_1, t_2)$ and $q < 0$ for $t \in t_2, t_1 + \tau$. The variational equation (3) has form

$$\begin{aligned} \dot{\mathbf{Y}} &= \begin{pmatrix} 0 & 1 \\ 0 & 0 \end{pmatrix} \mathbf{Y}, \text{ if } R = 0, \\ \dot{\mathbf{Y}} &= \begin{pmatrix} 0 & 1 \\ -c^2 & -2k \end{pmatrix} \mathbf{Y}, \text{ if } R \neq 0. \end{aligned} \qquad (12)$$

These equations can be solved easily, and we obtain

$$\begin{aligned} \mathbf{Y}^+(t) &= \begin{pmatrix} 1 & t \\ 0 & 1 \end{pmatrix} \mathbf{Y}^+(0), \quad \delta = \sqrt{c^2 - k^2}, \\ \mathbf{Y}^-(t) &= \exp(-kt) \begin{pmatrix} \cos \delta t + \frac{k}{\delta} \sin \delta t & \frac{1}{\delta} \sin \delta t \\ -\frac{c^2}{\delta} \sin \delta t & \cos \delta t - \frac{k}{\delta} \sin \delta t \end{pmatrix} \mathbf{Y}^-(0). \end{aligned} \qquad (13)$$

where superscripts "plus" and "minus" indicate the first and second cases in (12).

The jump of variation takes place at $t = t_2$ and we have according to (4)

$$\operatorname{grad} f = \begin{pmatrix} 1 \\ 0 \end{pmatrix}, \quad \mathbf{F}^- = \begin{pmatrix} V \\ P(t_2) \end{pmatrix}, \quad \mathbf{F}^+ = \begin{pmatrix} V \\ P(t_2) - 2kV \end{pmatrix},$$

with $V = \dot{q}(t_2)$, hence

$$\mathbf{B} = \mathbf{E}_2 + \frac{1}{V} \begin{pmatrix} 0 \\ -2kV \end{pmatrix} (1 \quad 0) = \begin{pmatrix} 1 & 0 \\ -2k & 1 \end{pmatrix}. \qquad (14)$$

At the moment $t = t_1$ we have $q = 0, \dot{q} > 0$, hence $c^2 q + 2k\dot{q} > 0$. This inequality indicates that the reaction force vanishes at some time $t_3 < t_1$ (second scenario in theorem 3), hence no discontinuity accompanies slackening of the constraint. Thus in (12) the first case is realized for $t \in (t_3, t_2)$ while the second case for $t \in (t_2, \tau + t_3)$. We set t_3 as initial time, then the monodromy matrix looks so:

$$\mathbf{Y}(t_3, t_3 + \tau) = \mathbf{Y}^-(\tau + t_3 - t_2)\mathbf{B}\mathbf{Y}^+(t_2 - t_3). \tag{15}$$

Stability conditions (6) in the considered case $n = 2$ is equivalent to the following inequalities

$$|a_1| < 1 + a_0 < 2, \quad a_1 = \mathrm{Tr}\,\mathbf{Y}(t_3, t_3 + \tau), \ a_0 = \det \mathbf{Y}(t_3, t_3 + \tau). \tag{16}$$

We put formulas (13)-(15) together to obtain

$$a_1 = 2\exp\left(-k(t_3 + \tau - t_2)\right)\left(\cos\Omega - \frac{k}{\delta}\sin\Omega + (t_2 - t_3)\left(\frac{c^2}{2\delta} - \delta\right)\sin\Omega - \right.$$

$$\left. -k(t_2 - t_3)\cos\Omega\right), \quad a_0 = \exp\left(-2k(t_3 + \tau - t_2)\right), \ \Omega = \delta(t_3 + \tau - t_2). \tag{17}$$

The second inequality (16) is satisfied automatically while the first one is to be checked for given parameter values.

Remark. If the coefficients c, k tend to infinity consistently (so that their ratio is constant), elastic constraint transforms to rigid one, where collisions are characterized by Newtonian coefficient $e = -\exp\left(-2\frac{k}{\delta}\mathrm{arctg}\,\frac{\delta}{k}\right)$ [11]. It can be shown that stability conditions of elastic and rigid system (cf.[12]) are consistent, but more subtle analysis than that in [11] is needed.

4 Conservative systems with periodic excitation and several impact pairs

In this section we discuss briefly linear systems

$$\ddot{\mathbf{q}} + \mathbf{A}\mathbf{q} = \mathbf{Q}(t), \quad \mathbf{x}, \mathbf{Q} \in R^m, \ q_k \geq 0 \ (k = 1, \ldots, m) \tag{18}$$

with τ - periodic excitation $\mathbf{Q}(t)$ (the stiffness matrix $\mathbf{A}$ is positive definite). The collision in k - th impact impact pair is characterized by restitution coefficient e_k. An example of such system is an elastic string with concentrated masses and rigid stops: it is known (see [12–14]) that there exist periodic motions with simultaneous collisions in the impact pairs.

Suppose $\mathbf{q}^*(t)$ is $N\tau$ - periodic solution of (18) with one multiple collision per period. The impact matrix $\mathbf{B}_k$ describing variation jumps for each impact pair has according to (5) the following form:

$$\mathbf{B}_k = \begin{pmatrix} \mathbf{E}_n - (1 + e_k)\mathbf{D}_k & 0 \\ \Phi_k \mathbf{D}_k & \mathbf{E}_n - (1 + e_k)\mathbf{D}_k \end{pmatrix}, \quad \Phi_k = \frac{1 + e_k}{\dot{q}_k(t' - 0)} Q_k(t'),$$

where t' is the impact time, $\mathbf{D}_k$ is the matrix with all elements equal zero except unity at the intersection of k - th row with k - th column. It is easy to show that the matrices $\mathbf{B}_i$ and $\mathbf{B}_j$ commute and the result of multiple collision does not depend on the order of simple impacts. Therefore, the multiple collision can be described by the following jump matrix:

$$\mathbf{B} = \begin{pmatrix} -\sum_{k=1}^{m} e_k \mathbf{D}_k & 0 \\ \sum_{k=1}^{m} \Phi_k \mathbf{D}_k & -\sum_{k=1}^{m} e_k \mathbf{D}_k \end{pmatrix}. \tag{19}$$

The monodromy matrix has form

$$\mathbf{Y}(t', t' + N\tau) = \mathbf{B} \exp(N\tau\mathbf{A}). \tag{20}$$

Verification of stability conditions (6) is connected with calculation of eigenvalues of the matrix (20). This problem is rather cumbersome for large n, and it is useful to obtain more simple sufficient conditions of stability. For such purpose we introduce a norm in R^{2m} by the formula

$$||\mathbf{x}||^2 = (\dot{\mathbf{q}}, \dot{\mathbf{q}}) + (\mathbf{A}\mathbf{q}, \mathbf{q}). \tag{21}$$

Obviously, the value $L = ||\mathbf{x}(t) - \mathbf{x}^*(t)||$ is preserved during smooth motion (18) but it changes at impacts. Keeping theorem 1 in mind, we arrive to the inequality

$$(\mathbf{A}\mathbf{z}, \mathbf{z}) + (\mathbf{w} - \dot{\mathbf{z}}, \mathbf{w} - \dot{\mathbf{z}}) \le (\dot{\mathbf{q}}, \dot{\mathbf{q}}) + (\mathbf{A}\mathbf{q}, \mathbf{q}) \tag{22}$$

with $z_k = e_k q_k$, $w_k = \Phi_k q_k$ $(k = 1, \dots, m)$. If relation (22) holds $\forall \mathbf{x} \in R^{2m}$, then L is Lyapunov function, and given motion is stable. Verification of this property (Silvester criterion) is much easier then direct evaluation of eigenvalues (Schur inequalities).

Let for instance $m = 1, A = \omega^2$. The sufficient condition (22) looks so:

$$\left|\frac{\Phi}{\omega}\right| < 1 - e^2 \tag{23}$$

while criterion (6) is equivalent to

$$\left|\frac{\Phi}{\omega} \sin N\omega\tau - 2e \cos N\omega\tau\right| < 1 + e^2. \tag{24}$$

Note that the condition (23) does not include the values N, τ. Moreover, if (24) is valid for any τ, then it coincides with (23). Therefore, the proposed analytical technique can be useful.

The work is supported in part by the Russian Foundation for Basic Researches (96-01-01440) and by the International association for the promotion of cooperation with scientists from the former Soviet Union (INTAS 96-2138).

References

1. Filippov AF, Differential equations with discontinuous right-hand side, Nauka, Moscow, 1985 (in Russian).

2. Ivanov AP, Impact oscillations: linear theory of stability and bifurcations, Journal of Sound and Vibrations, 178(3): 361–378, 1994.

3. Gayewskii K, Radzijevskii B, On the stability of impact systems. Bull. Polish Acad.Sci., Tech.Sci., 35(3–4): 183–188, 1987.

4. Akhmetov MU, Perestjuk NA, Periodic and quasi-periodic solutions of strongly nonlinear impulsive systems. Russian Journal of Applied Mathematics and Mechanics, 56(6): 926–934, 1992.

5. Ivanov AP, On the stability in the systems with unilateral constraints, Proc. Int.Symp. Analysis and Synthesus of Nonlinear Dynamical Systems in Mechanics, Riga, Latvia, pp.50–58, 1996.

6. Izerman MA, Gantmacher FR, Stability by linear approximation of periodic solutions of ODE systems with discontinuous right-hand sides. Russian Journal of Applied Mathematics and Mechanics, 21(5): 658–669, 1957

7. Müller PC, Calculation of Lyapunov exponents for dynamic system with discontinuities, Chaos, Solitons & Fractals, 5(9): 1671–1681, 1995

8. Ivanov AP, Bifurcations associated with grazing collisions, Nonlinear Dynamics: New theoretical and applied results. Akademie Verlag, 67–91, 1995.

9. Ivanov AP, Stability of periodic solutions in discontinuous systems in the presence of multiple intersections with jump surfaces, Russian Journal of Applied Mathematics and Mechanics, 62(5): 726–731, 1998

10. Ivanov AP, On the problem of multiple collision, Russian Journal of Applied Mathematics and Mechanics, 59(6): 930–946, 1995

11. Kozlov VV, Treshchev DV, Billiards, American Mathematical Society, Providence, 1991.

12. Babitsky VI, Theory of Vibro-impact Systems with Applications, Nauka, Moscow, 1978 (Revised English edition: Springer Verlag, 1998).

13. Babitsky VI, Veprik AM, Krupenin VL, Vibro-impact phenomena due to limited oscillations of one-dimensional elasto-connected particles, Proc. USSR Academy of Sciences 300(3) (in Russian).

14. Krupenin VL, To the analysis of vibro-impact processes in systems with a large number of impact pairs, Journal of Machine Manufacturing and Reliability (2): 85–92, 1994.

On the calculation of resonance oscillations of the vibro-impact systems by the averaging technique

V Sh Burd[*] *and V L Krupenin*[**]

The averaging technique is applied to calculation of resonance processes in the vibro-impact systems. In distinction from the method given in [1,2], based on use of outcomes [3], this technique based on use some results given in [4]. The small periodic perturbations by time of conservative vibro-impact systems with one degree of freedom are considered. The theorems of existence and stability the periodic solutions are proved.

The linear oscillator with the equation of motion

$$x''+\Omega^2 x = 0 \qquad (\Omega = const)$$

is considered. The fixed limiter is installed at the point $x = \Delta$. It is suggested that on this system elastic impact occurs when coordinate x takes the value Δ. If $x(t_\alpha) = \Delta$ at the moment t_α then

$$x'(t_\alpha - 0) = -x'(t_\alpha + 0).$$

We shall call this system vibro-impact system. For clarity $\Delta > 0$ will be considered. Now we shall state some facts set forth in [1] which will be necessary later on. In the presence of impacts the motion of the system will be periodic. The frequency of oscillations ω satisfies the inequality

$$\Omega < \omega < 2\Omega.$$

Let $t_\alpha = 0$. The law of motion has the form

---

[*] Yaroslavl State University, Sovetskaya ul. 14, Yaroslavl 150000, Russia , e-mail: burd@univ.uniyar.ac.ru.
[**] Blagonravov Mechanical Engineering Research Institute RAS, M. Haritonievskii per. 4, Moskow 101830, Russia, e-mail: krupenin@glasnet.ru.

$$x(t) = -J\left[\frac{\omega}{2\pi\,\Omega^2} + \frac{\omega}{\pi}\sum_{k=1}^{\infty}\frac{\cos k\omega t}{\Omega^2 - k^2\omega^2}\right] = -J\kappa\ (\omega\ t), \tag{1}$$

where $J = 2x'(-0) = -2\Omega\Delta\tan\frac{\pi\,\Omega}{\omega}$ is the impulse of impact. The solution of vibroimpact system is the periodic function (1) which when being substituted in the equation of motion converts it in to a true (in the sense of theory of distribution) equality of the form

$$x'' + \Omega^2 x + \Phi(x, x') = 0, \tag{2}$$

Here, $\Phi(x(t), x'(t)) = -J\delta_{\omega}(t)$, where $\delta_{\omega}(t)$ is periodic delta-function with period $\frac{2\pi}{\omega}$. Equation (2) is the equation of motion of vibroimpact system in temporal representation.

We consider perturbed vibroimpact system

$$x'' + \Omega^2 x + \Phi(x, x') = \varepsilon\ g(x, x', t), \tag{3}$$

Here ε is a small positive parameter, function $g(x, x', t)$ is sufficiently smooth with respect to the variables x, x' and periodic over t with period $\frac{2\pi}{\nu}$. Now convert the variables x, x' into new variables impulse-phase $J,\ \psi$ given by the formulas

$$x = -J\kappa(\psi), \quad x' = -J\omega_0(J)\kappa_{\psi}(\psi), \tag{4}$$

where

$$\psi = \omega t, \quad \kappa(\psi) = \omega_0^{-1}\left[\frac{1}{2\pi\,\Omega_0^2} + \frac{1}{\pi}\sum_{k=1}^{\infty}\frac{\cos k\psi}{\Omega_0^2 - k^2}\right], \quad \Omega_0 = \Omega[\omega_0(J)]^{-1},$$

$$\omega_0(J) = -\frac{\pi\,\Omega}{\pi - \arctan\dfrac{J}{2\Omega\Delta}}. \tag{5}$$

Change (4) is not smooth. Function κ_{ψ} has finite discontinuities with respect to ψ at the points $\psi = 2\pi l$ (l is an integer). The variables $J,\ \psi$ describe the system uniquely. The change (4) transforms the system (3) into a system

$$\frac{dJ}{dt} = -4\varepsilon\,\omega_0(J)\,g(-J\kappa(\psi), -J\omega_0(J)\kappa_\psi(\psi), t)\kappa_\psi(\psi) = \varepsilon\,X(J,\psi,t),$$
$$\frac{d\psi}{dt} = \omega_0(J) - 4\varepsilon\,\omega_0(J)J^{-1}g(-J\kappa(\psi), -J\omega_0(J)\kappa_\psi(\psi), t)(-J\kappa(\psi))_J =$$
$$= \omega_0(J) + \varepsilon\,Y(J,\psi,t) \qquad (6)$$

The function $X(J,\psi,t)$ has finite discontinuities with respect to ψ at the points $\psi = 2\pi l$ (l is an integer).

We will say that in the system (6) has a resonance if

$$\omega_0(J) = \frac{q}{p}\nu, \qquad (7)$$

where q, p are mutually prime numbers. The solution of the equation (7) denotes J_{pq} and is called resonance point. The transformation of variable

$$\psi = \phi + \frac{q}{p}\nu\, t$$

leads the system (6) to the following form

$$\begin{aligned} \frac{dJ}{dt} &= \varepsilon\,X(J, \phi + \frac{q}{p}\nu t, t), \\ \frac{d\phi}{dt} &= \omega_0(J) - \frac{q}{p}\nu + \varepsilon\,Y(J, \phi + \frac{q}{p}\nu t, t). \end{aligned} \qquad (8)$$

We will consider behavior of solutions of system (8) in $\mu = \sqrt{\varepsilon}$ - neighborhood of resonance point J_{pq}. Substitute

$$J = J_{pq} + \mu\, z$$

and represent the right-hand side of the transformed system as a power series expansion with respect to μ

$$\frac{dz}{dt} = \mu\,X(J_{pq}, \phi + \frac{q}{p}\nu t, t) + \mu^2 X_J(J_{pq}, \phi + \frac{q}{p}\nu t, t)z + O(\mu^3),$$
$$\frac{d\phi}{dt} = \mu\omega_{0J}(J_{pq})z + \frac{1}{2}\mu^2\omega_{0JJ}(J_{pq})z^2 + \mu^2\,Y(J_{pq}, \phi + \frac{q}{p}\nu t, t) + O(\mu^3).$$

(9)

The system (9) contains only one fast variable t. The right-hands side of the system (9) are periodic on t with period $\frac{2\pi\, p}{\nu}$. Let's make in the system (9) a standard transformation of the method of averaging, which will allow us to exclude the fast variable from the right-hand side of the system (9) up to the terms μ^2 exactly. We shall search this transform in the following form

$$z = \xi + \mu u_1(\eta,\, t) + \mu^2\, u_2(\eta,\, t)\xi, \qquad \phi = \eta + \mu^2\, v_2(\eta,\, t), \tag{10}$$

where $u_1(\eta,\, t)$, $u_2(\eta,\, t)$, $v_2(\eta,\, t)$ are defined as periodic functions over t with null average value from equations

$$\frac{\partial u_1}{\partial t} = X(J_{pq},\, \eta + \frac{q}{p}\nu t, t) - X_0(\eta),$$

$$\frac{\partial u_2}{\partial t} = X_J(J_{pq},\, \eta + \frac{q}{p}\nu t, t) - u_{1\eta}(\eta, t)\, \omega_{0J}(J_{pq}) - X_1(\eta),$$

$$\frac{\partial v_2}{\partial t} = Y(J_{pq},\, \eta + \frac{q}{p}\nu t, t) - u_1(\eta, t)\, \omega_{0J}(J_{pq}) - Y_0(\eta).$$

Functions $X_0(\eta)$, $X_1(\eta)$, $Y_0(\eta)$ are defined by formulas

$$X_0(\eta) = \frac{\nu}{2\pi\, p} \int_0^{\frac{2\pi p}{\nu}} X(J_{pq},\, \eta + \frac{q}{p}\nu\, t, t)\, dt,$$

$$X_1(\eta) = \frac{\nu}{2\pi\, p} \int_0^{\frac{2\pi p}{\nu}} X_J(J_{pq},\, \eta + \frac{q}{p}\nu\, t, t)\, dt,$$

$$Y_0(\eta) = \frac{\nu}{2\pi\, p} \int_0^{\frac{2\pi p}{\nu}} Y(J_{pq},\, \eta + \frac{q}{p}\nu\, t, t)\, dt.$$

An easy computations shows that functions $X_0(\eta)$, $X_1(\eta)$, $Y_0(\eta)$ are periodic over η with period $\frac{2\pi}{p}$ and functions $X_0(\eta)$, $X_1(\eta)$ have finite discontinuities with respect to η at the points $\eta = \frac{2\pi\, l}{p}$ (l is an integer). The transformation (10) leads to a system

$$\frac{d\xi}{dt} = \mu\, X_0(\eta) + \mu^2\, X_1(\eta)\,\xi + O(\mu^3)$$
$$\frac{d\eta}{dt} = \mu\,\omega_{0J}(J_{pq})\xi + \mu^2 Y_0(\eta) + \frac{1}{2}\mu^2\omega_{0JJ}(J_{pq})\xi^2 + O(\mu^3). \tag{11}$$

Let's assume a number η_0 $(0 < \eta_0 < \frac{2\pi}{p})$ exists such that

$$X_0(\eta_0) = 0 \tag{12}$$

and an η_0 is a simple root of equation (12). Then the averaged system of the first approximation

$$\frac{d\xi}{dt} = \mu\, X_0(\eta), \qquad \frac{d\eta}{dt} = \mu\,\omega_{0J}(J_{pq})\xi \tag{13}$$

has a solution

$$\xi = 0, \qquad \eta = \eta_0. \tag{14}$$

Next linearize the right-hand side of the system (13) on the solution (14) and obtain the matrix

$$A_0(\mu) = \begin{pmatrix} 0 & \mu\, X_{0\eta}(\eta_0) \\ \mu\,\omega_{0J}(J_{pq}) & 0 \end{pmatrix}.$$

It follows from the formula (5) that $\omega_{0J}(J_{pq}) > 0$. Therefore, if

$$X_{0\eta}(\eta_0) > 0 \tag{15}$$

then the matrix $A_0(\mu)$ has real eigenvalues of different signs. The classical theorem of Bogolyubov [5] about averaging on infinite interval gives that there exists an periodic solution with period $\frac{2\pi\, p}{\nu}$ of system (9) for sufficiently small μ and this solution is unstable. We apply the obtained result to system (3).

Theorem 1 *Assume that there exists a number η_0 satisfying the equality (12) and inequality (15). Then in $\sqrt{\varepsilon}$-neighborhood of the resonance point J_{pq} if ε is sufficiently small there exists a unique periodic solutions of the system (3) with period $\frac{2\pi\, p}{\nu}$ which is unstable.*

Now, assume that instead of inequality (15) the following inequality holds

$$X_{0\eta}(\eta_0) < 0 \tag{16}$$

In this case eigenvalues of the matrix $A_0(\mu)$ are pure imaginary. For investigation of the problem about the existence and stability of periodic solutions in μ-neighborhood of resonance point J_{pq} it is necessary to consider averaged equations of the second approximation. These equations have the following form

$$\begin{aligned} \frac{d\xi}{dt} &= \mu\, X_0(\eta) + \mu^2\, X_1(\eta)\,\xi, \\ \frac{d\eta}{dt} &= \mu\,\omega_{0J}(J_{pq})\xi + \mu^2 Y_0(\eta) + \frac{1}{2}\mu^2\omega_{0JJ}(J_{pq})\xi^{\,2}. \end{aligned} \tag{17}$$

If there is an η_0 such that $X_0(\eta_0) = 0$ and $X_{0\eta}(\eta_0) < 0$, by the implicit function theorem, for small μ, there is a unique $w(\mu) = (\xi(\mu), \eta_0)$ such that $\xi(0) = 0$ and $w(\mu)$ is the equilibrium point of the system (17). The matrix of linear approximation of system (17) will have the form

$$A_1(\mu) = \begin{pmatrix} \mu^2\, X_1(\eta_0) & \mu\, X_{0\eta}(\eta_0) \\ \mu\,\omega_{0J}(J_{pq}) & \mu^2 Y_{0\eta}(\eta_0) \end{pmatrix}$$

The eigenvalues of matrix $A_1(\mu)$ have a negative real parts if

$$X_1(\eta_0) + Y_{0\eta}(\eta_0) < 0 \tag{18}$$

and a positive real parts if

$$X_1(\eta_0) + Y_{0\eta}(\eta_0) > 0. \tag{19}$$

It follows from theorem 3.1 [6] that there exists a unique periodic solution with period $\dfrac{2\pi\, p}{\nu}$ of system (9) for sufficiently small μ. This solution is asymptotically stable if inequality (18) is satisfied and is unstable of inequality (19) is satisfied.

We apply the obtained statement to system (3).

Theorem 2. *Let number η_0 to satisfy the equality (12) and inequality (16). Let the inequality*

$$X_1(\eta_0) + Y_{0\eta}(\eta_0) \neq 0$$

holds. Then if ε *sufficiently small the system (3) has a unique periodic solution with period* $\dfrac{2\pi p}{\nu}$ *in* ε*-neighborhood of the resonance point* J_{pq}. *This solution is asymptotically stable if*

$$X_1(\eta_0) + Y_{0\eta}(\eta_0) < 0$$

and unstable if

$$X_1(\eta_0) + Y_{0\eta}(\eta_0) > 0.$$

For example consider the equation

$$x'' + \Omega^2 x + \Phi(x, x') = \varepsilon\left[-\gamma\, x' + a\sin(\nu t + \delta)\right], \tag{20}$$

where $\gamma > 0$, Ω, a, ν, δ are real constants. Making the transformation (4), we obtain the system

$$\frac{dJ}{dt} = -4\varepsilon\,\omega_0(J)\left[a\sin(\nu t + \delta) + \gamma\, J\omega_0(J)\kappa_\psi(\psi)\right]\kappa_\psi(\psi),$$
$$\frac{d\psi}{dt} = \omega_0(J) - 4\varepsilon\,\omega_0(J) J^{-1}\left[a\sin(\nu t + \delta) + \gamma\, J\omega_0(J)\kappa_\psi(\psi)\right](-J\kappa(\psi))_J. \tag{21}$$

Let a solution of equation (7) exists. The substitution $\psi = \phi + \dfrac{q}{p}\nu t$ leads the system (21) to the following form

$$\frac{dJ}{dt} = -4\varepsilon\,\omega_0(J)\left[a\sin(\nu t + \delta) + \gamma\, J\omega_0(J)\kappa_\psi(\phi + \frac{q}{p}\nu t)\right]\kappa_\psi(\phi + \frac{q}{p}\nu t),$$
$$\frac{d\phi}{dt} = \omega_0(J) - \frac{q}{p}\nu -$$
$$-\,4\varepsilon\,\omega_0(J) J^{-1}\left[a\sin(\nu t + \delta) + \gamma\, J\omega_0(J)\kappa_\psi(\phi + \frac{q}{p}\nu t)\right](-J\kappa(\phi + \frac{q}{p}\nu t))_J. \tag{22}$$

For the investigations of resonance regimes, apply Theorems 1 and 2. Average value of the first term in the right-hand side of the first equation from system (22) is different from 0 if, and only if, $q=1, p=n\ (n=0,1,\ldots)$. If $q=1, p=n$, this average value is equal to

$$X_0(\eta)=\frac{2a\nu^2}{\pi n(\Omega^2-\nu^2)}\cos(n\eta-\delta).$$

An average value of the second term in the right-hand side of the first equation of system (22) is equal to

$$-\frac{\gamma J_{pq}}{2}\left(1+\frac{4\Omega^2\Delta^2}{J_{pq}^2}\right)$$

Then the number η_0 is defined by equation

$$\cos(n\eta-\delta)=\frac{\gamma J_{pq}\pi n}{4a\nu^2}(\Omega^2-\nu^2)\left(1+\frac{4\Omega^2\Delta^2}{J_{pq}^2}\right)=A_n. \tag{23}$$

Since $A_n\to\infty$ as $n\to\infty$ then equation (23) can have a solution for the finite number of values of n. If equation (23) has a solution then

$$\eta_{0l}=\frac{\delta}{n}\pm\frac{\arccos A_n}{n}+\frac{2\pi l}{n}, \qquad l=0,\ldots,n-1.$$

By computing a derivative of the function $X_0(\eta)$ in the point η_{0l}, we obtain

$$X_{0\eta}(\eta_{0l})=\pm\frac{2a\nu^2}{\pi(\Omega^2-\nu^2)}\sqrt{1-A_n^2}$$

and, therefore, $X_{0\eta}(\eta_{0l})$ has n values with positive signs and n values with negative signs. Then an easy computation shows that $X_1(\eta_{0l})+Y_{0\eta}(\eta_{0l})<0$. From Theorems 1 and 2 the next results follows. If the resonance point J_{n1} is a solution of the equation

$$\omega(J_{n1})=\frac{\nu}{n},$$

then if ε sufficiently small the equation (20) has n solutions which are unstable resonance periodic with period $\frac{2\pi n}{\nu}$ in $\sqrt{\varepsilon}$-neighborhood of resonance point

J_{n1}, and n solutions which are asymptotically stable resonance periodic with period $\frac{2\pi n}{\nu}$ in $\sqrt{\varepsilon}$-neighborhood of the resonance point J_{n1}.

This research was supported by RFBR (project 98-01-00925).

References

1. Babitsky V I , Krupenin V L, Oscillations in Strongly Nonlinear Systems, Nauka, Moscow, 1985 (in Russian).

2. Babitsky V I, Kovaleva A S, Krupenin V L Investigation of Quasiconservative Vibroimpact Systems by the Method of Averaging, MTT (Mechanics of solid), 1, 41-50, 1982 (in Russian).

3. Moiseev N N Asymptotic Methods of Nonlinear Mechanics, Nauka, Moscow, 1981 (in Russian).

4. Volosov V M, Morgunov B I, Method of Averaging in Theory of Nonlinear Oscillatory Systems, MGU, Moscow, 1971 (in Russian).

5. Guckenheimer J, Holmes P J, Nonlinear Oscillations, Dynamical Systems, and Bifurcations of Vector Fields, Springer-Verlag, New York, NY, 1983.

6. Hale J K, Pavlu L C, Dynamic behavior from asymptotic expansions, Quart. of Appl. Math., 41, 161 -168, 1983.

Systems with high degree of nonlinearity as vibro-impact ones - asymptotic approaches

Igor V. Andrianov[*]

Method of "small δ" is proposed and tested. It gives asymptotic for large n ($n \to \infty$) of non-linear dynamical equation with power nonlinearity x^n. Method is based on the nontrivial changing of variables and on the method "small δ", proposed by C.M.Bender et. al.

1 Introduction

For non-linear systems with higher degree of nonlinearity quasi-linear approach is not adequate tool. A method, proposed in [1], is based on reducing governing problems to the vibro-impact ones and using iteration process. For generalisation and justification it will be very useful to construct an asymptotic approach. Below we propose this procedure, based on the changing of variables and so-called "small δ approach" [2].

2 Asymptotics: integral

We are searching periodical solution of the equation

$$\ddot{x} + x^n = 0, \; n = 2k+1, \; k \in N \tag{1}$$

with boundary conditions

$$\dot{x}(0) = 0; \; x = 1 \Rightarrow \dot{x} = 0. \tag{2}$$

Boundary value problem (1), (2) has exact solution in implicit form

$$\left(\frac{2}{n+1}\right)^{\frac{1}{n+1}} t = \int_0^{0 \le \xi \le 1} \frac{d\xi}{\sqrt{1-\xi^{n+1}}}. \tag{3}$$

By substituting

$$\xi = \sin^{\varepsilon}\theta, \; \varepsilon = 2/(n+1)$$

equation (3) may be reduced to the following form

$$\varepsilon^{2\varepsilon^{-1}} t = \varepsilon^{-1} \int_0^{0 \le \theta \le \pi/2} A \, d\theta, \tag{4}$$

where $A = \sin^{\varepsilon-1}\theta$. For asymptotic expansions of fright side in (4) we use the following relations:

[*] Prydniprovskaya State Academy of Civil Engineering and Architecture

$$A = \theta^{\varepsilon-1} + \left[\sin^{\varepsilon-1}\theta - \theta^{\varepsilon-1}\right],\ \int_0^\theta A d\theta = \int_0^\theta \theta^{\varepsilon-1} d\theta + O(1) = \theta^\varepsilon \varepsilon^{-1} + O(1).$$

Finally, one obtains in governing variables

$$x \approx \varepsilon^{-2\varepsilon} \sin^\varepsilon\left(\varepsilon^{2\varepsilon} t^{\varepsilon^{-1}}\right). \tag{5}$$

Period of this solution is

$$T(n) = 4\left[\frac{\pi}{2}\varepsilon^{-2\varepsilon^{-1}}\right]^\varepsilon. \tag{6}$$

Naturally, we have to use solution (5) on the quarter of period, then it may be periodically continued. Expanding expressions (5) in ε, one obtains in the first approximation solution from [1].

Formula (6) for $n = 1$ (linear problem) gives exact value 2π; for $n \to \infty$ one has $T(n) \to 4$. For various values of n one obtains:

n	Formula (6)	Exact value	Error, %
3	5.504	5.464	0.7
5	4.886	4.904	0.4

So, formulas (5), (6) give approximate solution for the problem of incomplete Beta-function inversion (Ateb-function construction) [3].

3 Asymptotic: equation

Analysis of above solution leads to the following conclusions: solution of boundary value problem (1), (2) contains branching point (root $\sqrt[n]{\ldots}$); natural independent variable is t^n.

Now let us assume model equation (1) with boundary conditions

$$x(0) = 0;\ \dot{x}(0) = 1. \tag{7}$$

Let us introduce changing of variables

$$x = y^{1/n};\ \tau = t^n. \tag{8}$$

Then boundary value problem (1), (7) may be reduced to the following one

$$\tau^{2(1-\varepsilon)} y \frac{d^2 y}{d\tau^2} + \tau^{1-2\varepsilon} y \frac{dy}{d\tau} - (1-\varepsilon)\tau^{2(1-\varepsilon)}\left(\frac{dy}{d\tau}\right)^2 + \varepsilon y^{3-\varepsilon} = 0,$$

$$y(0) = 0,\ \tau^{1-\varepsilon} y^{\varepsilon-1}\frac{dy}{d\tau} = 1. \tag{9}$$

Let us search periodic solution on quarter of period. For $\varepsilon << 1$ one has

$$\tau^{1-\varepsilon} = \tau - \varepsilon\tau \ln \tau + \dots ,$$
$$y^{3-\varepsilon} = y^3 - \varepsilon y^3 \ln y + \dots .$$

We search solution in the following form

$$y = y_0 + \varepsilon y_1 + \dots$$

In the first approximation one obtains:

$$\tau y_0 \ddot{y}_0 + y_0 \dot{y}_0 - \tau^2 \dot{y}_0^2 = 0, \tag{10}$$

$$y(0) = 0, \ \tau \frac{dy_0(0)}{d\tau} = y_0(0). \tag{11}$$

Solution of boundary value problem (10), (11) is

$$y_0 = \tau. \tag{12}$$

Solution (12) coincides with solution in [1].

Second approximation may be obtained from the following boundary value problem

$$\tau \ddot{y}_1 - \tau \dot{y}_1 + y_1 = -\tau^3, \tag{13}$$

$$y_1(0) = \dot{y}_1(0) = 0. \tag{14}$$

Let us suppose

$$y_1 = \sum_{i=3}^{\infty} \alpha_i \tau^i . \tag{15}$$

After substituting (15) in (13), (14) one obtains $\alpha_3 = -1/5$, then we may use, for example, "Mathematica".
So,

$$y \approx \tau \sqrt[2n]{1 - 0.2\tau^{2n}} . \tag{16}$$

Formula (16) is in good agreement with numerical solution for various values of n.

Literature

1. Manevitch L.I., Mikhlin Yu.V., Pilipchuk A.N. Method of normal modes in non-linear mechanics. Moscow: Nauka, 1989.
2. Bender C.M., Miltpon K.A., Pinsky S.S., Simmonds L.M. Nonperturbative approach to non-linear problems // J. Math. Physics, 1989, V.30, No.7, P.1447-1455.
3. Rosenberg R. The Ateb(h)-functions and their properties // Quart. Appl. Math., 1963, V. 21, No.1, P.37-47.

Simulation of liquid sloshing impact in moving stractures

V N Pilipchuk* and R A Ibrahim♣

This paper examines the nonlinear modal interaction between liquid hydrodynamic impacting with an elastic support structure. The liquid impact is modeled based on a phenomenological concept by introducing a power nonlinearity with a higher exponent. A special saw-tooth time transformation (STTT) technique is used to analytically describe the in-phase and out-of-phase strongly nonlinear periodic regimes. Based on explicit forms of analytical solutions all basic characteristics of the nonlinear free and forced response regimes such as time history, amplitude-frequency dependencies and nonlinear parametric resonance curves are estimated. The response behavior reveals that high frequency out-of-phase nonlinear mode takes place with a relatively small tank amplitude and is more stable than the in-phase oscillation mode under small perturbations. The in-phase mode has relatively large tank amplitudes and does not preserve its symmetry under the periodic parametric excitation.

1 Introduction

Liquid containers constitute major components in a number of dynamical systems such as aerospace vehicles, road tankers, and liquefied natural gas carriers. Liquid sloshing in a moving container creates a broad class of problems of practical safety importance, including tank trucks on highways, liquid tank cars on railroads, and liquid cargo in oceangoing vessels. In populated cities, gasoline and other flammable liquid tankers are prone to rollover accidents while entering and exiting highways.

Analytical techniques for predicting large-amplitude sloshing are not fully developed. Such loads are extremely important in designing the support sloshing technology developed for space applications [1] is not applicable to road tankers, because emphasis has been placed on frequencies and effects of local peak impact pressure on structural requirements have not been - studied to any extent. Furthermore, the excitation amplitudes considered in space applications are too small for road vehicle simulation.

* Professor of the Applied Mathematics Department, State Technology and Chemistry University, 320005 Dnepropetrovsk, Ukraine (Current address: ME, Wayne State University, Detroit, MI 48202; pylyp@me1.eng.wayne.edu)

♣ Professor of the Mechanical Enfineering, Wayne State University, Detroit, MI 48202; raouf_ibrahim@eng.wayne.edu

Sloshing phenomena are governed by the liquid fill depths. For the case of low fill depth, the liquid free surface motion is characterized by the formation of hydraulic jumps and traveling waves for excitation periods around resonance. At higher fill depths, large standing waves are usually formed in the resonant frequency range. When hydraulic jumps or traveling waves are present, extremely high impact pressures can occur on the tank walls. Typical pressure traces recorded under this sloshing condition were reported in [2]. Impact pressures can also occur on the tank top when tanks are filled to the higher fill depth, and the pressure variation is neither harmonic nor periodic, since the magnitude and duration of the pressure peaks vary from cycle to cycle.

Some of many possible formulations for solid body dynamics interacting with liquid can be found in [1,3]. It is clear that the liquid pressure impacts are one of the sources of strong nonlinearity in the liquid tank supported on an elastic structure. From the mathematical point of view, one has to deal with strongly nonlinear coupled sets of the partial differential equations of motion provided that physical features of pressure impacts have been clarified. Then different numerical techniques based on either direct discretization or modal analysis of the linearized problem can be applied. Any analytical estimation can be done after an appropriate modes truncation, in principal, reducing the system to a one or may be two degrees of freedom model. That is why a more direct way of the phenomenological modeling becomes quite reasonable in order to get the possible analytical estimation. Parameters of the model can be taken from an auxiliary calculation and experimental testing. In present paper the hydro-impact phenomenon is going to be modeled by phenomenological introducing a power nonlinearity with high exponent into a two degrees freedom rigid body model of appropriate properties [4].

2 Liquid Sloshing Modeling

The mathematical modeling of liquid containers supported on elastic structures can be developed by considering a liquid container supported by four massless rods of length L which are restrained by four torsional springs k at the base as shown in Figure1.

Let M be the total mass of the container including liquid, and m be the equivalent sloshing mass of the first asymmetric mode of the liquid. The fluid free surface is modelled as a pendulum of length l. The pendulum can reach the walls of the tank if its angle with the vertical axis is $\theta = \pm\theta_0$. One can phenomenologically describe the interaction between the pendulum and the tank walls by means of the following function of the force of interaction [5] $F_i = b(\theta/\theta_0)^{2n-1}$, where $n >> 1$ is a positive integer, and b is a positive constant parameter. One has a limit of absolutely rigid bodies interaction, if $n \longrightarrow \infty$. If the exponent $2n-1$ is large and finite then the interaction field is not absolutely localized at the points $\theta = \pm\theta_0$. It is assumed that the energy dissipation of the pendulum basically results from the pendulum interaction with the container walls. It means that the dissipation is spatially localized around the points $\theta = \pm\theta_0$ and is approximated by the expression $F_d = d(\theta/\theta_0)^{2p}\dot{\theta}$, where d is a constant parameter, $p >> 1$ is a positive integer (generally, $p \neq n$), and a dot denotes differentiation with respect to time t.

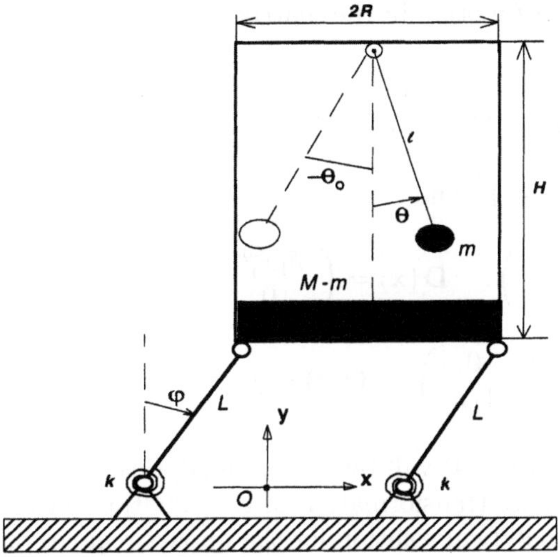

Figure 1. Schematic diagram of the model.

Keeping the "interaction" nonlinear terms but linearize the rest of terms in equations one obtains

$$l^2 m\ddot{\theta} + lLm\ddot{\varphi} + lm\left[g - F_y(t)\right]\theta + d\left(\frac{\theta}{\theta_0}\right)^{2p}\dot{\theta} + b\left(\frac{\theta}{\theta_0}\right)^{2n-1} = lmF_x(t) \quad (1)$$

$$lLm\ddot{\theta} + L^2 M\ddot{\varphi} - LM\left[g - F_y(t)\right]\varphi + c\dot{\varphi} + k\varphi = LMF_x(t) \quad (2)$$

where $F_x(t)$ and $F_y(t)$ are the components of an external force per unit mass along the x- and y-axes respectively. Introducing the following dimensionless variables

$$x_1 = \frac{\theta}{\theta_0}, \quad x_2 = \frac{\varphi}{\theta_0}, \quad \bar{t} = \sqrt{\frac{g}{l}}t, \quad \mu = \frac{m}{M}, \quad \lambda = \frac{l}{L}$$

$$2\delta_1 = \frac{d}{Lmg}\sqrt{\frac{g}{l}}, \quad 2\delta_2 = \frac{c}{Lmg}\sqrt{\frac{g}{l}}, \quad \beta = \frac{b}{Lmg\theta_0} \quad (3)$$

$$f_x(\bar{t}) = \frac{F_x(t)}{g\theta_0}, \quad f_y(\bar{t}) = \frac{F_y(t)}{g}, \quad \nu = \frac{\omega_L}{\omega_l} = \sqrt{\lambda\left(\frac{k}{MLg} - 1\right)}$$

$$\omega_L^2 = \frac{k - LMg}{ML^2}, \quad \omega_l^2 = \frac{g}{l}$$

The last parameter ν is impotent for future analysis and denotes the ratio of two local frequencies, i.e., ω_L is frequency of the tank in the absence of the pendulum motion, and ω_l is frequency of the pendulum if the tank is standstill. The frequency ω_l is equivalent to the fluid first antisymmetric sloshing mode and depends on the tank geometry and fluid depth (see [1]).

Taking into account the notations listed in (3) gives the following dimensionless equations of motion in the matrix form

$$\mathbf{B}\frac{d^2\mathbf{x}}{d\bar{t}^2} + 2\mathbf{D}(\mathbf{x})\frac{d\mathbf{x}}{d\bar{t}} + [\mathbf{K} + \mathbf{P}(\bar{t})]\,\mathbf{x} + \beta\mathbf{N}(\mathbf{x}) = \mathbf{Q}(\bar{t}) \tag{4}$$

where $\mathbf{x} = \begin{pmatrix} x_1 & x_2 \end{pmatrix}^T$ and

$$\mathbf{B} = \begin{pmatrix} \lambda & 1 \\ 1 & 1/(\lambda\mu) \end{pmatrix}, \quad \mathbf{D}(\mathbf{x}) = \begin{pmatrix} \delta_1 x_1^{2p} & 0 \\ 0 & \delta_2 \end{pmatrix}, \quad \mathbf{K} = \begin{pmatrix} \lambda & 0 \\ 0 & \nu^2/(\lambda\mu) \end{pmatrix}$$

$$\mathbf{P}(\bar{t}) = f_y(\bar{t})\begin{pmatrix} -\lambda & 0 \\ 0 & 1/\mu \end{pmatrix}, \quad \mathbf{Q}(\bar{t}) = f_x(\bar{t})\begin{pmatrix} \lambda \\ 1/\mu \end{pmatrix}, \quad \mathbf{N}(\mathbf{x}) = \begin{pmatrix} x_1^{2n-1} \\ 0 \end{pmatrix} \tag{5}$$

The non-linear system (4) will be studied analytically (for periodic regimes) by means of a special transformation of the equations of motion. Note that to get a correct analytical solution for the impact regimes one must keep the strongly non-linear term $\beta\mathbf{N}(\mathbf{x})$, starting from the first step of analysis.

3 Analytical Study

Consider the free undamped oscillation of the nonlinear system

$$\mathbf{B}\frac{d^2\mathbf{x}}{d\bar{t}^2} + \mathbf{K}\mathbf{x} + \beta\mathbf{N}(\mathbf{x}) = 0 \tag{6}$$

To consider the strongly non-linear situation including power terms with a higher exponent a sawtooth *temporal* transformations (STTT) will be applied. Details of the derivations represented below can be found in [6]. The idea of STTT is similar to a great extent to the trigonometric generating functions $\{\sin\xi, \cos\xi\}$ frequently used in constructing solutions of linear and weakly nonlinear systems. Similarly, one can consider a pair of non-smooth functions which have relatively simple forms and will be termed as the saw-tooth sine

$$\tau(\xi) = \begin{cases} \xi, & -1 \le \xi \le 1 \\ 2-\xi, & 1 \le \xi \le 3 \end{cases}; \qquad \tau(\xi+4) \overset{\forall\xi\in R}{=} \tau(\xi)$$

and the rectangular cosine, $e(\xi)$, which is the generalized derivative of $\tau(\xi)$. The functions $\{\tau(\xi), e(\xi)\}$ and $\{\sin\xi, \cos\xi\}$ describe the motions of the two simplest vibrating models, namely, the motion of a particle between two rigid barriers and a mass-spring oscillator, respectively.

We seek a family of periodic solutions of equations (6) in the following form

$$\mathbf{x} = \mathbf{X}(\tau), \quad \tau = \tau(\bar{t}/a) \tag{7}$$

where a is unknown scaling factor which is equal to quarter of the period $T = 4a$ and must be defined for the autonomous case.

Thus the solution is going to be constructed as a function of the saw-tooth function τ, which varied in the region $-1 \le \tau \le 1$. The symmetry of equation (6) admits the solution be constructed as an odd function: $\mathbf{X}(-\tau) \equiv -\mathbf{X}(\tau)$. It is a special feature of the STTT that solution is expressed as a function of the non-smooth argument τ. This, however, does not imply function $\mathbf{x}$ be non-smooth with respect to old independent variable $\bar{t}$. For example, function $x(\bar{t}) = \tau(\bar{t}/a) - \tau^3(\bar{t}/a)/3$ possesses first two continuous derivatives. One can verify that if using the following differentiation scheme for more general case (7). Due to equality $e^2\ (\bar{t}/a) = 1$ one has

$$\frac{d\mathbf{x}}{d\bar{t}} = \frac{1}{a}\frac{d\mathbf{X}}{d\tau}e, \qquad \frac{d^2\mathbf{x}}{d\bar{t}^2} = \frac{1}{a^2}\frac{d^2\mathbf{X}}{d\tau^2} + \frac{1}{a^2}\frac{d\mathbf{X}}{d\tau}\frac{de(\bar{t}/a)}{d(\bar{t}/a)} \tag{8}$$

The last term in (8) contains the series of Dirac's delta functions

$$\frac{de(\bar{t}/a)}{d(\bar{t}/a)} = 2\sum_{j=-\infty}^{j=\infty}\left[\delta\left(\frac{\bar{t}}{a}+1-4j\right) - \delta\left(\frac{\bar{t}}{a}-1-4j\right)\right] \tag{9}$$

Note that, the delta functions (9) are "localized" at points $\{\bar{t} : \tau(\bar{t}/a) = \pm 1\}$. It means that under the condition

$$\frac{d\mathbf{X}}{d\tau} = 0 \text{ for } \tau = \pm 1 \tag{10}$$

all delta functions of the series will be eliminated, and as a result the second derivative in (8) becomes continuous function. Roughly speaking condition (10) provides all of the delta functions in (8) with zero "coefficients."

Substituting (8) into the equations of motion (6) one obtains the boundary value problem

$$\mathbf{B}\frac{d^2\mathbf{X}}{d\tau^2} = -a^2\left[\mathbf{K}\mathbf{X} + \beta\mathbf{N}(\mathbf{X})\right] \tag{11}$$

$$\frac{d\mathbf{X}}{d\tau}|_{\tau=1} = 0, \quad \mathbf{X}(-\tau) \equiv -\mathbf{X}(\tau) \tag{12}$$

One of the advantages brought by the transformation is that the new time parameter is bounded, and hence one can seek solutions in power series form by making use successive approximation techniques. This does not require the equations involves any small parameter explicitly. For example, following the STTT algorithms one obtains solution of the boundary value problem (11) and (12) in the power series form with respect to the saw-tooth function τ as

$$x_1 = A_{01}\tau - \frac{h_0\left(\lambda A_{01} - \nu^2 A_{02}\right)}{\lambda(1-\mu)}\frac{\tau^3}{6} - \frac{\beta h_0 A_{01}^{2n-1}}{\lambda(1-\mu)}\frac{\tau^{2n+1}}{2n(2n+1)} + \cdots \quad (13)$$

$$x_2 = A_{02}\tau + \frac{h_0\left(\lambda\mu A_{01} - \nu^2 A_{02}\right)}{(1-\mu)}\frac{\tau^3}{6} + \frac{\mu\beta h_0 A_{01}^{2n-1}}{(1-\mu)}\frac{\tau^{2n+1}}{2n(2n+1)} + \cdots \quad (14)$$

where $\tau = \tau\left(\bar{t}/a\right)$, $a^2 = h_0\left(1 + \gamma_1 + \cdots\right)$, and

$$\gamma_1 = [-\frac{\beta^2 h_0^2 (1-2n) A_{01}^{4n-2}}{8\lambda(1-\mu) n^2 (1+2n)} + \frac{\beta h_0^2 \left(3 - 4n + 4n^2\right)\left(\lambda A_{01} - \nu^2 A_{02}\right) A_{01}^{2n-1}}{12\lambda(1-\mu) n (1+2n)} +$$

$$+\frac{h_0^2\left(\lambda^2 \mu A_{01}^2 - 2\lambda\mu\nu^2 A_{01}A_{02} + \nu^4 A_{02}^2\right)}{24\lambda(1-\mu)\mu}]/\left(2A_{01}A_{02} + \lambda A_{01}^2 + \frac{A_{02}^2}{\lambda\mu}\right) \quad (15)$$

In expressions (13) and (14), the triple dots stay for higher order terms in sense of the successive approximation algorithm, and the series convergence can be estimated by a geometric progression. More details regarding the estimations and physical reason of these algorithm can be found in [6]. Note that the series are not asymptotic by Poincare, however some asymptotic estimations of the series terms can be done based on parameter $1/n$, if it is small. Solution (13) and (14) satisfies boundary condition (12) if vector $\mathbf{A}_0 = \begin{pmatrix} A_{01} & A_{02} \end{pmatrix}^T$ is chosen by solving the nonlinear eigenvector problem

$$\left(\mathbf{K} - (2/h_0)\,\mathbf{B}\right)\mathbf{A}_0 + (\beta/n)\,\mathbf{N}\left(\mathbf{A}_0\right) = 0 \quad (16)$$

(the solution is given in the next subsection). In the linear case ($\beta = 0$) the vector amplitude remains indefinite, and its direction gives the linear normal modes. If the nonlinearity exists ($\beta \neq 0$) then the vector length depends on the parameter h_0. Thus, solution (13)-(15) includes one constant parameter h_0. Since the equation of motion (6) admits a group of time shift $\bar{t} \longrightarrow \bar{t} + \alpha$, another arbitrary parameter α can be always introduced into the solution. Finally the solution will contain two parameters h_0 (or the length of eigenvector can be chosen), and phase α. Using these parameters one can satisfy a definite type of initial conditions, which provide the existence of the periodic regimes.

Consider the nonlinear eigenvalue problem (16), which can be written in terms of the vector components using the notations (5)

$$\lambda\left(1 - \Omega_0^2\right) A_{01} - \Omega_0^2 A_{02} + (\beta/n)\, A_{01}^{2n-1} = 0 \quad (17)$$

$$\Omega_0^2 A_{01} - (\lambda\mu)^{-1}\left(\nu^2 - \Omega_0^2\right) A_{02} = 0 \quad (18)$$

where $\Omega_0^2 = 2/h_0$.

Equation (18) gives

$$A_{02} = \lambda\mu\Omega_0^2 \left(\nu^2 - \Omega_0^2\right)^{-1} A_{01} \tag{19}$$

Substituting (19) into (17) one obtains

$$\beta A_{01}^{2n-2} = n\lambda (1-\mu) \left(\Omega_0^2 - \omega_1^2\right) \left(\Omega_0^2 - \omega_2^2\right) \left(\Omega_0^2 - \nu^2\right)^{-1} \tag{20}$$

where ω_1 and ω_2 are the two linear normal mode natural frequencies.

If $\beta = 0$ one has the linear case, and equation (20) should be considered as an equation for linear frequencies with respect to Ω_0. In the nonlinear case ($\beta \neq 0$) equation (20) defines an amplitude A_{01} as a function of Ω_0. Having this function one can find another amplitude A_{02} by means of the expression (19). Taking Ω_0 as a variable parameter, one can say that expressions (19) and (20) define two curves on the plane $A_{01}A_{02}$ (indicated by solid curves in Figures 2a,b). Figure 2a is obtained for the case $n = 2$ while Figure 2b is for $n = 6$. For a given point (A_{01}, A_{02}) belonging to one curve we can find the amplitudes of the pendulum and the tank $(\hat{x}_1, \hat{x}_2)$ using expressions (13) and (14), for $\tau = 1$. Figure 3 shows the estimated amplitudes for the out-of-phase and in-phase motions. For the out-of-phase mode the relationship is represented by almost straight line both for small and large amplitudes. In the case of strongly nonlinear regime, when the pendulum amplitude is $\hat{x}_1 \approx 1$, the tank amplitude is relatively small, $\hat{x}_2 \approx 0.2$. Another situation is observed for the in-phase mode, when both the pendulum and the tank simultaneously move in the same direction. The tank amplitude is large enough compared to the pendulum, and the curve slope is not a constant in the nonlinear region (Figure 3, the top curve). Numerical simulations also show that due to impact nonlinearity the in-phase periodic regime is not stable with respect to small perturbations of out-of-phase mode, as a result the analytical and numerical solutions for strongly nonlinear in-phase regime (Figure 4b) is not as good as for out–of-phase one (Figures 5a,b). Figures 4a,b show a comparison between the numerical estimation (shown by dashed curves) and analytical result (shown by solid curves in the time domain for two values of the nearness of the nonlinear natural frequency to the linear natural frequency, i.e., for $\Omega_0^2 - \omega_1^2 = 0.001$ and 0.005 and for $n = 6$. Figures 4 are taken for the in-phase mode and it shows that for large difference between the nonlinear natural frequency and the linear one, the matching between numerical and analytical solutions is relatively poor. On the other hand, Figures 5a,b are obtained for the out-of-phase mode and reveal good agreement between numerical and analytical solutions.

Under parametric excitation of the first mode and the second mode , it is found [4] that if the excitation frequency ω_f is close to $2\omega_{p1}$ or to $2\omega_{p2}$ then oscillations of large amplitudes appear. Asymmetry of the curves indicates that the low frequency oscillations involve both in-phase and out-of-phase motions. On the other hand, the out-of-phase mode is mainly excited about the higher frequency.

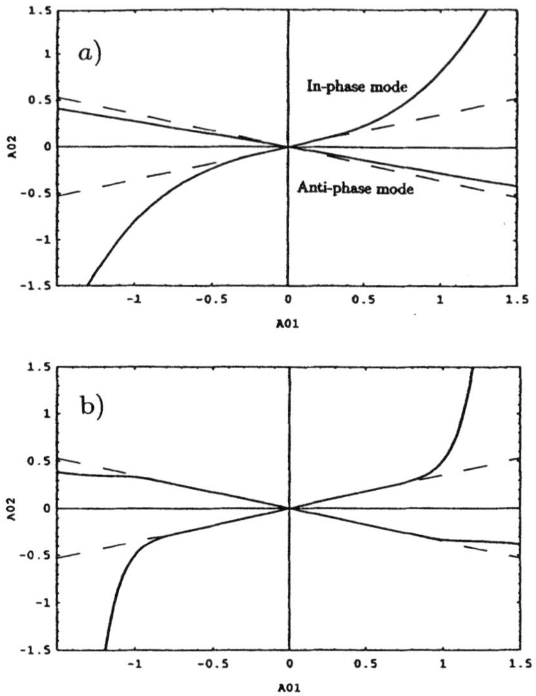

Figure 2.Solution for the nonlinear eigenvector problem: $\beta = 1$, $\lambda = 0.5$, $\mu = 0.5$, $\nu = 1$, a)$n = 2$, b)$n = 6$ (− − − linear normal modes, — nonlinear normal modes) for parameters

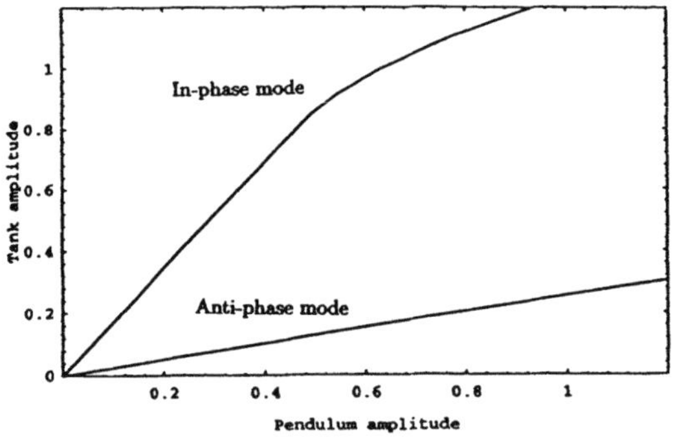

Figure 3.Plane of amplitudes for nonlinear in-phase (top line) and anti-phase (bottom line) periodic free oscillations regimes. The model parameters are: $n = 6$, $\beta = 1$, $\lambda = 0.5$, $\mu = 0.5$, $\nu = 0.5$.

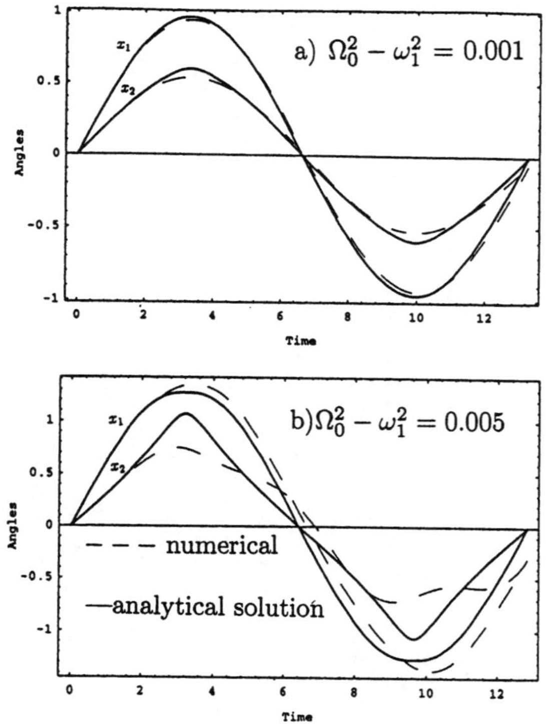

Figure 4.Time history for one period of in-phase mode

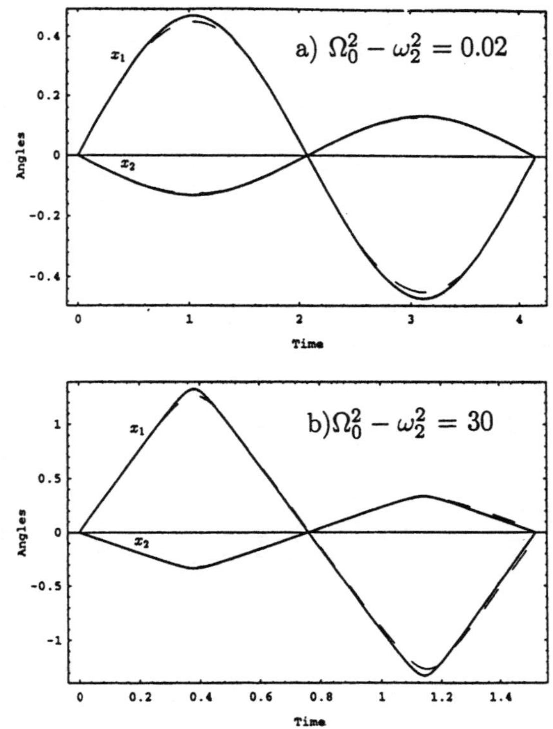

Figure 5.Time history for one period of anti-phase mode

Under lateral periodic loading it is found that for small amplitude level the curves behave like a typical linear oscillator at the points of linear natural frequencies. In the nonlinear region, i.e. for sufficiently large A_{01}, the response behavior essentially depends on the impact power nonlinearity n.

4 Conclusions

The nonlinear interaction the liquid free surface impact motion with the support structure dynamics has been considered. The impact is modeled based on a phenomenological concept by introducing a power function with a higher exponent. A special saw-tooth time transformation (STTT) technique has been applied to analyze the strongly nonlinear periodic regimes of the in-phase and out-of-phase modes. Based on an explicit forms of analytical solutions all basic nonlinear free and forced response characteristics such as in terms of time and frequency domains are obtained. The results have shown that high frequency out-of-phase nonlinear mode appears with a relatively small tank amplitude and preserves stability with respect to small perturbations of the in-phase oscillation mode. The in-phase mode has relatively large tank amplitudes and unstable behavior. The in-phase mode is excited by a low frequency forcing function and then may be realized first during a transient period when the model starts its motion.

References

1 Abramson H N (Editor), The Dynamic Behavior of Liquids in Moving Containers. NASA SP 106, 1966.
2 Cox P A, Bowels E B and Bass R L, Evaluation of Liquid Dynamic Loads in Slack LNG Cargo Tanks, Southwest Research Institute, San Antonio, Technical Report SR-1251, 1980.
3 Moiseev N N, and Rumiantsev V V, Rigid Body Dynamics with Empties Filled by Liquid, Nauka, Moscow, 1965.
4 Pilipchuk V N and Ibrahim R A, The Dynamics of a Non-Linear System Simulating Liquid Sloshing Impact in Moving Structures, Journal of Sound and Vibration 205(5):593-615, 1997.
5 Vedenova E G, Manevich L I, and Pilipchuk V N, Normal oscillations of a string with concentrated masses on nonlinear supports, Pricladnaya Matematika Mekhanika (PMM) 49 (2):153-159, 1985.
6 Pilipchuk V N, Analytical study of vibrating systems with strong nonlinearities by employing saw-tooth time transformations, Journal of Sound and Vibration, 192 (1):43-64, 1996.

Exact solutions for a discrete systems undergo-ing free vibro-impact oscillations

L.I.Manevitch[*], M.A.F. Azeez[**] and A.F.Vakakis[**]

Non-smooth transformations are applied to study strongly nonlinear, periodic free oscillations of a two degree-of-freedom system undergoing purely elastic vibro-impacts. The imposition of suitable boundary conditions at the instants of impacts eliminates the singularities introduced by the transformations, and the smoothened problem is solved numerically. Numerical analysis reveals vibro-impact solutions of complicated form and interesting bifurcation structures of nonlinear modes of three types. Because the system supports a reach structure of periodic motions the study was focused on solutions with periods smaller than twice the period of in-phase normal mode. Higher-period solutions can be studied similarly.

Vibro-impact oscillations represent a very interesting field of study since they often occur in engineering practice. Systems with clearances or contacting parts are examples. Various techniques exist for analyzing vibro-impacting systems, including Poincare' maps [1,2], reduction to boundary value problems [3], purely numerical methods [4,5], perturbation theory and other analytical techniques [6,14].

In this work we study exact free vibro-impact solutions of a two degree-of-freedom system using nonsmooth transformations of the type used in [8] and [9]. The normalized equations of motion are given by

$$\frac{d^2u_1}{dt^2}+u_1+\varepsilon(u_1-u_2)+P\left(u_1,\frac{du_1}{dt}\right)=0$$

$$\frac{d^2u_2}{dt^2}+u_2+\varepsilon(u_2-u_1)+P\left(u_2,\frac{du_2}{dt}\right)=0, \qquad (1)$$

where $P\left(u_i,\frac{du_1}{dt}\right)=2\frac{d\ u_i}{dt}[\delta(u_i+1)-\delta(u_i-1)]$, i=1,2 and the normalized clearances occur at $u_1,u_2=\pm 1$.

We examine periodic vibro-impact solutions introducing the nonsmooth transformations of the time variable,

---

[*] *Institute of Chemical Phisycs RAS, Russia, Moscow*

[**] *University of Illinois, USA, Urbana*

$$\tau(t) = \frac{T}{2\pi}\arcsin[\sin 2\pi\, t/T]\,;\; e(t) = \overset{\circ}{\tau}(t)\; (\overset{\circ}{*}) = \frac{d}{dt}, \tag{2}$$

where represents the period of solution. Expressing the derivatives of u_1 and u_2 (in terms of τ), we obtain:

$$\frac{du_i}{dt} = \frac{du_i}{d\tau}e\,;$$

$$\frac{d^2u_i}{dt^2} = \frac{d^2u_i}{d\tau^2}e^2 - 2\frac{du_0}{d\tau}\sum_{k=-\infty}^{\infty}\delta\,(t - T/4 - kT/2)(-1)^{k+1t} \tag{3}$$

Here $e^2 = 1$.

We note the second derivative contains singular terms that are localized at time instants $t = T/4 + kT/2$ and the equal zero everywhere besides these points. These singularities are non-essential if the derivative $\frac{du_i}{d\tau}$ equal exactly to zero at same instants; these instants correspond to impacts of one of the masses. So the nonlinear terms in the equations of motion can be written as

$$P\left(u_i, \frac{du_1}{dt}\right) = 2\frac{du_i}{d\tau}\sum^{\infty}\delta(t - T/4 - kT/2)(-1)^{k+1}.$$

Hence, assuming the periodic vibro-impact motion involves impacts in u_1 and u_2 only at time instants $t = T/4 + kT/2$, $-\infty < k < +\infty$, the terms with singularities in the second derivatives o u_1 and u_2 are eliminated and the equations of motion (1) are transformed to

$$\frac{d^2u_1}{d\tau^2} + u_1 + \varepsilon\,(u_1 - u_2) = 0$$

$$\frac{d^2u_2}{d\tau^2} + u_2 + \varepsilon(u_2 - u_1) = 0 \tag{4}$$

These equations are linear and free of singularities, and possess τ as independent variable. The solution of (4) for initial conditions:

$$u_1(0) = 1;\quad u_2'(0) = 0;\quad (*)' \equiv d/d\tau \tag{5}$$

are given by

$$\begin{Bmatrix} u_1(\tau) \\ u_2(\tau) \end{Bmatrix} = \left(\frac{1+u_2(0)}{2}\right)\cos\tau + \begin{Bmatrix} +1 \\ -1 \end{Bmatrix} \times \left(\frac{1-u_2(0)}{2}\right)\cos\alpha\tau + \\ + \left(\frac{u_1^1(0)}{2}\right)\sin\tau + \begin{Bmatrix} +1 \\ -1 \end{Bmatrix} \times \left(\frac{u_1^1(0)}{2\alpha}\right)\sin\alpha\tau \tag{6}$$

where $\alpha = \sqrt{1 + 2\varepsilon}$ is the coupling parameter.

We note that, in contrast to (1), the solutions of (6) are only in the range $|\tau| \leq 1$ [see definition (2)]; In essence, we will construct the vibro-impact solution for duration of a period and then rely on the periodicity properties of τ to periodically continue the solution over large interval of time. We will consider three types of periodic vibro-impact oscillations corresponding to following boundary conditions:

$$\begin{aligned}&\text{Type I:} \quad u_1'(T/2) = 0; \quad u_2(T/2) = -1\\ &\text{Type II:} \quad u_1(T/2) = -1; \quad u_2'(T/2) = 0\\ &\text{Type III:} \quad u_1'(T/2) = 0; \quad u_2(T/2) = 1\end{aligned} \tag{7}$$

In addition the boundary conditions (5) are enforced. Substituting (6) in to (7) we obtain the following algebraic relations:

$$\begin{aligned}&\left(\frac{1+u_2(0)}{2}\right)\cos\left(T/2\right)+I_u\left(\frac{1-u_2(0)}{2}\right)\cos\left(\alpha T/2\right)+\\ &+\left(\frac{u_1^1(0)}{2}\right)\sin\left(T/2\right)+I_u\left(\frac{u_1^1(0)}{2\alpha}\right)\sin\left(\alpha T/2\right)=I_\varphi\\ &-\left(\frac{1+u_2(0)}{2}\right)\sin\left(T/2\right)-I_u\left(\frac{1-u_2(0)}{2}\right)\alpha\sin\left(\alpha T/2\right)+\\ &+\left(\frac{u_1^1(0)}{2}\right)\cos\left(T/2\right)+I_u\left(\frac{u_1^1(0)}{2}\right)\cos\left(\alpha T/2\right)=0\end{aligned} \tag{8}$$

where depending on the type of vibro-impact oscillation the various coefficients in (8) take the following values:

$$\begin{aligned}&\text{Type I:} \quad I_u = -1, \quad I_{u'} = +1, \quad I_\varphi = -1\\ &\text{Type II:} \quad I_u = +1, \quad I_{u'} = -1, \quad I_\varphi = -1\\ &\text{Type III} \quad I_u = -1, \quad I_{u'} = +1, \quad I_\varphi = +1\end{aligned} \tag{9}$$

Expressing $u_1'(0)$ as a function $u_2(0)$ from the second of the equations (8) and substituting the result into the first equation, we obtain the following expression relating the period T and $u_2(0)$:

$$\begin{aligned}&u_2(0)\left[\cos(T/2)-I_u\cos(\alpha T/2)\right]+\left[\cos(T/2)+I_u\cos(\alpha T/2)\right]+\\ &+u_1^1(0)\left[\sin(T/2)+(I_u/\alpha)\sin(\alpha T/2)\right]-2I_\varphi=0\end{aligned} \tag{10}$$

which must be solved for $\max_{0\leq\tau\leq 1}\left\{|u_1(\tau)|_1, |u_2(\tau)|\right\} \leq 1$.

The numerical solution of equation (10) for varying $u_2(0)$ reveals complicated structures of vibro-impact solutions of the types I, II and III. The system supports the rich structure of periodic solutions. In the following study we are focusing on solutions with periods in the range $0 \leq T \leq 4\pi$. Higher period solutions can be study similarly. The stability of the numerical orbits was studied by numerical integrating the equations of motion (1) with the theoretically predicted initial conditions.

The simplest case corresponds to smallness of parameter ($\varepsilon<<1$) [11]. In Figure 1 we present the solution branches in total energy $u_2(0)$ for the moderate value of coupling $\varepsilon = 1$ and corresponding vibro-impact modes. The solution is restrained by the requirement $|u_2(0)|<<1$. Solution 1 is stable whereas solution 2 is unstable; These are typical Type-I vibro-impact motions involving impacts in both u_1 and u_2 directions. Solution 3 is unstable and solution 4 is stable, solution 4 is a Type-II localized vibro-impact motion involving impacts only in the u_1 direction. The Type-III solution 3 is unstable.

To show the effect on the solutions of an increase in the coupling, in Figure 2 we present the vibro-impact solutions for $\varepsilon = 10.0$. Solutions 1 (stable), 2 (stable), 3 (unstable), 4 (unstable) are Type-I oscillations. Two Type-I branches of periodic solutions are shown, one corresponding to periods in the range 4< <6 (1,2,3), and the other to periods 3< <4. Three Type-II branches of vibro-impact oscillations were found, the first one (with periods in the range 5< <7) containing solutions 10 (unstable), 11 (stable) and 12 (unstable), the second one (2< <5) containing solutions 6 and 7 (unstable), and the third one (0< <2) containing solutions 8 (unstable) and 9 (stable); these last solutions is the only localized motion. Finally, a single Type-III branch was detected with periods in the range 11< <4 π, containing solutions 13 (unstable) and 14 (stable).

These results indicate that as the coupling increases more complicated classes of solutions are generated. Additional higher-period vibro-impact solutions of the aforementioned types were detected for $T > 4\pi$, but these are not shown here. Finally, we note that by changing the boundary conditions (7) more classes of vibro-impact periodic motions can be studied.

Acknowledgements. This work were supported in part by NSF International supplement to young investigated Award CMS-94-57750 (of A.F.Vakakis). Dr. Devendra Garg and Mrs. Cassandra Dudca are Grant monitors.

References

1. Shaw, S., Holmes, P. 1983. A periodically forced linear oscillator with impacts: chaos and long period motions // Physical Review Letters. V.51. P. 623-626.

2. Shaw, S. 1985. The dynamics of a harmonically excited system having rigid amplitude constraints, Part I: Subharmonic motions and local bifurcations, Part II: Chaotic motions and global bifurcations //Journal of Applied Mechanic. V. 52. P. 453-464.

3. Mikhlin, Yu.V., Vakakis, A.F., Salenger, G. Direct and inverse problems encountered in vibro-impact oscillations of a discrete system // Journal of Sound and Vibration (submitted).

4. Azeez, M.A.E., Vakakis, A.F. 1997. Numerical and experimental analysis of the nonlinear dynamics due to impacts of a continuous overhung rotor // 16th Bi-

ennial ASME Conference on Mechanical Vibration and Noise, Symposium on Impact. Sacramento, California.

5. Emaci, E., Nayfeh, T.A., Vakakis, A.F. 1997. Numerical and experimental stidy of nonlinear localization in a flexible structure with vibro-impacts // Zeitschrift fur Angewandte Mathematik und Mechanik, ZAMM. V.77. N.7. P.527-541.

6. Babitsky V.I. 1978. Theory of Vibro-Impact Systems (in Russian), Moscow, Nauka.

7. Ivanov A.P. 1993. Analitical Methods in the theory of Vibro-Impact Systems// PMM. V. 57(2). p.5-21.

8. Peterka, F., Vacik, J. 1992. Transition to chaotic motion in mechanical systems with impacts // Journal of Sound and Vibration. V.154. N.1. p.95-115.

9. Zuravlev V.F. 1977. Investigation of Certain Vibro-Impact Systems by the Method of Non-smooth Transformations. // Izv. AN SSSR. MTT. V.12(6). pp.24-28.

10. Pilipchuk V.N. 1988. A Transformation for Vibrating Systems Based on a Non-smooth Periodic Pair of Functions.// Doclady AN Ukr. SSR. Ser.. 4. pp. 37-40.

11. Vedenva E.G., Manevitch L.I., Pilipchuk V.N. 1985. Normal Oscillations of a String with Concentrated Masses on Nonlinearly Elastic supports. // PMM. V. 49.(2). pp.203-211.

11. Vedenva E.G., Manevitch L.I. 1985. Periodic and Localized Waves in vibro-Impact Systems of regular Configurations. // Mashinovedenie, V. 4. pp.21-39.

12. Manevitch L.I., Mikhlin Yu.V. and Pilipchuk V.N. 1989. Method of Normal Vibrations for Essentially Nonlinear Systems. Moscow. Nauka.

13. Vakakis, A.F., Manevitch L.I., Mikhlin Yu.V., Pilipchuk V.N., Zevin A.A. 1996. Normal modes and localization in nonlinear systems. New-York: Wiley, 552c.

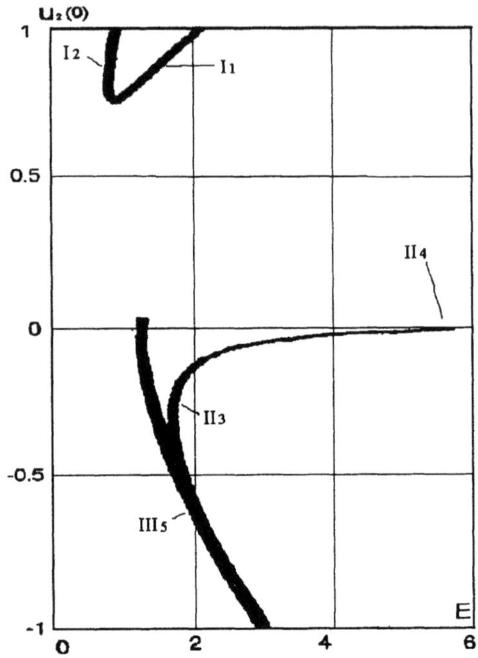

Fig. 1, a

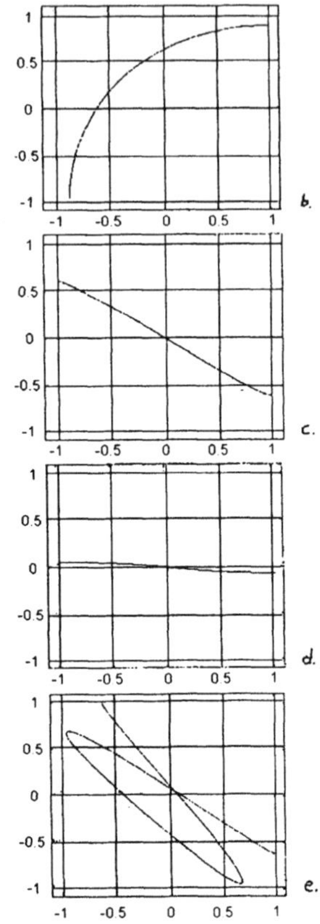

Fig. 1, (b-e)

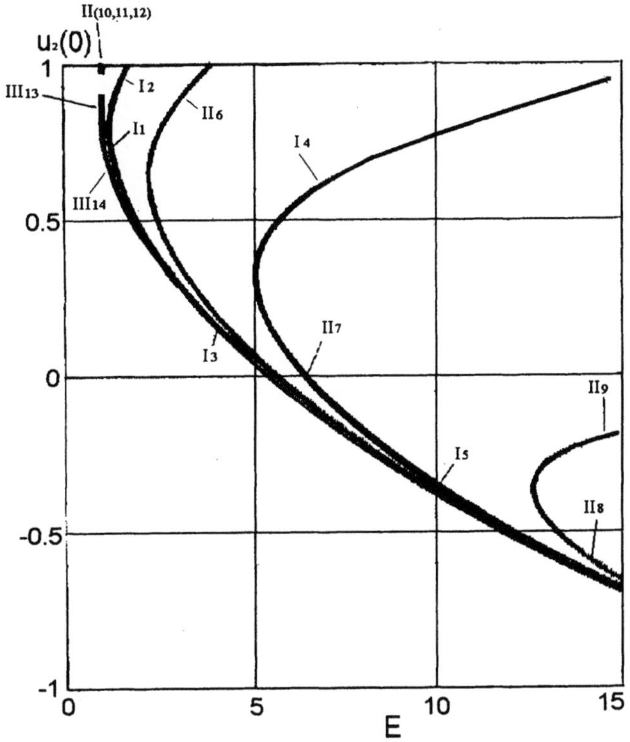
II(10,11,12)
u₂(0)
1
0.5
0
-0.5
-1
0
5
10
15
E
III13
I2
II6
I1
I4
III14
II7
I3
II9
I5
II8

Fig. 2, a

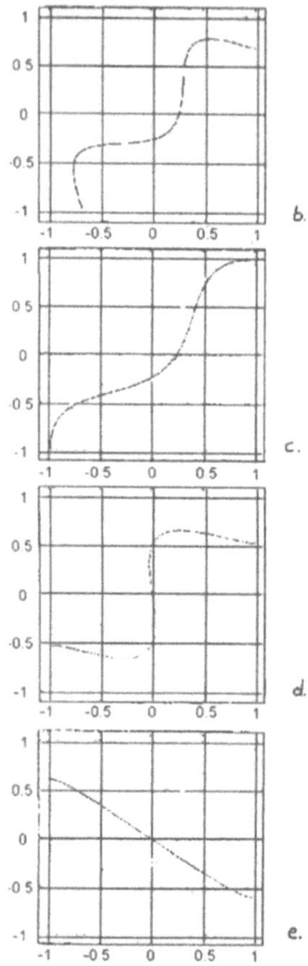

Fig. 2, (b-e)

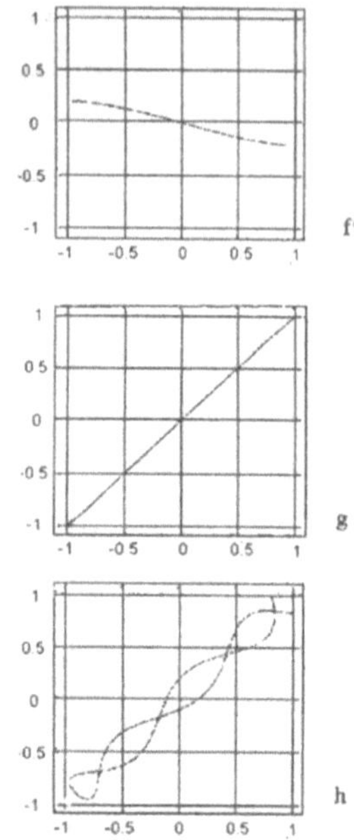

Fig. 2, (f-h)

Asymptotic study of damped one-dimensional oscillator with close to impact potential

O.V.Gendelman and L.I.Manevitch**

Dynamics of linearly damped oscillator with strongly nonlinear elastic potential is ivestigated. The description of the oscillator dynamics is provided by means of matching asymptotic expansions for strongly nonlinear and quasilinear vibrations. The quasilinear vibrations are investigated by standard methods whereas the strongly nonlinear vibrations require special asymptotics based on exactly integrable partial case. The coincidence of the results with numerical simulation data is satisfactory for any combination of the regimes.

1 Introduction and the exactly integrable case.

The impact interaction can be efficiently approximated by elastic force with odd power nonlinearity of high order [1]. In this connection a theoretical analysis of the oscillator with strong nonlinearity and dissipation of the energy seems to be significant for the theory of vibro-impact systems.

Let us consider the following equation describing the dynamics of one-dimensional oscillator with strong nonlinearity and linear dissipation:

$$G = \ddot{y} + \frac{1}{\tau_0}\dot{y} + \omega_0^2 y + Cy^m = 0 \tag{1}$$

where y denotes the displacement, τ_0 - characteristic time of linear relaxation, ω_0 - the frequency of linear vibrations, C is the intensity of nonlinear potential, m is odd number more than unity.

This equation cleary cannot be integrated for arbitrary set of parameters. However fortunately it is possible to reveal the set of parameters which enables the explicit integration of eq. (1).

In order to proceed we search for the symmetry groups of eq. (1), i.e. for the transformations remaining it invariant. These groups of transformations are completely determined by their infinitesimal operator acting on the equation. In the most general case it may be written in the following form:

$$X = \varepsilon\frac{\partial}{\partial t} + \eta\frac{\partial}{\partial y} + X_1 + X_2 \tag{2}$$

where ε and η are the functions of y and t, X_1 and X_2 are the first and the second continuations of the operator [2]. The explicit expressions of these continuation

* Polymer Physics Department, Institute of Chemical Physics RAS, Kosygin str.4, Moscow, 117977, Russia

operators may be computed according to standard procedure [3] and are explicitly expressed as

$$X_1 = (\frac{\partial\eta}{\partial t} + p(\frac{\partial\eta}{\partial y} - \frac{\partial\varepsilon}{\partial t}) - p^2\frac{\partial\varepsilon}{\partial y})\frac{\partial}{\partial p} \tag{3}$$

$$X_2 = (\eta_{tt} - p^3\varepsilon_{yy} + p(2\eta_{yt} - \varepsilon_{tt}) + p^2(\eta_{yy} - 2\varepsilon_{yt}) + r(\eta_y - 2\varepsilon_t) - rp\varepsilon_y)\frac{\partial}{\partial r} \tag{4}$$

where $p = \dot{y}$, $r = \ddot{y}$. The condition of invariance of eq.1 with respect to the transformation group determined by the infinitesimal operator (2) is expressed as

$$XG = 0 \tag{5}$$

While expressing the value r via y, t and p with the help of equation (1) as

$$r = -\frac{1}{\tau_0}p - \omega_0^2 y - Cy^m \tag{6}$$

the condition (5) is transformed to the form:

$$- p^3\varepsilon_{yy} + p^2(\eta_{yy} - 2\varepsilon_{yt}) + p(2\eta_{yt} - \varepsilon_{tt} + \frac{\varepsilon_t}{\tau_0} + \varepsilon_y(\omega_0^2 y + Cy^m)) +$$

$$+ \eta_{tt} + \frac{\eta_t}{\tau_0} + \eta(\omega_0^2 + Cmy^{m-1}) - (\omega_0^2 y + Cy^m)(\eta_y - 2\varepsilon_t) = 0$$

(7)

All variables in this equation do not depend on *p*, therefore the coefficients of all powers of p shoul be equal to zero and the following system of equations holds:

$$\begin{cases} \varepsilon_{yy} = 0 \\ \eta_{yy} - 2\varepsilon_{yt} = 0 \\ 2\eta_{yt} - \varepsilon_{tt} + \dfrac{\varepsilon_t}{\tau_0} + \varepsilon_y(\omega_0^2 y + Cy^m) = 0 \\ \eta_{tt} + \dfrac{\eta_t}{\tau_0} + \eta(\omega_0^2 + Cmy^{m-1}) - (\omega_0^2 y + Cy^m)(\eta_y - 2\varepsilon_t) = 0 \end{cases}$$

(8)

It is easy to derive from the system (8) the expressions for the functions η and ε:

$$\eta = yf(t)$$
$$\varepsilon = g(t) \tag{9}$$
$$2f - g_{tt} + \frac{\varepsilon_t}{\tau_0} = 0$$

These relationships are to be substituted in the last equation of system (8). If recalling that the functions mentioned above depend solely on t it is possible to obtain the following solutions:

$$g = \exp\left(\frac{m-1}{m+3}\frac{t}{\tau_0}\right)$$
$$f = \frac{-2}{(m+3)\tau_0}\exp\left(\frac{m-1}{m+3}\frac{t}{\tau_0}\right) \tag{10}$$

These solutions are valid when certain relationship imosed on the parameters of eq.(1) holds. This relationship is written as

$$\omega_0^2\tau_0^2 = \frac{2(m+1)}{(m+3)^2} \tag{11}$$

and constitutes the single case when the equation (1) may be exactly solved analytically due to its nontrivial group of symmetry. The appropriate solution is expressed as follows:

$$y(t) = Y_0 \exp\left(-\frac{2}{m+3}\frac{t}{\tau_0}\right) CAM_m\left(\sqrt{\frac{2C}{m+1}}\frac{\tau_0(m+3)}{m-1} Y_0^{\frac{m-1}{2}} \exp\left(-\frac{m-1}{m+3}\frac{t}{\tau_0}\right) + \phi_0\right) \tag{12}$$

where the special function $CAM_m(u)$ is defined as [5]:

$$du = \int \frac{dCAM_m}{\sqrt{1 - CAM_m^{m+1}}}$$

and Y_0 is the initial magnitude of the solution, ϕ_0 is the phase additive. The typical form of this solution is presented at Fig.1. It is easy to check that the solution (12) of equation (1) with condition (11) for any appropriate values of m describes the nonlinear vibrations of large amplitude which turn to nearly linear relaxation in the case of low amplitudes.

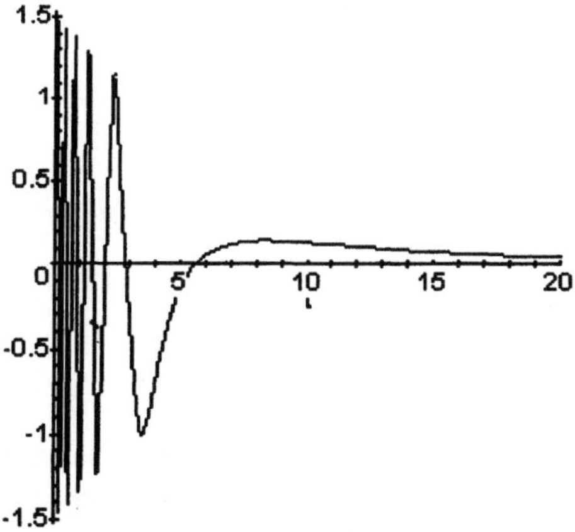

Figure 1. The solution for the integrable case.

$m = 15, C = 8, y(0) = 0, y(0) = 25.6, \tau = 1$

As it was mentioned before, the condition (11) fixes very special case when the solution of eq. (1) may be expressed in explicit analytical form. However the physical meaning of this solution is doubtful. One never can suppose that the values of the characteristic time of relaxation, linear vibration frequency of the oscillator and the power of the nonlinear force are connected by exact mathematical relation because they physical origins usually are different and not related. Such situation is common for significant number of nonlinear equations which can be solved exactly if certain conditions relating their parameters hold. However we are going to demonstrate that in the system considered the integrable case constitutes important approximation for any set of parameters and may be used for computing of approximate solutions.

2 Asymptotic procedure for arbitrary set of parameters.

Let us consider the equation (1) with arbitrary set of parameters (the fact that the parameters of the equation become arbitrary is denoted by omitting of subscript 0 which was above related to the integrable case):

$$\ddot{y} + \frac{1}{\tau}\dot{y} + \omega^2 y + Cy^m = 0 \tag{13}$$

The change of variables "inspired" by the expressions for the infinitesimal operator (2)-(10)

$$z = \exp\left(-\frac{m-1}{m+3}\frac{t}{\tau}\right)$$

$$X = y\exp\left(\frac{2}{m+3}\frac{t}{\tau}\right) \tag{14}$$

reduces equation (13) to the following form:

$$X_{zz} + \frac{C\tau^2(m+3)^2}{(m-1)^2}X^m + \Delta\frac{X}{z^2} = 0 \tag{15}$$

where

$$\Delta = \frac{(m+3)^2}{(m-1)^2}\left(\omega^2\tau^2 - \frac{2(m+1)}{(m+3)^2}\right) \tag{16}$$

The condition Δ=0 obviously corresponds to the integrable case considered above. However it is clear from equation (15) that for sufficiently large value of z, the last term (which «spoils» the integrability) becomes small compared to other terms of the equation regardless actual values of parameters. The condition of this term being small is expressed via the initial variables as follows:

$$CY_0^{m-1} >> \omega^2 - \frac{2(m+1)}{(m+3)^2\tau^2} \tag{17}$$

where Y_0 denotes characteristic magnitude of the solution. The fact that the condition (17) does not contain the time dependence is of great importance. It means that for any set of parameters of eq. (13) in the case of sufficiently large amplitudes the solution is close to appropriate integrable case, determined by the relaxation time and power of nonlinearity. The actual value of ω however effects the regime of low amplitudes and determines the transition point where the solution «leaves» the trajectory of the intergable case. The integrable case constitutes the «attractor» for the case of large amplitudes.

Now the idea of asymptotic approach to the investigation of equation (13) becomes clear. The case of large amplitudes always may be described by the solution (12). It has to be matched with the solution for the case of small amplitudes which may be described in simple quasilinear approximation. The matching point may be determined with the help of condition (17).

In order to proceed with the matching process it is desirable to investigate the behavior of the solutions of eq.(13) in the non-integrable case. The initial form of equation is more convenient for this purpose.

Three main possible cases should be distinguished. They are:

a) Damped vibrations: $1-4(\omega\tau)^2 < 0$

b) Fast relaxation: $1-4(\omega\tau)^2 > 0$

c) Slow relaxation: $\omega = 0$.

The first approximation for the cases a) and b) constitutes simply the solution of dynamical equation of linear oscillator with damping and may be written as follows:

$$y_{small}(t) = D_1 \exp(\lambda_1 t) + D_2 \exp(\lambda_2 t)$$
$$\lambda_{1,2} = -\frac{1}{2\tau} \pm \sqrt{\left(\frac{1}{2\tau}\right)^2 - \omega^2} \tag{18}$$

In the case c) the situation becomes essentially more complicated. One of the roots of the secular equation (18) turns to zero and it means that the «slow» damping is governed by nonlinear term rather than by linear force as in the former case. According to [4] the appropriate solution may be presented in the following form:

$$y_{small}(t) = D_1 \exp(-\frac{t}{\tau}) + ((m-1)C\tau t + D_2)^{\frac{1}{1-m}} \tag{19}$$

The constants D_1, D_2 in equations (18,19) have to be determined from the conditions of matching. The initial conditions of the equation (13) should be satisfied with the help of the constants Y_0 and ϕ_0.

The calculations necessary for computing the appropriate approximations were performed numerically with the help of MAPLE software, because the transcendent equations determining the constants hardly can be solved in other way. The matching point was chosen in order to preserve the continuity of the function and its first and second derivatives. The results for all regimes mentioned above are presented at Fig. 2, a-c. The solid line represents the rsults of matching and the dots reprsent the numerical simulation data. The coincidence seems to be rather satisfactory.

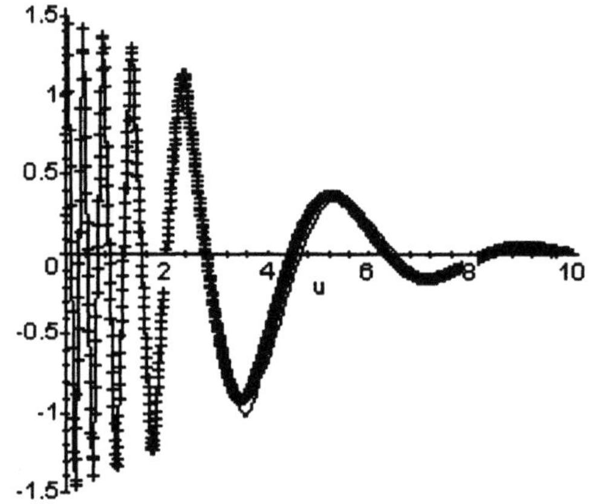

Figure 2a - the case of damped vibrations.

$\omega = 1.73, m = 15, C = 8, y(0) = 0, y(0) = 25.6, \tau = 1$

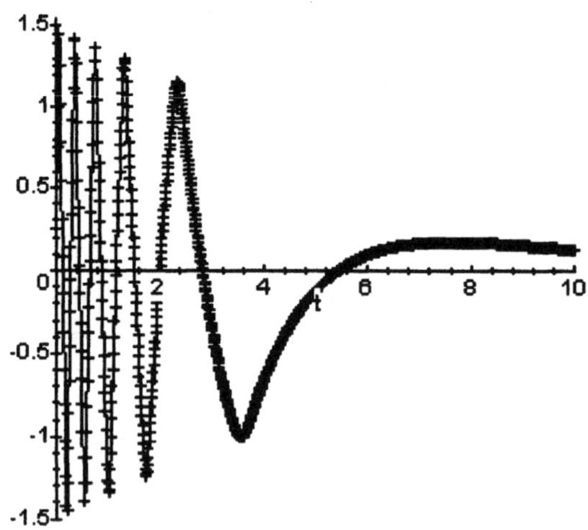

Figure 2b - the case of fast relaxation.

$\omega = 0.45, m = 15, C = 8, y(0) = 0, y(0) = 25.6, \tau = 1$

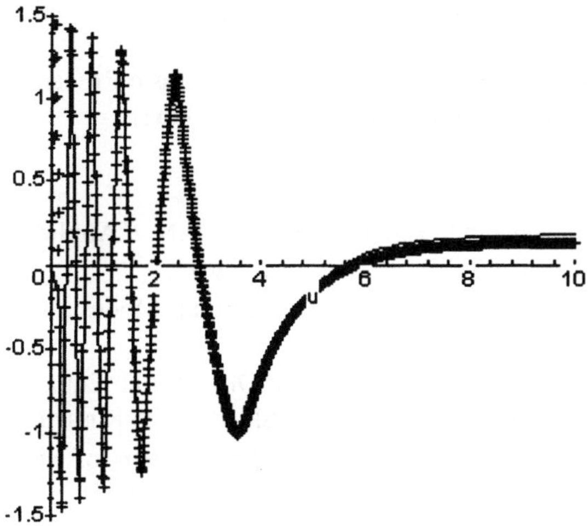

Figure 2c - the case of slow relaxation.

$\omega = 0, m = 15, C = 8, y(0) = 0, y(0) = 25.6, \tau = 1$

The problem of computing higher approximations seems rather complicated from technical point of view. It needs explicit account of the solution (12) modification due to the last term in eq. (15) as well as the effects of weak nonlinearity in expressions (18) for the low-amplitude regime. These investigations are beyond the scope of present paper.

3 Conclusive remarks.

The investigation presented here demonstrates that efficient asymptotic analysis of the dynamics of one-dimensional oscillator wits strong nonlinearity may be performed analytically based on the integrable case. The latter one is characterized by unique combination of parameters of the initial equation. This integrable case serves as «attractor» for all possible combinations of parameters in the case of large amplitudes, which seems to be the most difficult case for regular asymptotic methods. The opposite case of low amplitudes may be easily analyzed by regular quasilinear approximation. The matching point also may be determined with the help of the integrable case solution.

The situation described is rather regular for wide class of nonlinear equations because for certain combination of parameters there exists additional symmetry and the equation may be solved explicitly. The present research gives an idea how this «exotic» exact solution may serve a basis for the asymptotic analysis of more general cases.

The authors are grateful to Russian Foundation of Basic Research for the financial support of their investigations (grant No. 98-03-33366a).

References.

1. Vedenova E.G., Manevitch L.I. and Pilipchuk V.N. // Applied Mathematics and Mechanics, v.49, n.2, p.203-211, 1980 (in Russian)

2. Vakakis A.F., Manevitch L.I. et al,. Normal Modes and Localization in Nonlinear Systems, Wiley, NY, 1996

3. Ovsiannikov L.V. Group Properties of Differential Equations, Academy of Sciences Publishers, Sibir branch, Novosibirsk, 1962 (in Russian)

4. Vasilyeva A.B., Butuzov V.F., Asymptotic Analysis of Singular Differential Equations, Moscow, Vyschaya Shkola, 1990 (in Russian)

5. Rosenberg R., The Ateb(h)-Functions and Their Properties, Quarterly of Applied Mathematics, v.21(1),37-47, 1963,

Interaction between impulses and impacts in the nonlinear dynamics of an impacting system: a non-classical bifurcation

***Stefano Lenci*[†] *and Giuseppe Rega*[‡]**

This work is aimed at studying a non-classical bifurcation arising in the dynamics of mechanical systems subjected to impulses and impacts. It is shown that when the impulses (which are applied at known times but at unknown positions) and the impacts (which are applied at known positions but at unknown times) synchronize, i.e., occur simultaneously, then a previously existing periodic solution is suddenly destroyed. These local bifurcations are detected analytically and it is discussed how they contribute to the determination of the overall bifurcation scenario of an investigated class of periodic subharmonic solutions. The theoretical results are illustrated and checked by some numerical simulations.

1 Background and motivations

In recent and past years, the study of impacting dynamical systems arising in many fields of engineering drew the attention of many researchers ([1-4]). The interest suggested by practical applications was paralleled by the interest of specialists in dynamical systems theory, which found in impacting systems very powerful prototypes of complex dynamics ([5], [6]). In this area, referring to impacting systems often entails the possibility to obtain closed form solutions of the governing equations [7].

In this work, some aspects of the nonlinear dynamics of an inverted pendulum with rigid unilateral constraints are analyzed under the hypothesis that the system is forced by a base T-periodic excitation. We assume the Newton law ($\dot{x}(t^+) = -r\dot{x}(t^-)$, $r \leq 1$) for the instantaneous impacts on the lateral barriers and we consider the case of two (equal and opposite) impulses excitation applied after constant time intervals $T/2$. This excitation arises in the field of controlling nonlinear dynamics ([8], [9]), and it is an approximation of the best excitation permitting to avoid chaos for large values of the excitation amplitude.

We are interested in studying a class of nT (subharmonic) periodic motions bouncing one time in each period, which seems to play an important role in the system response [7, 9]. The hypothesis of piecewise linear behavior of the pendulum allows obtaining analytical criteria for the existence and stability of these solutions. These criteria are discussed and they are graphically illustrated in the period-amplitude parameter plane, paying attention to the possible coexistence of solutions with different n. Furthermore, the ranges of stability are investigated

[†] Istituto di Scienza e Tecnica delle Costruzioni, Università di Ancona, via Brecce Bianche, Monte D'Ago, 60131, Ancona, Italy.
[‡] Dipartimento di Ingegneria Strutturale e Geotecnica, Università di Roma "La Sapienza", via A. Gramsci 53, 00197, Roma, Italy.

in detail. It is shown that they initiate and terminate either by classical bifurcations (saddle-node, *SN*, and period doubling, *PD*) or by *non-classical* (*NC*) bifurcations, which will be analyzed in detail.

The system is subjected to two different kinds of pulsed forces: the impulses of the external excitation and the impulses causing the rebounding of the hitting mass at the impacts. The formers depend on time, and are applied at *a priori* unknown positions in the phase space. The latters, on the other hand, depend on the position of the pendulum, and are applied at *a priori* unknown times. It is shown that the synchronization of these two kinds of impulses entails the sudden destroying of a previously existing periodic solution. This fact may occur only for determined values of the impulses amplitude, which correspond to the occurrence of the mentioned *non-classical* local bifurcation.

Some numerical simulations based on time histories, bifurcation diagrams and basins of attractions, and aimed at confirming and illustrating the theoretically results previously obtained, end the work. Some typical aspects of nonlinear systems, such as coexistence and competition of attractors and fractal basin boundaries, are evidenced.

2 A class of periodic solutions and their stability

The equations of motion of the inverted pendulum with rigid unilateral constraints subjected to base periodic impulsed excitation are

$$\ddot{x}+2\delta\dot{x}-x=f(t),\ |x|<1, \qquad \dot{x}(t^+)=-r\dot{x}(t^-),\ |x|=1, \tag{1}$$

where $x=\vartheta/\vartheta_{\max}$, ϑ is the deflection angle measured from the vertical, $\delta\in[0,1[$ is the damping coefficient of free motion, $r\in]0,1]$ is the coefficient of restitution for the *instantaneous* impacts, $f(t)=\gamma[-\delta(t-nT)+\delta(t-T/2-nT)]$, $n\in\mathbb{N}$, is the T-periodic excitation force ($\delta(t)$ is the Dirac delta function) and γ the magnitude of the equally spaced in time impulses.

Let $x_1=x(t)$, $y_1=\dot{x}(t)$, $x_0=x(0)$ and $y_0=\dot{x}(0)$. During the free motion, the initial condition $\mathbf{x}_0=[x_0,y_0]^T$ is mapped into $\mathbf{x}_1=[x_1,y_1]^T$ by the linear transformation $\mathbf{x}_1=\mathbf{M}_t\mathbf{x}_0$, where $\mathbf{M}_t$ is given in [10].

The impulses are characterized by the (affine) map $\mathbf{x}_1=\mathbf{x}_0+\mathbf{g}$, where $\mathbf{g}=[0,\gamma]^T$, which increases the velocity of the point by an amount equal to γ. Finally, the impacts occurring at $x_0=x_1=\pm1$ are described by $\mathbf{x}_1=\mathbf{R}\mathbf{x}_0$, where $\mathbf{R}$ is also given in [10].

The stroboscopic Poincaré map of the system is obtained by an orderly repeated application of the previously illustrated maps. For example, if there are no impacts, the point $\mathbf{x}_0$ is mapped after one period of the excitation into the point $\mathbf{x}_1=\mathbf{M}_T\mathbf{x}_0-\mathbf{M}_T\mathbf{g}+\mathbf{M}_{T/2}\mathbf{g}$.

In this work we study a class of *n*-periodic solutions corresponding to the pendulum bouncing repeatedly at one lateral constraint, say the right $x=+1$, with free flight time of nT. Thus, in the investigated solutions there is only one impact in each motion period. These correspond to Type I solutions in the Shaw and Rand

[7] classification, to type $z=1/n$ motions in the Peterka [2] classification and to $(n,1)$-type orbits in the Ivanov [4] classification.

After some algebra (see [10] for more details), we obtain that the condition of existence of these periodic motions is given by

$$\mathbf{R}^{-1}\mathbf{M}_{\beta-T/2}\mathbf{x}_0 = \mathbf{M}_{nT}\mathbf{M}_{\beta-T/2}\mathbf{x}_0 - (\mathbf{I}-\mathbf{M}_T)^{-1}(\mathbf{I}-\mathbf{M}_{nT})(\mathbf{M}_{T/2}-\mathbf{I})\mathbf{M}_\beta\mathbf{g}, \quad (2)$$

where $\beta \in [0,T/2]$ (unknown) is the time between the impact and the foregoing impulse. Accordingly, $T/2$-β is the time between the impact and the subsequent impulse. Let us define $\mathbf{z} = \mathbf{M}_{\beta-T/2}\mathbf{x}_0 = [1,\xi]^T$, where $\xi<0$ is the velocity after the rebounding. With this definition, equation (2) can be rewritten as

$$(\mathbf{I}-\mathbf{M}_T)(\mathbf{R}^{-1}-\mathbf{M}_{nT})\mathbf{z} = (\mathbf{I}-\mathbf{M}_{nT})(\mathbf{I}-\mathbf{M}_{T/2})\mathbf{M}_\beta\mathbf{g}. \quad (3)$$

These are two equations in the two unknowns β and ξ, where T, n, r, δ and γ are parameters. To solve this system, we observe that it is linear in ξ and γ and can easily be solved with respect to these two variables:

$$\gamma = \frac{A}{Be^{s_2\beta}+Ce^{s_1\beta}}, \qquad \xi = \frac{De^{s_2\beta}+Ee^{s_1\beta}}{Be^{s_2\beta}+Ce^{s_1\beta}}, \quad (4)$$

where $s_{1,2} = -\delta \pm \sqrt{1+\delta^2}$ and where $A\geq 0$, $B>0$, $C>0$, $D<0$ and $E<0$ are functions of T, n, r and δ and are given in [10]. Since B and C have the same sign, γ and ξ are well defined, $\forall\beta\in\mathbb{R}$, and γ is positive. Inverting the function $(4)_1$ we get $\beta=\beta(\gamma)$, which, substituted in $(4)_2$, gives the second unknown $\xi=\xi(\gamma)$.

The inverse of $(4)_2$ is $\beta = (s_2-s_1)^{-1}\ln[-(C\xi-E)/(B\xi-D)]$. This shows that we can arbitrarily consider β or ξ as "independent" variable. Contrarily to $(4)_2$, on the other hand, the function $(4)_1$ has not a unique inverse. It tends to zero for $|\beta|\to\infty$ and it has a local maximum γ_{max}. It follows that for $\gamma>\gamma_{max}$ there are no periodic solutions, while for $\gamma<\gamma_{max}$ there are two oscillations. Thus, γ_{max} actually corresponds to an *inverse* saddle-node (*SN*) bifurcation.

The *SN* bifurcation is characterized by the condition $d\gamma/d\beta=0$, which yields

$$\gamma_{SN} = A\left[B\left(-\frac{Cs_1}{Bs_2}\right)^{(s_2/(s_2-s_1))} + C\left(-\frac{Cs_1}{Bs_2}\right)^{(s_1/(s_2-s_1))}\right]^{-1}. \quad (5)$$

The previous computations have been performed under the hypothesis of rebounding on the right wall $x=+1$. However, according to the symmetry of the system and the antisymmetry of the excitation, there exist also "specular" oscillations in the opposite potential well. To consider the bouncing on the left constraint, we must assume that the impact occurs for $x=-1$, *i.e.*, we must use $\mathbf{z} = [-1,\xi]^T$, $\xi>0$, in (3). It is easily seen that in this case the solution is obtained by simply changing the sign of γ and of ξ given in (4).

To study the stability of the periodic solutions, we calculate the eigenvalues of the stroboscopic Poincarè map associated to the system. They are the solutions of the second order equation $\lambda^2 - \text{tr}\lambda + \det = 0$ [10], where

$$\det = r^2 e^{-2\delta nT}, \ \text{tr}=\frac{(s_1 r + s_2)e^{ns_1 T} - (s_2 r + s_1)e^{ns_2 T}}{s_2 - s_1} \text{-}(1+r)\frac{Fe^{s_2\beta} + Ge^{s_1\beta}}{He^{s_2\beta} + Ie^{s_1\beta}}, \quad (6)$$

and where F, G, $H<0$ and $I<0$ are functions of T, n, r and δ and are given in [10].

If we want to detect classical local bifurcations, we must set $\lambda=+1$ (*SN* or more degenerate bifurcations) and $\lambda=-1$ (*PD* or more degenerate bifurcations) in the previous equation. Inserting these values and rearranging, we obtain [10] the bifurcation parameters values. A substantial calculation verifies that in the case of the *SN* bifurcation we reobtain the value given in (5), while in the case of the *PD* bifurcation we have

$$\gamma_{PD} = A\left[B\left(-\frac{I\zeta_{PD} - G}{H\zeta_{PD} - F}\right)^{(s_2/(s_2-s_1))} + C\left(-\frac{I\zeta_{PD} - G}{H\zeta_{PD} - F}\right)^{(s_1/(s_2-s_1))}\right]^{-1}, \quad (7)$$

where the value of ζ_{PD} is reported in [10].

3 Non-classical bifurcations and bifurcation scenario

So far, we have not explicitly taken into account the limiting values $\beta=0$ and $\beta=T/2$ of the physically admissible range for β. They correspond to the sudden disappearing of the periodic solutions (both of the stable and of the unstable path) by a mechanism which involves "collision" between impulses and impacts.

To fix ideas, let us suppose that all the parameters except the excitation amplitude γ are fixed, for example, T=2.91, n=1, r=0.8 and δ=0.2 (with this values, γ_{SN}=0.607 and β_{SN}=0.798). Furthermore, let us assume that for a given $\gamma<\gamma_{SN}$, say γ=0.55, there exists a periodic solution, either stable or unstable (the stability does not seem to play a role in this section), which corresponds to some admissible β, β=0.3585 in the considered case. The projection of the orbit on the phase space is depicted in Fig. 1a.

When γ decreases, according to $(4)_1$ either β decreases (if $\beta<\beta_{SN}$) or it increases (if $\beta>\beta_{SN}$). In the case of Fig. 1 it decreases, as confirmed by Fig. 1b. So, the distance (both in time and in space) between the second impulse (denoted by $\mathbf{x}_n^{II}$ in Fig. 1) and the impact (denoted by $\mathbf{x}_n^{IV}$ in Fig. 1) shrinks, until for a determined value of γ (γ=0.448 in the illustrative example), it vanishes and the impact and the foregoing positive impulse *synchronize* and occur simultaneously. After this threshold, since β further decreases, the periodic solution *no longer exists*, because negative values of β are inadmissible.

The conclusion is that when the impulse and the impact synchronize, the periodic solution *suddenly disappears*. This is *not* a classical bifurcation, because it is not related to the eigenvalues of the stability matrix, but to a completely different mechanism. In any case, it can certainly be classified as a bifurcation (a *non-classical* bifurcation, *NC* henceforth), because it leads to sudden qualitative change in the system response. It is worth remarking that the *NC* bifurcation completely destroys the solution, without affecting the local stability.

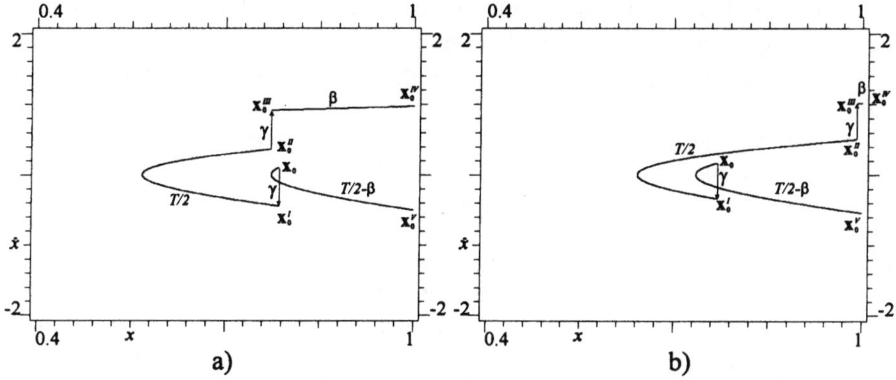

Figure 1: The orbits for T=2.91, n=1, r=0.8 and δ=0.2. a) γ=0.55 and b) γ=0.50.

The previous arguments can be repeated word by word for the other periodic solution ensuing from the *SN* bifurcation. The only difference is that in this case β increases when γ decreases. The distance $T/2$-β between the impact and the subsequent negative impulse decreases until they synchronize when β becomes equal to $T/2$, where another *NC* bifurcation takes place.

Mathematically, the two *NC* bifurcations are characterized by the conditions β=0 and $\beta=T/2$. One is related to the stable solution (β=0 in the illustrative example), and the other to the unstable solution. Inserting these values in $(4)_1$ we obtain the formulae

$$\gamma_{NC0} = \frac{A}{B+C}, \qquad \gamma_{NC1} = \frac{A}{Be^{s_2 T/2} + Ce^{s_1 T/2}}. \tag{8}$$

So far, we have seen that there are four bifurcations which characterize the bifurcation scenario of the investigated periodic solutions. Initially, we recall that the *SN* bifurcation, which generates the stable and the unstable basic paths, is of reverse kind, so it certainly occurs for the highest value of γ. Secondarily, we note that the *PD* bifurcation may occur only on the stable path (because for $\lambda_1 = -1$ we have $\lambda_2 = -r^2 e^{-2\delta nT} > -1$).

The four bifurcations curves are illustrated in the (T,γ) excitation control parameter space of Fig. 2 for the case n=1, r=0.8 and δ=0.2. For T<1.134, both *NC*0 and *NC*1 break the stable path on different sides. This means the *SN* bifurcation does *not* actually exist in this range, and the stable periodic solution appears by the *NC*0 bifurcation with increasing amplitude and disappears by the *NC*1 one. The unstable periodic solution does not exist. A numerical bifurcation diagram corresponding to this situation is represented in Fig. 6a.

When T crosses the value 1.134, the SN bifurcation enters the actual dynamics. The $NC1$ now breaks the unstable path, which exists in the interval $[\gamma_{NC1}, \gamma_{SN}]$ and which appears for higher (with respect to the stable one) values of the excitation amplitude. The stable solution, on the other hand, exists in the interval $[\gamma_{NC0}, \gamma_{SN}]$. This case is reported in Fig. 6b.

Fig. 2 shows that the PD bifurcation can theoretically exists only for T>2.9958, although initially it certainly does not break the stable solution, which was previously destroyed by the $NC0$ bifurcation with decreasing amplitude. The $NC0$ and the PD bifurcations exchange at T=3.347. After this threshold, the stable solution loses its stability by the PD bifurcation before disappearing through the $NC0$ bifurcation, as depicted in Figs. 6c and 6d. It is thus evident that the NC bifurcations have no practical meaning in this range.

Anyway, for T>3.347 there are two other qualitative changes in the bifurcation scenario. The first occurs when γ_{PD} becomes greater than γ_{NC1} and the other when γ_{NC0} becomes greater than γ_{NC1}. Though having no effects on the attractor scenario, these two modifications may considerably modify the basins of attraction organization.

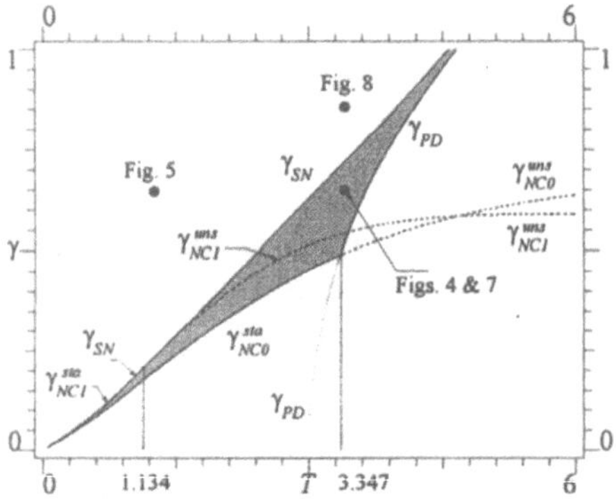

Figure 2: The four bifurcation curves γ_{SN}, γ_{PD}, γ_{NC0} and γ_{NC1} for n=1.

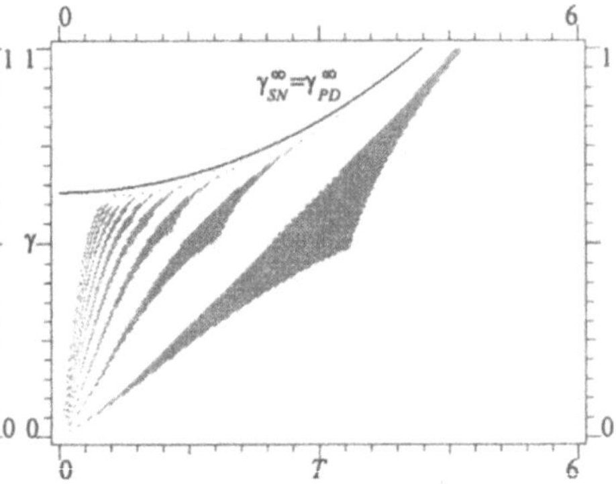

Figure 3: The stable regions for n=1 (the right one) to n=10 (the left one).

In the (T,γ) parameters space, the region of existence of the stable periodic solution is the oblique tongue delimited by the curves γ_{NC0} and γ_{NC1} for T<1.134, by γ_{NC0} and γ_{SN} for 1.134<T<3.347, and by γ_{PD} and γ_{SN} for T>3.347. It is the grey region in Fig. 2. It is very tiny for low values of the excitation period, then it increases in magnitude up to reaching the maximum width in correspondence of T=3.347 ($NC0 \rightarrow PD$ transition), above which it shrinks to zero with $T \rightarrow \infty$.

These qualitative features of the bifurcation chart do not depend on the particular choice of the parameters, but are general. In fact, by varying the system parameters, the bifurcation charts change only quantitatively and not qualitatively.

For example, by increasing the period n of the investigated solutions the region of existence moves to the left and decreases in width, as shown in Fig. 3. Therein, the right stable region corresponds to n=1, the subsequent to n=2 and so on until the left one which is depicted for n=10.

Fig. 3 emphasizes that the different stable regions do not overlap, at least for low values of n, so that there is no coexistence of different periodic attractors with one impact per period. We note that the stable regions seem to shrink and to accumulate onto the vertical axis when $n \to \infty$. Thus, high periodic solutions exist only for low excitation period and are more and more difficult to observe.

For a fixed T, stable segments corresponding to increasing n-periods occur for increasing values of the excitation amplitude γ (however, with decreasing magnitude of the existence interval) and the regions accumulate onto the curves $\gamma_{SN}^{\infty} = \gamma_{SN}(n \to \infty)$ and $\gamma_{PD}^{\infty} = \gamma_{PD}(n \to \infty)$, which actually coincide. These properties also occur for the periodic solutions of the system under harmonic excitation [7], and therefore we guess they are intrinsic properties of the system dynamics not depending on the shape of the excitation.

4 Numerical simulations

In this section we report some numerical simulations aimed at illustrating the results previously obtained. A Runge-Kutta-Fehlberg's method is used for the numerical integrations, and the analysis is based on a combined use of time histories, bifurcation diagrams and basins of attraction. We fix r=0.8 and δ=0.20 and we allow the period and the amplitude of the external excitation to vary.

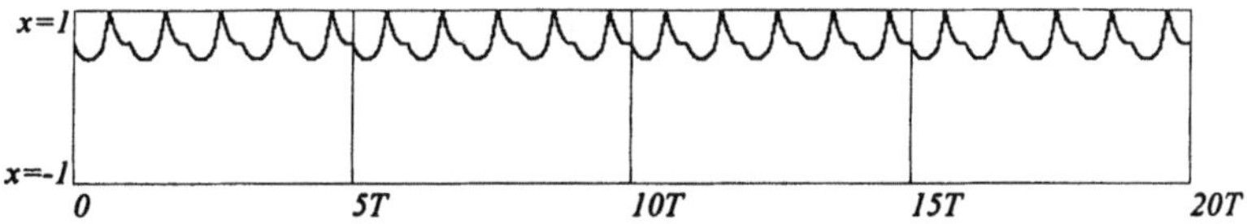

Figure 4: The time history of the n=1 periodic solution for T=3.40 and γ=0.65.

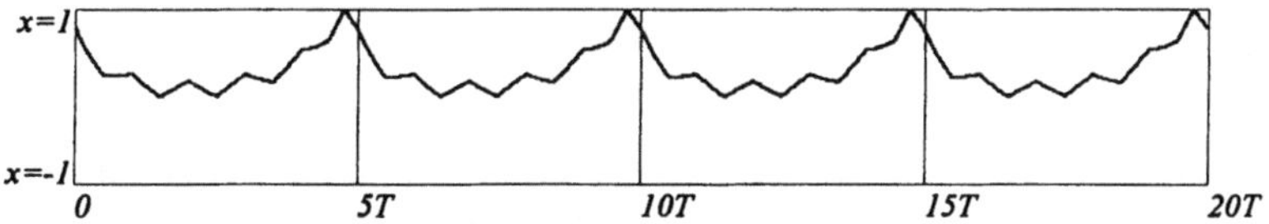

Figure 5: The time history of the n=5 periodic solution for T=1.257 and γ=0.647.

In Fig. 4, the time history of the n=1 periodic solution for T=3.40 and γ=0.65 is reported. The picture clearly emphasizes the free motions, the two impulses and the single impact in each period of the oscillation. Fig. 5, on the other hand, illustrates the n=5 periodic solution for T=1.257 and γ=0.647.

Although the periodic solutions corresponding to large n values have very narrow stable regions (see Fig. 3) and strongly intertwined basins of attraction (see Fig. 8), thus being very difficult to be realized in practice, they have an important practical feature with respect to the low-period solutions. In fact, as highlighted by the comparison between Figs. 4 and 5, they are associated with considerable reduction of the number of impacts in time and, as a consequence, of the progressive damage of the system.

In the following pictures (Figs. 6a-d), we report the bifurcation diagrams of the n=1 periodic solution obtained for different values of the excitation period. The continuous (dotted) line corresponds to stable (unstable) solution. As a consequence of the particular choice of the Poincaré section (we sample the points just before the negative impulse), the $NC1$ bifurcation is clearly recognizable in the mentioned Figures because it corresponds to the condition x=1.

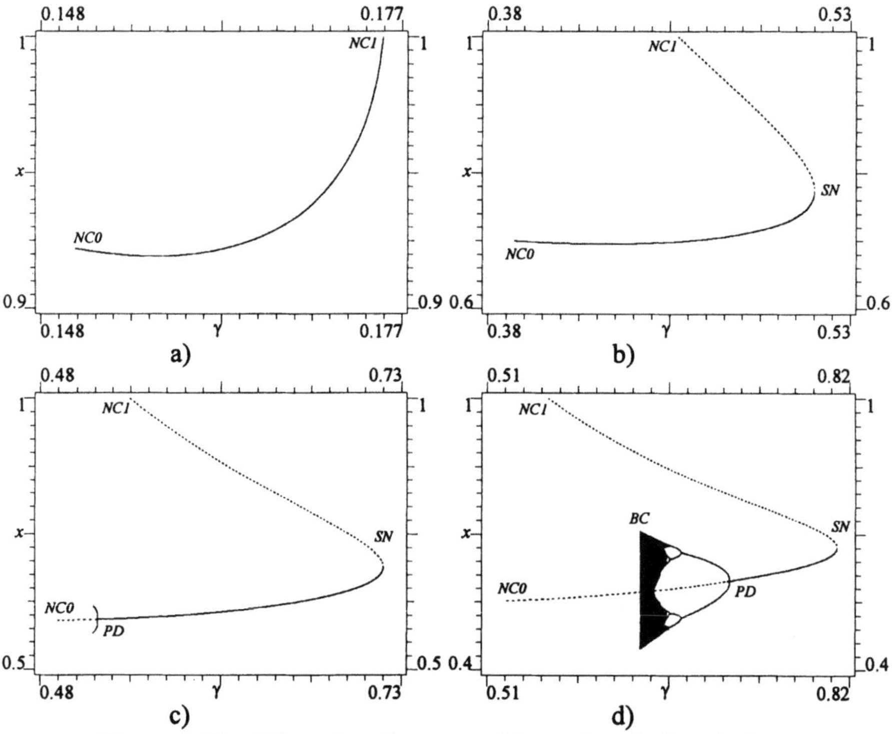

Figure 6: The bifurcation diagrams of the n=1 periodic solutions for a) T=1.0, b) T=2.5, c) T=3.4, d) T=3.8.

Figs. 6a and 6b have been discussed in the previous section and do not deserve further comments. Figs. 6c and 6d, on the other hand, possess some specific characteristics we wish to analyze. In particular, we focus on the stable branch which undergoes the PD bifurcation by decreasing γ. In the case of Fig. 6c, the PD and the $NC0$ are quite close, and the classical PD cascade which initiates at the PD bifurcation is suddenly interrupted, likely as a consequence of a NC bifurcation of the period 2 solution ensuing from the PD bifurcation.

In Fig. 6d, a different situation is observed. In this case the *PD* and the *NC*0 are very distant and the *PD* cascade is not interrupted by *NC* bifurcations. The cascade is complete and, according to the classical theory, it leads to the creation of a chaotic attractor, which is successively enlarged by interior crises and suddenly disappears likely as a consequence of a boundary crisis (*BC*). There is strong numerical evidence that this situation is common in the system response [9], but this fact will not further be considered here.

Finally, we want to discuss the possible coexistence of the investigated periodic solutions with other attractors, periodic and/or chaotic. This property clearly depends on the values of the system parameters. An example of occurrence of a unique attractor in each of the two wells of the system is presented in Fig. 7, which reports the attractor-basin phase portrait for T=3.40 and γ=0.65, and shows how in this case there are only the two "specular" n=1 cycles. However, this is neither a general nor a common situation, because usually there is coexistence of different attractors. An example is depicted in Fig. 8, which is obtained by increasing the amplitude with respect to the previous case (T=3.40 and γ=0.8564). This picture reports the basins of attraction of the n=2 periodic solutions (greys) and of other two attractors (black and white).

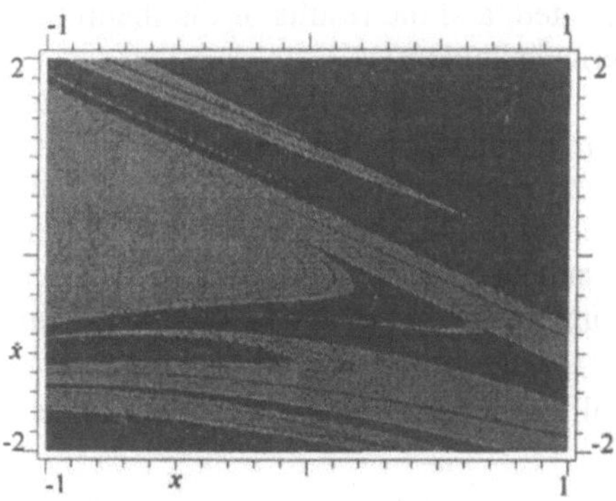

Figure 7: The basins of attraction for T=3.40 and γ=0.65.

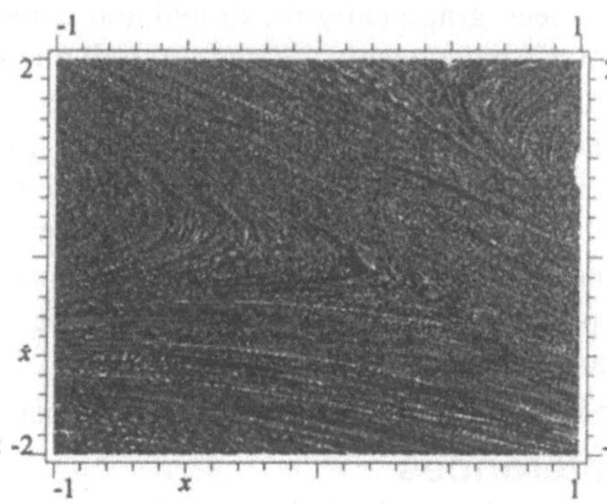

Figure 8: The basins of attraction for T=3.40 and γ=0.8564.

Comparison of Figs. 7 and 8 illustrates an important aspect induced by the growth of amplitude of the excitation in the global dynamics of the system: the coexistence of the periodic solutions with other (chaotic as a rule) attractors frequently entails spread over the whole phase portrait of the fractal nature of the boundaries of the basins of attraction, which implies sensitivity to initial conditions, already observable in Fig. 7. The ensuing overall fractalization of the basins correspond to strong – and very undesirable – reduction of the robustness of the attractors. This is a very general property of the systems response, that has been studied in detail in [9].

We have observed that the chaotic attractors which coexist and compete with the periodic solutions are usually created and/or destroyed by classical *PD* cascades (an example is reported in Fig. 6d), by boundary crises and by subductions. However, a detailed analysis of these arguments is out of the scope of the present paper and is reported in [9]. We only remark that, as a rule, the periodic solutions corresponding to low values of n are often the unique attractors in the whole phase space, while those related to large values of n usually coexist with other (chaotic as a rule) attractors. This is due to the fact that these last cycles exist for large values of the excitation amplitude (see Fig. 3).

5 Conclusions

Periodic motions and non-classical bifurcations of an inverted pendulum with lateral constraints have been analyzed under the hypothesis that the system is forced by impulsed T-periodic excitation. A class of bouncing nT-periodic solutions has been detected analytically. The occurrence of classical *SN* and *PD* bifurcations of these periodic solutions has been established theoretically.

A non-classical mechanism leading to the *sudden* appearance or disappearance of the periodic solutions has been discussed and it has been shown that it involves simultaneous occurrence (synchronization) of impulses and impacts.

The *SN*, *PD* and *NC* critical curves have been investigated in the period-amplitude (of the excitation) parameters plane, emphasizing how they modify the system response. The stability regions of the investigated periodic solutions have also been graphically illustrated and commented, and the results of the theoretical analysis have been validated through numerical simulations.

In this work we have considered a simple (piece-wise linear) inverted pendulum. This choice is motivated by the interest in obtaining closed-form expressions for the main events, but the phenomena investigated, in particular, the *NC* bifurcations, do not depend on the properties of this simple model and are very general. They require only the presence and synchronization of impulses and impacts, so we guess they should be common also to more general systems sharing these two basic properties.

Acknowledgments: this work was partially supported by MURST 1997 funds.

References

1 Kobrinskii A A, Kobrinskii A E, *Two-dimensional Vibro-impact Systems*, Nauka, Moscow, 1981.

2 Peterka F, Bifurcations and Transition Phenomena in an Impact Oscillator, *Chaos, Solitons & Fractals* **7(10)**, 1635-1647, 1996.

3 Brach R, *Mechanical Impact Dynamics. Rigid Body Collisions,* Wiley, New York, 1991.

4 Ivanov A, Bifurcations in Impact Systems, *Chaos, Sol. & Fract.* **7(10)**, 1615-1634, 1996.

5 Holmes P, The dynamics of Repeated Impacts with a Sinusoidally Vibrating Table, *J. Sound Vibr.* **84(2)**, 173-189, 1982.

6 Moon F C, Shaw S W, Chaotic Vibrations of a Beam With Non-linear Boundary Conditions, *Int. J. Non-linear Mech.* **18(6)**, 465-477, 1983.

7 Shaw S W, Rand R H, The Transition to Chaos in a Simple Mechanical System, *Int. J. Non-linear Mech.* **24**, 41-56, 1989.

8 Lenci S, Rega G, A Procedure for Reducing the Chaotic Response Region in an Impact Mechanical System, *Nonlinear Dynamics*, **15**, 391-409, 1998a.

9 Lenci S, Rega G, Controlling Nonlinear Dynamics in a Two-Well Impact System. Part I & Part II, in press on *Int. J. Bif. Chaos*, 1998b.

10 Lenci S, Rega G, Theoretical analysis of periodic solutions and bifurcations in an impact system under impulsive excitation, submitted for publication, 1998c.

Dynamics of an impact unit for percussion machines

S A Kember and V I Babitsky**

A percussion machine impact unit has been studied as a vibro-impact system. The development of this concept involved the application of various techniques in a co-operative manner. The techniques included are periodic Green's functions, Matlab-Simulink models and the use of laser vibrometry. The dynamic characteristics revealed by the analytical and numerical studies are compared with those found by experiments on a simple mechanical model and a laboratory model of an impact drill.

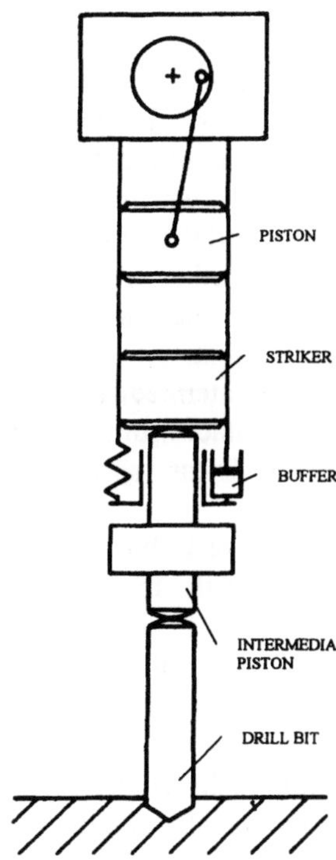

Figure 1

1 Introduction

There are several mechanical devices that employ a vibro-impact process in their mechanisms. Probably the widest known are the percussion machines. Some, like the hammer drill, contain one or more impacting pairs which act together as a force amplifier [1]. In order to understand the dynamical process of force conversion and to find the most effective form of motion, it is necessary to analyse the transformation of impact impulses by a nonlinear (vibro-impact) system. This can be achieved with the use of the periodic Green's function (PGF) method [2,3]. The exploitation of this method allows the system behaviour to be expressed and analysed in a general form, including the resonant and anti-resonant phenomena which provide the main synchronous behaviour of the converter. The timing and optimisation of the impact unit to achieve a given performance level by utilising these phenomena requires the development of more realistic models of impact using dynamic compliances. To this end simulation models were constructed in the Matlab-Simulink environment. The experimental validation of the

* Dept of Mechanical Engineering, Loughborough University, Loughborough, LE11 3TU, UK.

results of the analysis required the use of non-contacting methods of measurement. A special rig based on a real hammer drill was therefore developed and used with laser vibrometry instrumentation so that the dynamic perfomance of the rig was not affected.

A typical hammer drill is shown diagramatically in Figure 1. The exciting mechanism, in this case, is a rotating crank which drives a piston. The changes in air pressure caused by the piston motion excite a striker which impacts on an intermediate piston. The intermediate piston then strikes the drill bit then impacts upon the item to be drilled (for example concrete). On the rebound from striking the drill bit, the intermediate piston impacts against a buffer which acts as a shock absorber.
The aim of the study described here was to develop a simplified analytical model to provide an understanding of the physical phenomena, computer simulations to extend the simplified model, and experiments to validate both models.

2 Analytical study

The hammer action shown in Figure 1 is complicated and so it was decided to model simplified versions first. An initial simplified model is shown in Figure 2.

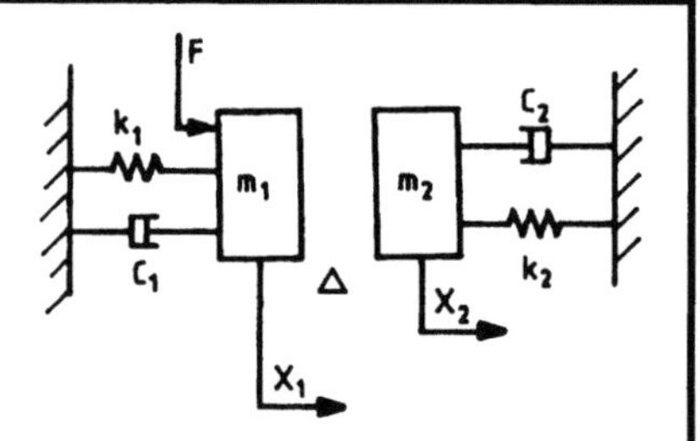

Figure 2

It consists of two linear single degree of freedom systems, one of which is excited by an external force and this is known as the active system, the other is known as the passive system. The assumptions behind this model are that:

1) Mass m_1 is the intermediate piston which is always in contact with the buffer (represented by a linear spring and damper),
2) Mass m_2 is the drill bit which is always in contact with the concrete (represented by a linear spring and damper),
3) Both masses are rigid compact bodies so that classical impact theory can be used and stress waves can be ignored,
4) The striker is replaced by a cyclic excitation force,
5) One impact between the two masses occurs in each cycle of the excitation.

This last assumption was chosen to provide the most efficient transfer of energy.
The method chosen to study this model is the periodic Green's function (PGF) method. Here only the results will be shown, as an extended explanation and the complete solution method will be published elsewhere [4].
The active system can be excited either by a sinusoidally varying force or, more realistically, by a series of repeated impacts (here only the impulse excitation will be considered). It will also receive excitation in the form of repeated impacts through its interaction with the passive system. As the active system is linear, its

responses to external excitation and interaction with the passive system can be added together. The steady state responses to impact excitation are calculated and given in the form of a PGF denoted by the symbol χ. The PGF is made up of the dynamic compliance of the system and the Fourier series for repeated Dirac δ functions.

$$\chi_1(t)=\frac{1}{Tm_1}\sum_{n=-\infty}^{\infty}\frac{1}{\Omega_1^2-(n\omega)^2+2\Omega_1\varsigma_1 in\omega}e^{in\omega t}$$

where $\omega=2\pi/T$, T is the period of excitation, Ω_1 is the first natural frequency and ς_1 is the damping ratio. The PGF can be transformed to cover just one period of motion [5] and this is carried out here.

The equation of motion for the active system is:

$$x_1(t)=F\chi_1(t+\tau)-J\chi_1(t)$$

The equation for the passive system is:

$$x_2(t)=J\chi_2(t)$$

There are two unknowns J and τ, these can be found with the help of the conditions of periodicity and impact. These are:

$$\Delta=x_1(0)-x_2(0) \text{ and } J=M(1+R)(\dot{x}_1(T)-\dot{x}_2(T))$$

where R is the coefficient of restitution, Δ is the gap between m_1 and m_2 and $M=m_1m_2/(m_1+m_2)$. In order to simplify the analytical solution for J, it was proposed that the damping coefficient c_1 equal zero and so J is given by the following expression:

$$J=\frac{\chi\Delta\Omega_1^2}{\left(\frac{1}{M(1+R)}+\dot{\chi}\right)^2+\Omega_1^2\chi^2}\left\{-1\pm\sqrt{1-\frac{\left[\left(\frac{1}{M(1+R)}+\dot{\chi}\right)^2+\Omega_1^2\chi^2\right](\Delta^2-2rF^2A_1^2)}{\chi^2\Delta^2\Omega_1^2}}\right\}$$

where $r=1-\cos\Omega_1 T$, $\chi=\chi_1(0)+\chi_2(0)$, $\dot{\chi}=\dot{\chi}_1(T)+\dot{\chi}_2(T)$, and $A_1=1/2rm_1\Omega_1$.

The value of τ can be found from these expressions:

$$sin\,\Omega_1\tau = \frac{r\Omega_1(\Delta + J\chi) - J\left(\frac{1}{M(1+R)} + \dot{\chi}\right)sin\,\Omega_1 T}{2rFA_1\Omega_1}$$

$$cos\,\Omega_1\tau = \frac{\Omega_1(\Delta + J\chi) - J\left(\frac{1}{M(1+R)} + \dot{\chi}\right)sin\,\Omega_1 T}{2rFA_1\Omega_1}$$

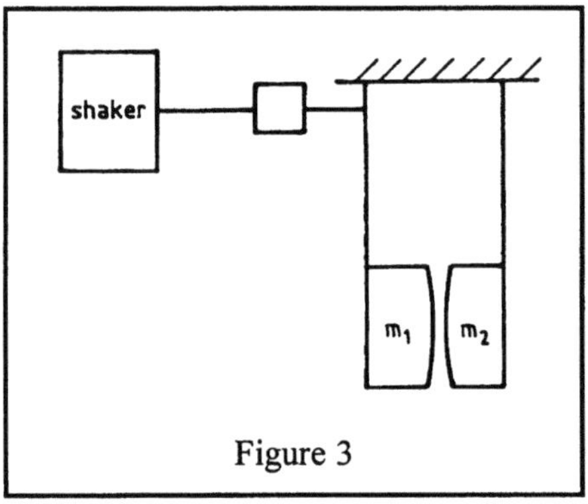

Figure 3

This solution was obtained algebraically without resort to numerical methods. Note that the equation for J has two possible solutions, only one of which is real. Following a stability analysis (given in [4]), the real solution is the one with the positive sign before the square root.

Now some numerical values can be inserted into the equations of motion. To give a realistic result the parameters were chosen from an experiment, which consisted of two steel masses attached to spring steel strips. One mass was excited by a shaker pressed up against its supporting beam at a point close to the base (see figure 3). The frequency of the sinusoidal excitation was slowly swept up and down. During the up sweep, at first only the active system moved, then impacts occurred between the two masses. At 5.6 Hz both masses moved in phase with large displacements, a *grazing* resonance. This motion declined and at the anti-resonance the active mass hardly moved while the passive mass moved with a larger amplitude like a dynamic absorber. As the frequency increased, the impacts became louder and the displacements increased. The two masses were moving out of phase, a *clapping* resonance. Suddenly at 9 Hz the

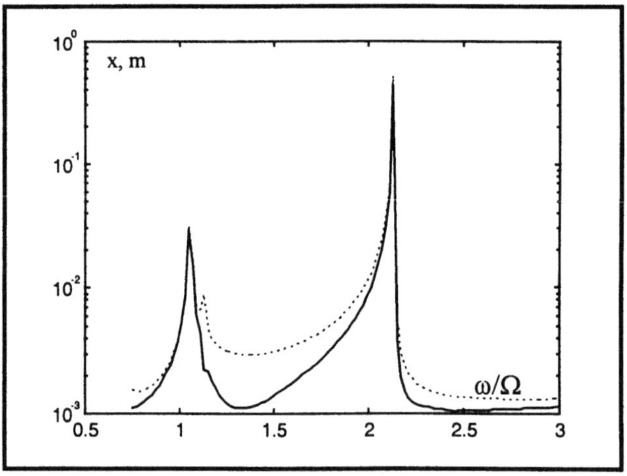

Figure 4

clapping ceased and no further impacts occurred. When the frequency was slowly reduced, the clapping resonance was not encountered and the grazing resonance occurred at 5.6 Hz.

The frequency at which the clapping resonance occurred was dependent upon the magnitude of excitation. By varying the amplitude level on the shaker amplifier, the clapping resonance occurred at 8.8 Hz, 9 Hz, 10.3 Hz, and with a larger shaker, 11.4 Hz. However the larger shaker caused the two masses to become magnetised which affected the results. Contact studies using a small electrical circuit demonstrated clearly that the clapping resonance had one impact per cycle.

Based on this experiment the following parameters were chosen for the PGF model: $m_1 = 0.125$ kg, $m_2 = 0.094$ kg, $k_1 = 135$ N/m, $k_2 = 128$ N/m, $R = 0.5$, and Δ=0.001m. The damping in the experiment was very low, so c_2 was set at a low value, 0.1 Ns/m and c_1 at zero. Although the chosen parameters come from a sinusoidally excited experiment, it is reasonable to use them for the impulse excitation model as the model gives similar results for both types of excitation [4].

There was some difficulty with the choice of force level for the model as the experimental force level was not constant due to feedback from the experimental rig to the shaker. As a consequence precise comparisons of amplitude are not possible at this stage.

The results from the PGF method show two peaks, one at $\omega/\Omega_1 = 1.05$ (5.5 Hz) and one at $\omega/\Omega_1 = 2.13$ (11.1 Hz) which is in close agreement with the experiment. Figure 4 shows the peak displacement against frequency with the solid line representing x_1 and the dotted line x_2. Note the behaviour at the antiresonance where m_2 has a greater displacement than m_1.

The behaviour at the lower natural frequency shows a grazing resonance (Figure 5) and at the higher natural frequency the behaviour is a clapping resonance (Figure 6).

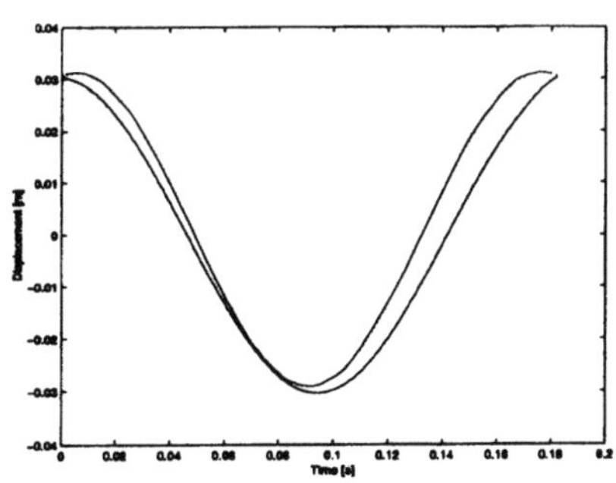

Figure 5

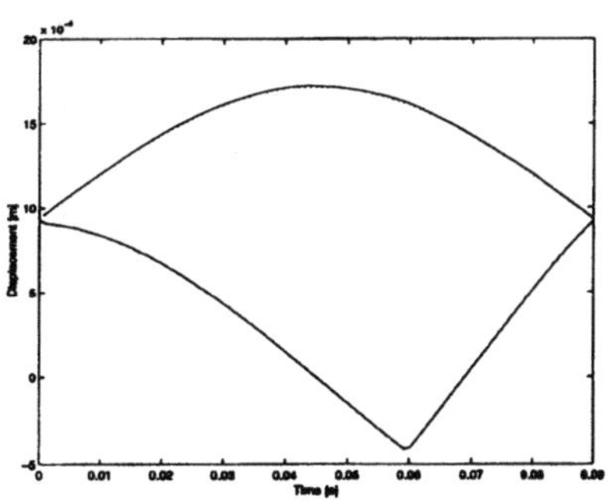

Figure 6

The PGF method results agree with the experimental results on the frequencies of the resonances and the behaviour of the resonances. In addition, the advantage of this method is that it gives a good understanding of the principles of resonant behaviour of the vibro-impact system, but additional features such as nonlinear springs would require a linearisation process to be applied to the formation of the PGF.

3 Computer simulation

The advantage of a computer simulation is that adding additional features is straightforward but often it is difficult to know if the result is reliable, or suffering from accumulated errors. The initial simulations were therefore of exactly the same model as used for the PGF method with the same parameters. The package used was Simulink. However the results were disappointing, the behaviour was irregular for the case with sinusoidal excitation and no results were achieved with impulse excitation. According to Press et al [6], the best numerical integrator for systems with sharp changes is Runge-Kutta but even Runge-Kutta may suffer from errors. As the integrator used in this model was 5th order Runge-Kutta with variable step size, the conclusion was that the errors due to the discontinuities had caused the poor results so a different impact model is required. The PGF method used a coefficient of restitution approach and so this was replaced with a compliant surface model [7] (Figure 7), where the surface was modelled with a spring and damper and the reaction force was monitored so that contact ceased when the reaction was zero. If instead the contact is considered to have ended after a particular distance, then a tensile force can exist in the compliant surface which is physically impossible.

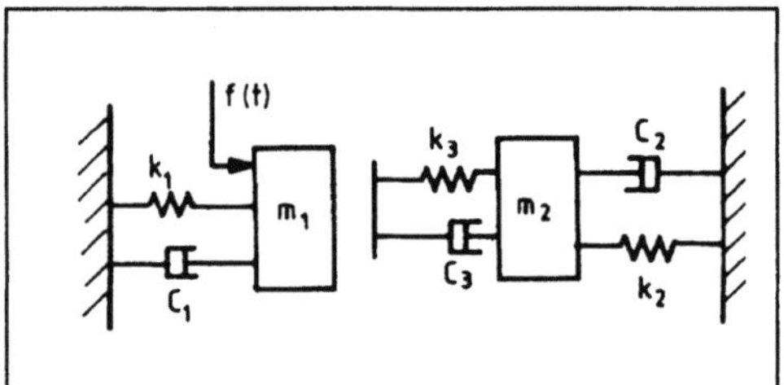

Figure 7

When the compliant surface is applied to the model in Figure 2, the results were much improved. Using a pulse sweep excitation, the response of the system was found. Again there are two resonances at 5.7 Hz and 10.7 Hz and they are clearly

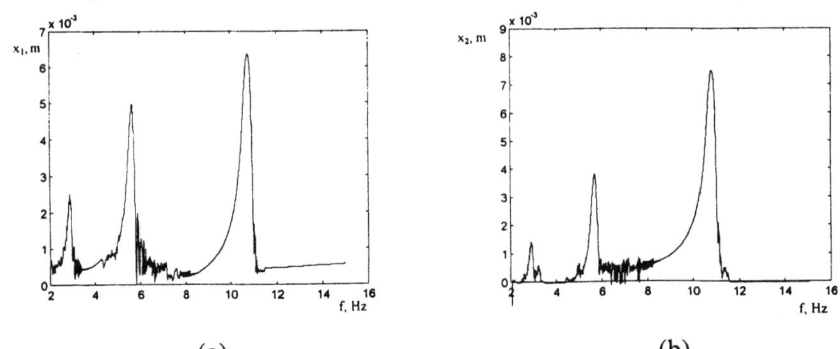

(a) (b)

Figure 8

the grazing and clapping resonances respectively. Figure 8a shows the rms displacement of m_1 against frequency and Figure 8b shows the displacement of m_2. The first peak is symmetric while the second peak is assymmetric. When the pulse sweep was reversed to sweep down through the frequencies only the first resonance was found (Figures 9 a and b). This agrees with the experimental results.

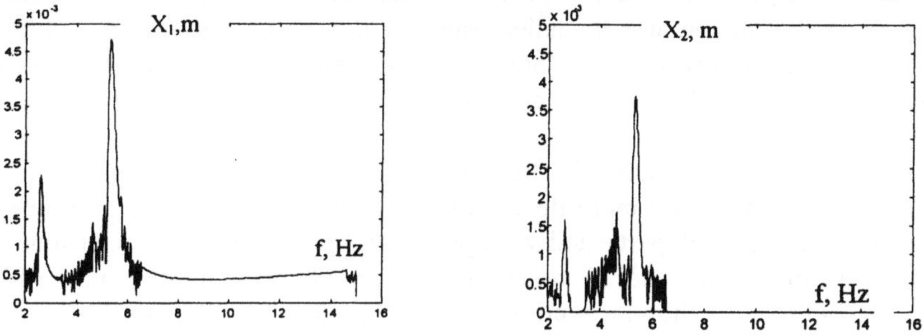

(a) (b)

Figure 9

The frequency response shows some additional peaks due to the presence of subharmonics and the results suffer from noise and some irregularities suggesting that there are errors in the integration process due to the pulse excitation.
These simulations agree with both the PGF method and the experimental results with regard to the frequencies and general behaviour. There is some discrepancy between the amplitudes and the relative sizes of the two peaks, this is likely to be due to the choice of excitation amplitude. Further features can now be added to the model to simulate an industrial drill more closely. However, these more advanced models would have to be validated with data from the drill itself.

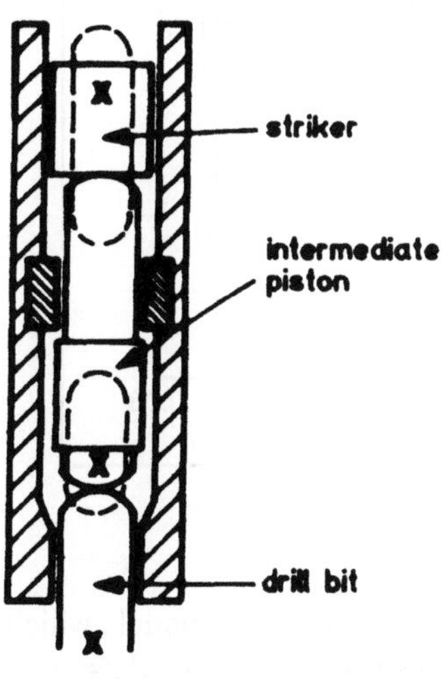

Figure 10

4 Experiment

An experiment rig was devised to provide data for comparison with the simulation models. The ideal experiment would collect data from a real hammer drill. However, the impact unit is buried deep within the drill and access is extremely difficult. The parts move inside a closely fitting tube which rotates so there is no room for wires and accelerometers. Contacting transducers are likely to affect the dynamic behaviour. The alternative is to use optical methods but these require windows cut into the outer casing and inner tube and any light beams would have to avoid a number of parts that exist between the two tubes. So the solution chosen was to replace the outer casing and inner tube with one tube with two acrylic windows (needed to maintain pressure levels) and connect this tube to the same motor to provide identical excitation. The tube was carefully bored to replicate the original as closely as possible so that the original internal components could be used. Concrete was modelled using a damper as suggested by ISO 8662-3:1992 so that the drill could be operated in chiselling mode (no rotation). The vibration levels at the handle were compared to the vibration levels for the original drill and were found to be similar.

A laser vibrometer was chosen as the means of measuring the movements of the hammer action parts - for more details of this technique see [8]. Accelerometers were used on external parts and an optical probe was used to determine the position of the piston and so relate the results obtained to the cycle of the excitation. Figure 10 shows a section of the experimental tube with the windows marked as dotted lines. The points measured by the laser vibrometer are marked by x. Figure 11 shows output from the vibrometer for the intermediate piston. The upper signal comes from the optical probe, the spikes indicate the point in each cycle when the piston is at top dead centre. Results from this experiment will be used to validate more complex models created in Simulink in the future.

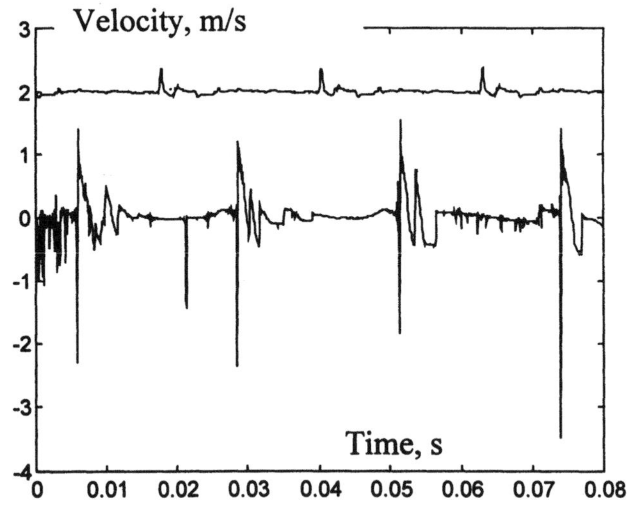

Fig. 11

5 Conclusion

The PGF method of studying vibro-impact processes is ideal for models which are difficult to study by other methods and provides results which give clear physical meanings. Here the PGF results have helped to provide a firm base for the numerical Simulink models which can now be further developed with the help of data from the laboratory hammer drill. It is hoped that in the future further models will allow optimisation of the hammer action so reducing vibration levels transmitted to the user.

References

1 Babitsky, VI, Hand-held percussion machine as discrete nonlinear converter, Journal of Sound and Vibration 214(1):165-182, 1998.

2 Babitsky , VI, Theory of Vibro-impact Systems and Applications, Nauka, Moscow, 1978 (Revised English edition: Springer Verlag, 1998).

3 Babitsky, VI & Krupenin, VL, Vibrations in Strongly Nonlinear Systems, Nauka, Moscow, 1985.

4 Kember, SA & Babitsky, VI, Excitation of vibro-impact systems by periodic impulses, to be published in the Journal of Sound and Vibration

5 Rosenwasser, EN, Vibrations of Nonlinear Systems, Nauka, Moscow, 1969. (in Russian)

6 Press, WH, Teukolsky, SA, Vetterling, WT & Flannery, BP, Numerical Recipes in Fortran, 2nd Ed, Cambridge University Press, Cambridge, 1992.

7 Babitsky VI, Veprik AM, Universal bumpered vibration isolator for severe environment, Journal of Sound and Vibration, 218(2):269-292, 1998.

8 Halliwell NA, Kember SA, Rothberg SJ, Laser vibrometry for impact measurements, Proceedings of EUROMECH 386.

The influence of asymmetries on the double impact oscillator dynamics

F Peterka and O Szőllős *

The dynamics of the double impact oscillator, as the model of a forming machine, is studied in the paper. The mechanical model is composed from two symmetrically arranged simple impact oscillators, the masses of which are periodically excited and can mutually impact. The influence of asymmetries of own frequencies and amplitudes of exciting forces in both impact oscillators is investigated. Asymmetries do not affect considerably the optimal periodic regime, which is useful for practical purposes. Results of theoretical analysis and numerical calculations are discussed.

1 Introduction

The one-degree-of-freedom impact oscillator is one of the simplest strongly non-linear mechanical systems. It consists of an elastically suspended and periodically excited mass impacting to a rigid stop. Its dynamics were thoroughly investigated experimentally, theoretically and by using simulation methods [1]. The influence of viscous and dry friction damping on the system behaviour was studied, with respect to the maximum before-impact velocities of the fundamental periodic impact motion. This motion is practically most important and is characterised by the repetition of one impact in every period of the excitation force [2]. There exist series of different periodic and chaotic impact motions and a certain region of existence in the space of system parameters correspond to every motion. Regions can overlap and create hysteresis regions where several regimes of the system motion can exist. It depends on initial motion conditions or other conditions, which determine definitely which motion will appear (see the set of papers, ref. 16. in [1]).

It has been shown, that the results, obtained for the mentioned oscillator are also applicable for two symmetrically arranged impact oscillators, which are identical to the aforesaid one. It can be considered as a model of the forming machine.

An advantage of such a system, compared to the single impact oscillator, consists in:

- Isolation of impact impulses from the neighbourhood of a forming machine and
- Doubling of before-impact velocities.

---

* Institute of Thermomechanics of the Academy of Sciences of the Czech Republic, Dolejškova 5, 182 00 Prague 8, CZ Fax: 00420-2-858 4695, E-mail: peterka@it.cas.cz

The second degree of freedom can introduce asymmetries into the system, which are, in general, undesirable for the application. Differences in initial conditions, in driving forces and in the eigenfrequencies of both systems were found to be the main sources of these asymmetries. Constructional or assembly imperfections can also lead to small asymmetries, however it has been shown via numerical simulations, that such asymmetries do not remarkably affect the optimal operation of this system.

This paper is the next in a series of contributions, investigating the behaviour of a system of two harmonic oscillators, which can mutually impact.

The mathematical model and the simulation analysis of regions of existence and stability of the typical kinds of motion were published in [3,4]. An optimisation of parameters for achieving maximum before-impact velocities was the subject of paper [5]. In the article [6], basins of attraction of two various types of motion, which occur in the area of two-fold response of the optimally tuned system, were investigated. At last, contribution [7] contains the stability analysis of the fundamental impact motion.

The effort of this paper is concentrated towards a determination of the influence of various types of asymmetries on the systems response.

2 Symmetric Case

The scheme of the system is in Fig. 1. The masses are driven by a periodic force with a period $T = 2\pi/\omega$. Their behaviour is described by the system of differential equations

$$X'' + \beta X' + X = \cos(\eta\tau + \varphi), \tag{1}$$

$$Y'' + \beta Y' + Y = \cos(\eta\tau + \varphi), \tag{2}$$

where $X = xk/F_0$, $Y = yk/F_0$, $\tau = \Omega t$, $\eta = \omega/\Omega$, $\beta = b/\sqrt{km}$.

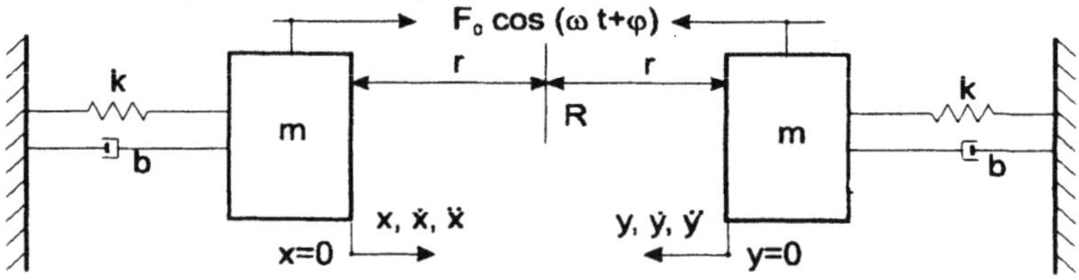

Figure 1 - *Scheme of the double oscillator*

Impacts occur, when the following condition is met

$$x + y \geq 2r, \tag{3}$$

where $2r$ is the static clearance between the masses. It is assumed that they can be described by the Newton's theory of impacts.

Let x'_-, y'_- and x'_+, y'_+ denote the before-impact and the after-impact velocity, respectively. Then:

$$x'_+ = \frac{1}{2}\left[(1-R)x'_- - (1+R)y'_-\right], \quad y'_+ = \frac{1}{2}\left[(1-R)y'_- - (1+R)x'_-\right], \tag{4}$$

where R represents the coefficient of restitution.

3 Asymmetric Cases

Asymmetries can, in general, considerably influence the systems behaviour. The following cases are considered:

1) The asymmetry of the initial conditions,
2) The asymmetry in the amplitudes of driving forces,
3) The asymmetry in eigenfreq.

ad 1) The oscillations symmetrize for asymmetric initial conditions when viscous friction is present. However, when the viscous friction is zero, then the initial asymmetry preserves in spite of a considerable amount of energy losses during impacts.

An explanation was found by a transformation of coordinates X, Y to anti-phase and in-phase components X_1 and X_2, respectively [8] :

$$X''_1 + \beta X'_1 + X_1 = 2\cos(\eta\tau + \varphi),$$
$$X'_{1+} = -RX'_{1-}, \tag{5}$$

$$X''_2 + \beta X'_2 + X_2 = 0,$$
$$X'_{2+} = -X'_{2-}, \tag{6}$$

where $X_1 = X + Y$ and $X_2 = X - Y$.

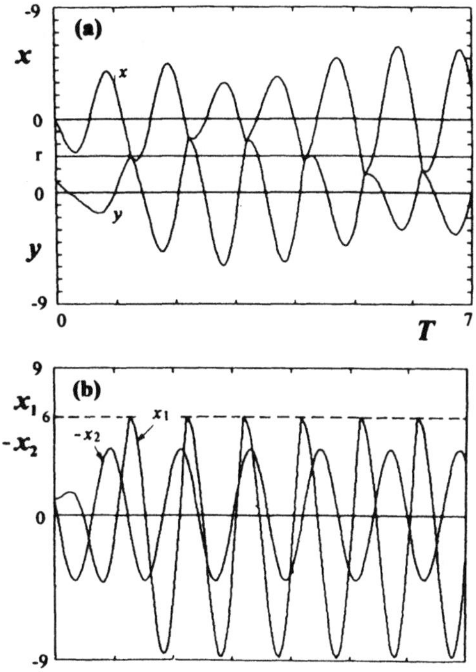

Figure 2 - *Sample time-trajectories of the system*

As it can be seen that X_1 corresponds to oscillations of the single impact oscillator, while X_2 is the usual linear (impactless) motion (Fig. 2(b)). In the case of nonzero viscous friction, the in-phase component gradually vanishes, which implies that only symmetric anti-phase impact component X_1 remains. Further details can be found in [5].

ad 2) By an asymmetry of the amplitudes of driving forces the asymmetric component of the system oscillation preserves. This follows again from the transformation of motion coordinates.

The equations, describing the asymmetry of the driving forces, are:

$$X'' + \beta X' + X = \cos(\eta\tau + \varphi), \quad Y'' + \beta Y' + Y = 0.95\cos(\eta\tau + \varphi). \tag{7}$$

and after the transformation :

$$X_1'' + \beta X_1' + X_1 = 1.95\cos(\eta\tau + \varphi), \; X_2'' + \beta X_2' + X_2 = 0.05\cos(\eta\tau + \varphi). \tag{8}$$

Equations (8) show, that the in-phase motion preserves, i.e. the asymmetry of motion preserve too.

ad 3) Equations of motion, expressing the asymmetry of eigenfrequencies, are:

$$X'' + \beta X' + X = \cos(\eta\tau + \varphi), \quad Y'' + \beta Y' + 0.95Y = \cos(\eta\tau + \varphi). \tag{9}$$

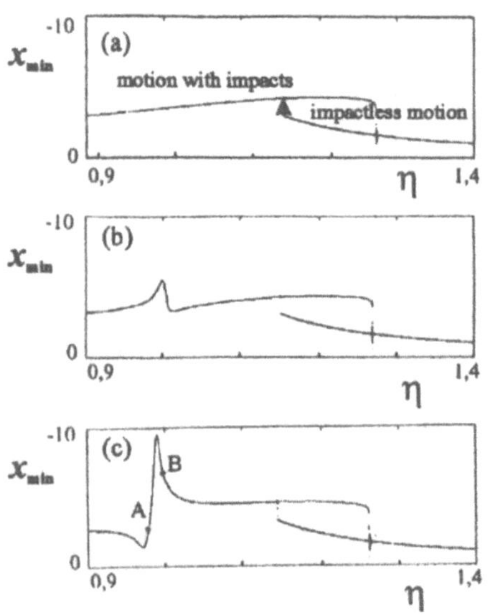

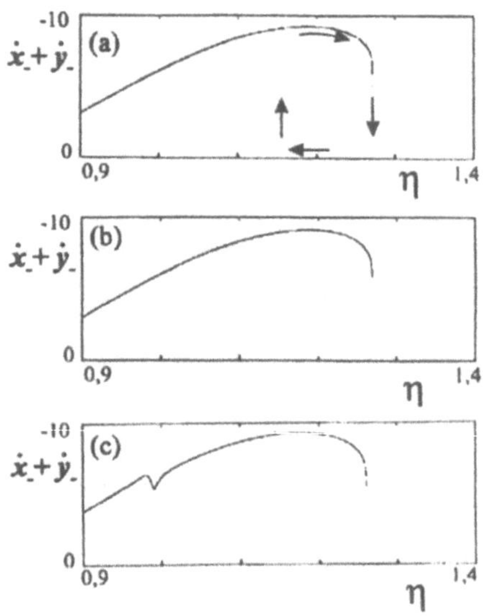

Figure 3 Figure 4

Frequency characteristics of the double impact oscillator

a) Symmetric case

b) Assymmetry of driving forces

c) Asymmetry of eigenfrequencies

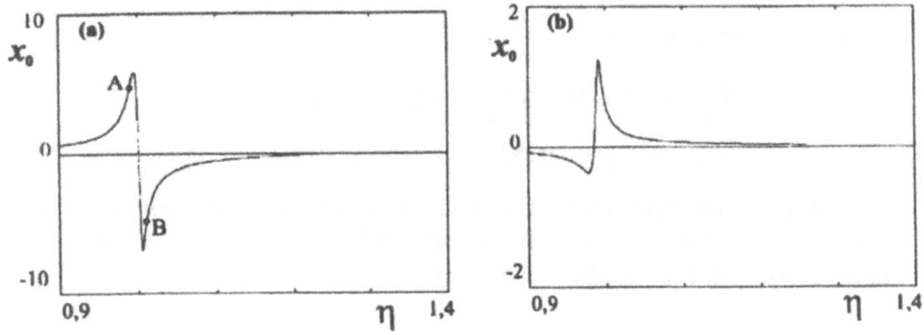

Figure 5 - *Frequency characteristics of the position of masses at impacts*
a) Asymmetry of driving forces
b) Asymmetry of eigenfrequencies

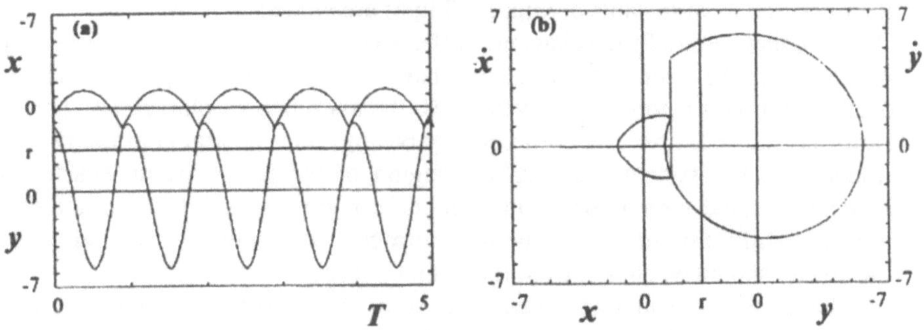

Figure 6 – *Motion trajectories in point A in Figs.* 3*(c),* 5*(a)*

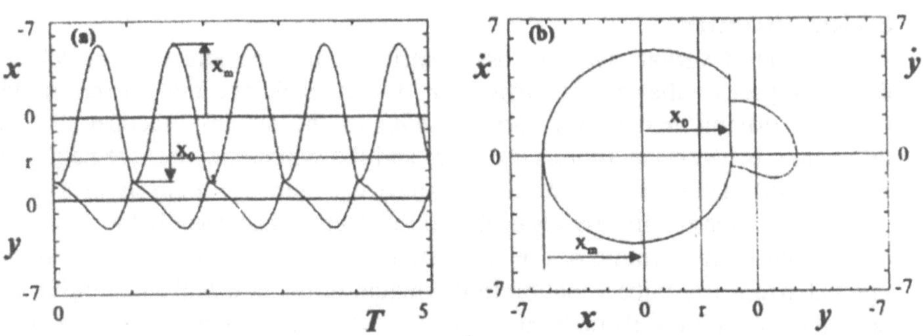

Figure 7 – *Motion trajectories in point B in Figs.* 3*(c),* 5*(a)*

After the transformation:

$$X_1'' + \beta X_1' + 0.975X_1 + 0.025X_2 = 2\cos(\eta\tau + \varphi),$$
$$X_2'' + \beta X_2' + 0.025X_1 + 0.975X_2 = 0 . \tag{10}$$

It follows from Equations (10) that anti-phase and in-phase components cannot be separated, similarly as before, and the asymmetry of the motion remains preserved, even in the presence of viscous damping.

4 Results of Numerical Simulations

The system was investigated in more detail by frequency characteristics in the range $0.9<\eta<1.4$ (Figs. 3-5). Minimum stationary amplitudes X_{min} of the displacement X (Fig.3), relative before-impact velocities of the bodies (Fig.4) and their position X_0 at instants of impacts (Fig.5) are plotted.

The comparison of the behaviour of the asymmetric system (Figs. 3(b),(c) and 4(b),(c)) with the symmetric one shows that when frequency is optimally tuned to obtain maximum before-impact velocities, the asymmetries play almost no role. However, the optimum value of frequency belongs into the region of the two-fold response of the system, i.e. into the region of hysteresis, where besides the optimum regime with impacts an impactless regime occurs too. To achieve a reliable stabilisation of the regime with impacts, the motion initial conditions of the system should be chosen from the basin of attraction of this motion.

The asymmetries manifest themselves through resonance phenomena near the resonance of natural impactless motion of partial systems ($\eta=1$, see Figs. 3(b), (c) and 5(a), (b)). The phase-trajectories of the bodies are also asymmetric around these resonance (see points A, B in Fig.3(c) and Figs.6 and 7).

5 Conclusion

An analysis of various types of asymmetries in the mathematical model of a double impact oscillator was carried out. It was explained that in the optimal range of driving frequencies corresponding to the system practical use, the differences between the asymmetric cases and the symmetric ones are negligible. It was also established that results obtained via thorough analysis of a single impact oscillator are directly applicable to the double impact oscillator.

Though small asymmetries of the system have no considerable effect on the optimal symmetric impact motion in the region of the practical importance, it is advisable to minimise them, because they can cause undesirable resonance effects near the eigenfrequency of partial systems.

Acknowledgement

This work was supported by grants No. 101/94/0126 and No. 101/97/0670 of the Grant Agency of the Czech Republic.

References

1 Peterka F, Vacík J, Transition to chaotic motion in mechanical systems with impacts, Journal of Sound and Vibration 154(1):95-115, 1992.

2 Peterka F, Introduction to Oscillations of Mechanical Systems with Internal Impacts. ACADEMIA, Prague, 1981 (in Czech).

3 Peterka F, Szőllős O, Mathematical model and numerical simulation of the motion of a two-mass-system with impacts and dry friction, Proc. Colloquium Dynamics of Machines'95, Institute of Thermomechanics AS CR, Prague, pp 161-168 (in Czech).

4 Peterka F, Szőllős O, Impact-oscillations of two colliding harmonic oscillators, Int. Czech-Polish-Slovak Colloquium-Workshop "Advanced Simulation of Systems", Zábřeh na Moravě, 1995, Sb., Vol.1, MARQ Ostrava 1995, pp 160-166 (in Czech).

5 Peterka F, Szőllős O, Dynamics of the Opposed Pile Driver. IUTAM Symposium "Interaction between Dynamics and Control in Advanced Mechanical Systems", Eindhoven, The Netherlands, April 21-26, 1996, Proc., Kluwer Academic Publishers.

6 Peterka F, Szőllős O, Zones of Attraction of Two Different Regimes of Motion of a Pile Driver. Proc. Conference IFToMM , Liberec, 1996, pp 447-452 (in Czech).

7 Peterka F, Szőllős O, The Stability Analysis of a Symmetric Two-Degree-of-Freedom System with Impacts. Proc. EUROMECH - 2nd European Nonlinear Oscillations Conference, Prague, Sept. 9.-13. 1996, ČVUT, pp 345-348.

8 Landa P S. Nonlinear Oscillations and Waves in Dynamical Systems, Kluwer Academic Publishers, Dordrecht, The Netherlands, 1996.

Dynamics of non-linear vibro-impact systems

*P.Z. Lugovoy** and V.I. Gouliaev* *

Dynamical systems with unilateral constraints, turning on and off in their motion, have peculiarities associated with successive variation of their kinematical structure and change of their number of degrees of freedom. Description of their periodical vibrations should be performed via joint consideration of constitutive differential equations and a set of inequalities transformed into equalities at the times of impact interactions unknown in advance. So it is proposed for the system solution to introduce additional unknown variables characterizing the time values when the impacts occur and magnitudes of the velocities discontinuities. For non-linear vibro-impact systems this approach permits to transform the starting two-boundary in time problem for a system of differential equations and inequalities into the multi-boundary in time problem for differential equations without inequalities. In this case, the intermediate boundaries are displacing with the vibration intensity change and additional unknown variables variation.

To construct periodical solutions to the gained systems of non-linear differential equations and to study their stability method of continuation by parameter is used jointly with Newton's method and Liapunov's and Floquet's theory of stability. The approach includes sequential linearization of the equations and construction of a transfer matrix at each step of the leading parameter variation. Applied problems are solved with the use of the elaborated technique.

1 Set up of the problem

Let T-periodical vibrations of a mechanical system be described by the non-linear vector differential equation

$$M\frac{d^2\mathbf{x}}{dt^2}+Q\frac{d\mathbf{x}}{dt}+C\mathbf{x}+\mathbf{G}(\mathbf{x})=\lambda\mathbf{P}(t), \qquad (1.1)$$

where $\mathbf{x}$ is the n-dimensional vector of dynamical displacements; M, Q, C the n×n constant matrices; $\mathbf{G}=(G_1, G_2, ..., G_n)^T$ the non-linear vector-function possessing sufficient number of derivatives with respect to x; $\mathbf{P}(t)$ the T-periodical vector-function featuring the external dynamical forces; $\omega=2\pi/T$ the cyclical frequency; t the time; λ the parameter of the external forces intensity.

Besides this equation the mechanical system is subjected to condition of unilateral constraint

$$g(\mathbf{x}, \dot{\mathbf{x}})\leq 0 \qquad (1.2)$$

where $g(\mathbf{x}, \dot{\mathbf{x}})$ is some functional defined in the class of T-periodical functions.

---

* Institute of Mechanics, Ukrainian Academy of Sciences, Kiev, Ukraine.

The considered set up takes place both for non-linear vibro-impact systems and mechanical systems with Coulomb`s friction. By far the most frequently assumed type of the functional represents the function $g(\mathbf{x}, \dot{\mathbf{x}})=g(\mathbf{x}(t), \dot{\mathbf{x}}(t))$. Below these occasions are examined. Firstly the algorithm for construction of a T-periodical solution to the equation (1.1) is considered for the case, when the constraint (1.2) is not applied. The solution is sought in the set N of the indicated functions determined by the correlations of T-periodicity

$$\mathbf{x}(0)=\mathbf{x}(T), \ \dot{\mathbf{x}}(0)=\dot{\mathbf{x}}(T). \tag{1.3}$$

Let $\mathbf{x}_{(k)}$ be a certain approximate T-periodical solution to the equation (1.1) at $\lambda=\lambda_{(k)}$. The equation T-periodical solution for the value $\lambda_{(k+1)}=\lambda_{(k)}+\Delta\lambda_{(k)}$ can be represented in the form

$$\mathbf{x}_{(k+1)}=\mathbf{x}_{(k)}+\Delta\mathbf{x}_{(k)}. \tag{1.4}$$

The increment vector $\Delta\mathbf{x}_{(k)}$ corresponding to the parameter λ small increment $\Delta\lambda_{(k)}$ must also belong to the set N.

If to substitute (1.4) into (1.1) and to linearize the derived equation in the vicinity of the state $(\mathbf{x}_{(k)}, \lambda_{(k)})$ then the linear differential equation with T-periodical coefficients will be gained for construction of the approximate increment of the vector-function $\Delta\mathbf{x}_{(k)}$:

$$M\frac{d^2\Delta\mathbf{x}_{(k)}}{dt^2}+Q\frac{d\Delta\mathbf{x}_{(k)}}{dt}+\left(C+G'_x(\mathbf{x}_{(k)})\right)\Delta\mathbf{x}_{(k)}=\Delta\lambda_{(k)}\mathbf{P}-\mathbf{r}_{(k)}. \tag{1.5}$$

Here the symbol G'_x denotes the matrix of corresponding partial derivatives

$$G'_x=\left\|\frac{\partial G_i}{\partial x_j}\right\|_1^n,$$

the residual vector $\mathbf{r}_{(k)}(t)$ is determined by the relation

$$\mathbf{r}_{(k)}(t)=M\frac{d^2\mathbf{x}_{(k)}}{dt^2}+Q\frac{d\mathbf{x}_{(k)}}{dt}+C\mathbf{x}_{(k)}+\mathbf{G}(\mathbf{x}_{(k)})-\lambda_{(k)}\mathbf{P}.$$

In general, the residual vector $\mathbf{r}_{(k)}(t)$ does not equal zero as $\mathbf{x}_{(k)}$ is approximate solution to the equation (1.1).

The solution to the equation (1.5) can be constructed as follows

$$\Delta\mathbf{x}_{(k)}(t)=\mathbf{y}_0(t)+Y(t)\alpha+Z(t)\beta \tag{1.6}$$

In this correlation the vector $\mathbf{y}_0(t)$ is the solution to the initial value problem for the equation (1.5) with zero initial values; the matrix $Y(t)=(\mathbf{y}_1(t), \mathbf{y}_2(t), \dots, \mathbf{y}_n(t))$ consists of n solutions to the initial value problem for the equation

$$M\frac{d^2\mathbf{y}}{dt^2}+Q\frac{d\mathbf{y}}{dt}+\left(C+G'_x(\mathbf{x}_{(k)})\right)\mathbf{y}=0 \tag{1.7}$$

with the initial values

$$\mathbf{y}_i(0)=\left(\delta_1^i,\delta_2^i,\dots,\delta_n^i\right)^T, \tag{1.8}$$

$\dot{\mathbf{y}}_i(0) = (0,0,\ldots,0)^T$ (i=1, 2, ..., n);

the matrix $Z(t)=(\mathbf{z}_1(t), \mathbf{z}_2(t), \ldots, \mathbf{z}_n(t))$ consists of n solutions to the initial value problem for the equation (1.7) with the initial values

$\mathbf{z}_i(0) = (0,0,\ldots,0)^T$,

$\dot{\mathbf{z}}_i(0) = (\delta_1^i, \delta_2^i, \ldots, \delta_n^i)^T$ (i=1, 2, ..., n);

the constant vectors $\alpha=(\alpha_1, \alpha_2, \ldots, \alpha_n)^T$, $\beta=(\beta_1, \beta_2, \ldots, \beta_n)^T$ are chosen in such a way as to satisfy the periodicity condition (1.3). In this manner the vectors α, β must obey the system of 2n linear algebraic equations

$$\begin{aligned} (Y(T)-E)\alpha + Z(T)\beta &= -\mathbf{y}_0(T), \\ \dot{Y}(T)\alpha + (\dot{Z}(T)-E)\beta &= -\dot{\mathbf{y}}_0(T), \end{aligned} \qquad (1.9)$$

where E is the unit n×n matrix; the matrices $\dot{Y}(t)$, $\dot{Z}(t)$ are the derivatives of the Y(t), Z(t) with respect to t correspondingly.

Hence, in order to find the equation (1.5) T-periodical solution there is a need to solve 2n+1 initial value problems for determination of vectors $\mathbf{y}_0(t)$, $\mathbf{y}_1(t)$, ..., $\mathbf{y}_n(t)$, $\mathbf{z}_1(t)$, ..., $\mathbf{z}_n(t)$ and thereafter to solve the system of 2n linear algebraic equations (1.9).

If the periodicity matrix

$$\begin{Vmatrix} Y(T)-E & Z(T) \\ \dot{Y}(T) & \dot{Z}(T)-E \end{Vmatrix} \qquad (1.10)$$

is not degenerated the set up problem for the selected $\lambda=\lambda_{(k)}$ may be solved. For the degenerated matrix (1.10), the solution $\mathbf{x}_{(k)}$ is singular and to realize the next step of the parameter λ variation it is necessary to employ the methods of bifurcation theory [2].

Upon determination of the solution $\Delta\mathbf{x}_{(k)}$, the approximate solution $\mathbf{x}_{(k+1)}=\mathbf{x}_{(k)}+\Delta\mathbf{x}_{(k)}$ to the equation (1.1) is constructed.

In a similar manner the T-periodical solutions to the equation (1.1) are found for subsequent values of the parameter λ.

The solution $\mathbf{x}_{(0)}\equiv 0$ at the corresponding value of the parameter $\lambda_{(0)}=0$ is usually taken to be the initial state of the considered system.

The involved approach also permits to investigate stability of the T-periodical motion. The issue of stability of the T-periodical solution $\mathbf{x}(t)$ to the equation (1.1) reduces to investigation of stability of the trivial solution $\mathbf{y}(t)\equiv 0$ to the linear differential equation (1.7) with periodical coefficients. This analysis may be performed on the basis of Floquet's theory [2].

The above outlined algorithm is applicable for study of steady periodical vibrations of a wide class of non-linear mechanical systems.

In the case when besides the conditions (1.3) the vibrational system endures other constraits, it is necessary to fulfill additional rearrangements. Consider the algorithm for derivation of the solution using an example of a vibro-impact mechanical system. The peculiarity of investigations of the system forced vibrations consists in the necessity to take into account the conditions of impacts between moving bodies and constraining bodies and in impossibility to predict a priori the impacts moments [1, 3, 4].

Let the vibro-impact system non-linear vibrations be described by the equation (1.1) and they are limited by the constraint

$$g(t)=g(\mathbf{x}(t))\leq 0. \qquad (1.11)$$

If for some value $\lambda=\lambda_{(k)}$ the T-periodical solution $\mathbf{x}_{(k)}(t)$ pertains to the admissible domain D, that is

$$g(t)<0 \ (0\leq t\leq T), \qquad (1.12)$$

the algorithm for the next step construction remains as before. But at entry of the solution in the domain D boundary when at some value $0<\tilde{t}<T$

$$g(\tilde{t})=0, \qquad (1.13)$$

the solution $\mathbf{x}(t)$ reduces its smoothness at the point $\tilde{t}$ and the algorithm should be properly transformed. In this case the solutions are constructed in the segments $(0, \tilde{t})$ and $(\tilde{t}, T)$ separately and they are joined at the point $\tilde{t}$. The conditions of the joint include: 1) condition of the solution $\mathbf{x}(t)$ discontinuity at $\tilde{t}$; 2) character of the velocity $\dot{\mathbf{x}}(\tilde{t})$ saltus.

So let the system enters in the constraint at some values $\lambda=\lambda_{(k)}$ and $\tilde{t}$, that is $g(\mathbf{x}_{(k)}(\tilde{t}))=0$.

Transitting to the new scales of time $\tau=t/\omega_1$ $(\omega_1=\tilde{t})$ before impact and $\tau=(t-\tilde{t})/\omega_2$ $(\omega_2=T-\tilde{t})$ after impact, the differential equation (1.1) may be reduced to the system

$$\frac{1}{\omega_1^2}M\frac{d^2\mathbf{x}^{(1)}}{d\tau^2}+\frac{1}{\omega_1}Q\frac{d\mathbf{x}^{(1)}}{d\tau}+C\mathbf{x}^{(1)}+\mathbf{G}(\mathbf{x}^{(1)})=\lambda\mathbf{P}(\omega_1\tau)$$

$(0\leq\tau\leq 1)$, (1.14)

$$\frac{1}{\omega_2^2}M\frac{d^2\mathbf{x}^{(2)}}{d\tau}+\frac{1}{\omega_2}Q\frac{d\mathbf{x}^{(2)}}{d\tau}+C\mathbf{x}^{(2)}+\mathbf{G}(\mathbf{x}^{(2)})=\lambda\mathbf{P}(\omega_1+\omega_2\tau$$

$(0\leq\tau\leq 1)$,

where $\mathbf{x}^{(1)}(\tau)=\mathbf{x}(\omega_1\tau)$, $\mathbf{x}^{(2)}(\tau)=x(\omega_1+\omega_2\tau)$ $(0\leq\tau\leq 1)$.

The vector-functions $\mathbf{x}^{(1)}(\tau)$ and $\mathbf{x}^{(2)}(\tau)$, by virtue of the periodicity conditions (1.3), satisfy the equalities

$$\mathbf{x}^{(1)}(0)=\mathbf{x}^{(2)}(1),\ \dot{\mathbf{x}}^{(1)}(0)=\dot{\mathbf{x}}^{(2)}(1). \qquad (1.15)$$

In addition, the condition of solution joining brings into existence the equality

$$\mathbf{x}^{(1)}(1)=\mathbf{x}^{(2)}(0) \qquad (1.16)$$

and a correlation between $\dot{\mathbf{x}}^{(1)}(1)$ and $\dot{\mathbf{x}}^{(2)}(0)$.

Gained in this manner the aggregate of the equations (1.14) - (1.16) describes the motion of the mechanical system of varying structure. We generalized the above proposed method for construction of periodical solution to the cases of non-linear differential equations with discontinuous characteristics. The generalization is based on the fact, that in variation of the leading parameter λ, the values of the velocities saltus and the parameters ω_1, ω_2, defining the moment of impact, are considered to be variable as well.

Let parameter $\lambda = \lambda_{(k_1)}$ gains the increment $\Delta\lambda_{(k_1+1)}$. In this case the solution $\mathbf{x}_{(k_1)}$ also gains increment $\Delta\mathbf{x}_{(k_1+1)}$. Linearize the equations (1.4) in the vicinity of the solution $\mathbf{x}_{(k_1)}$ taking into account variations of ω_1 and ω_2:

$$\frac{1}{\omega_1^2}M\frac{d^2\Delta\mathbf{x}_{(k_1)}}{d\tau^2}+\frac{1}{\omega_1}Q\frac{d\Delta\mathbf{x}_{(k_1)}}{d\tau}+C\Delta\mathbf{x}_{(k_1)}+G'_x(\mathbf{x}_{(k_1)})\Delta\mathbf{x}_{(k_1)}-$$

$$-\frac{2}{\omega_1^3}M\frac{d^2\mathbf{x}_{(k_1)}}{d\tau^2}\Delta\omega_1+\frac{1}{\omega_1^2}Q\frac{d\mathbf{x}_{(k_1)}}{d\tau}\Delta\omega_1=\Delta\lambda_{(k_1)}\mathbf{P}(\omega_1\tau)+ \qquad (1.17_1)$$

$$+\lambda_{(k_1)}\frac{d\mathbf{P}(\omega_1\tau)}{dt}\cdot\tau\Delta\omega_1+\mu\mathbf{r}_{(k_1)}(\omega_1 t),$$

$$\frac{1}{\omega_2^2}M\frac{d^2\Delta\mathbf{x}_{(k_1)}}{d\tau^2}+\frac{1}{\omega_2}Q\frac{d\Delta\mathbf{x}_{(k_1)}}{d\tau}+C\Delta\mathbf{x}_{(k_1)}+G'_x(\mathbf{x}_{(k_1)})\Delta\mathbf{x}_{(k_1)}-$$

$$-\frac{2}{\omega_2^3}M\frac{d^2\mathbf{x}_{(k_1)}}{d\tau^2}\Delta\omega_2+\frac{1}{\omega_2^2}Q\frac{d\mathbf{x}_{(k_1)}}{d\tau}\Delta\omega_2=\Delta\lambda_{(k_1)}P(\omega_2\tau+\omega_1) \qquad (1.17_2)$$

$$+\lambda_{(k_1)}\frac{d\mathbf{P}(\omega_2\tau+\tilde{t})}{dt}\cdot\tau\Delta\omega_2+\mu\mathbf{r}_{(k_1)}(\omega_2\tau+\omega_1),$$

where the multiplier μ=1 is introduced for convenience of representation.

The equation (1.17$_1$) solution is constructed in the form

$$\Delta\mathbf{x}^{(1)}_{(k_1)}(\tau)=Y_1(\tau)\alpha_1^{(1)}+Z_1(\tau)\beta^{(1)}+\Delta\lambda_{(k_1)}\cdot\mathbf{x}^{(1)}_p(\tau)+\Delta\omega_1\cdot\mathbf{x}^{(1)}_{\omega_1}(\tau)+\mathbf{x}^{(1)}_r(\tau). \qquad (1.18)$$

Here $\mathbf{x}^{(1)}_p(\tau)$, $\mathbf{x}^{(1)}_{\omega_1}(\tau)$, $\mathbf{x}^{(1)}_r(\tau)$ are the solutions to the initial problems for the equation (1.17$_1$) with nullified initial conditions and correspondingly for the values as follows:

$\Delta\lambda_{(k_1)}=1,\ \Delta\omega_1=0,\ \mu=0;$

$\Delta\lambda_{(k_1)}=0,\ \Delta\omega_1=1,\ \mu=0;$

$\Delta\lambda_{(k_1)}=0,\ \Delta\omega_1=0,\ \mu=1.$

In a similar manner the equation (1.17$_2$) solution is represented as

$$\Delta\mathbf{x}^{(2)}_{(k_1)}(\tau)=Y_2(\tau)\alpha^{(2)}+Z_2(\tau)\beta^{(2)}+\Delta\lambda_{(k_1)}\cdot\mathbf{x}^{(2)}_p(\tau)+\Delta\omega_2\cdot\mathbf{x}^{(2)}_{\omega_2}+\mathbf{x}^{(2)}_r. \qquad (1.19)$$

By this means the solution to the system (1.17_1), (1.17_2) is determined by the set of 4n+3 constants $\alpha^{(1)}$, $\beta^{(1)}$, $\Delta\omega_1$, $\alpha^{(2)}$, $\beta^{(2)}$, $\Delta\omega_2$, $\Delta\lambda_{(k_1)}$. It is necessary to construct system of 4n+2 equations for their calculation. The first group of 2n equations is evolved from periodicity conditions (1.15) in the form

$$\Delta\mathbf{x}^{(1)}_{(k_1)}(0)=\Delta\mathbf{x}^{(2)}_{(k_1)}(1),\ \Delta\dot{\mathbf{x}}^{(1)}_{(k_1)}(0)=\Delta\dot{\mathbf{x}}^{(2)}_{(k_1)}(1). \qquad (1.20)$$

The second group of n equations is deduced from condition (1.16) of continuity at the point $\tilde{t}$:

$$\Delta\mathbf{x}^{(1)}_{(k_1)}(1)=\Delta\mathbf{x}^{(2)}_{(k_1)}(0). \qquad (1.21)$$

The condition, determining the velocities discontinuity, is associated with n equations

$$\Delta\dot{\mathbf{x}}^{(1)}_{(k_1)}(1)=\mathbf{f}\left(\Delta\dot{\mathbf{x}}^{(2)}_{(k_1)}(0)\right) \qquad (1.22)$$

The linearized correlation of the constraint (1.13) results in equation

$$\left(\nabla g\left(\mathbf{x}_{(k_1)}(\tilde{t})\right)\cdot\Delta\mathbf{x}^{(1)}_{(k_1+1)}(1)\right)=0 \qquad (1.23)$$

The last equation ties ω_1 and ω_2:

$$\omega_1+\omega_2=T. \qquad (1.24)$$

Inasmuch as the number 4n+2 of the set up equations (1.20) - (1.24) is less by unit than the number 4n+3 of the unknown values, one of them, for example $\Delta\lambda_{(k_1)}$, is considered to be free. Its independent change determines evolution of the vibro-impact system motion.

The determination of the deduced algebraic equations matrix is also used for identification of the system critical states, which occur when its value equals zero.

Previously the case was analysed when the vibro-impact system entered on the constraint only one time in the period of its steady vibrations. The algorithm can be naturally extended to the case where l impacts occur in a period. In such an occasion the period T is divided into l+1 segments [0, $\tilde{t}^{(1)}$], [$\tilde{t}^{(1)}, \tilde{t}^{(2)}$], ..., [$\tilde{t}^{(l)}$, T]. In the ranges of every segment the vector differential equation of the (1.14) type and the linearized one of the (1.17) type are formulated. The solutions to the linearized equations are constructed in the form (1.18). The system of 2n(l+1)+l+1 linear algebraic equations for the 2n(l+1)+l+2 constants $\alpha^{(1)}$, $\beta^{(1)}$, $\Delta\omega_1$, ..., $\alpha^{(1)}$, $\beta^{(1)}$, $\Delta\omega_1$, $\Delta\lambda$ is constructed with the use of periodical conditions, discontinuity conditions, values of the velocity saltuses and conditions of enter on the constraints at the impact moments. Besides, one more equation

$$\omega_1+\omega_2+\ldots+\omega_l+\omega_{l+1}=T$$

is added.

In the events that the determinant of the 2n(l+1)+l+1 linear equations matrix differs from zero, the system has the unique solution. The states of the matrix degeneration are associated with the vibro-impact periodical process stability loss.

Application of the proposed approach to investigation of periodical processes in the system with unilateral constraints allows to reduce the specified problem to, as it is called, multi-point boundary value problem with moving boundaries.

2 Vibrations of oscillators with unilateral constraints

Consider an example. Let an oscillator non-linear vibrations be constrained by a rigid obstacle which is spaced at the distance Δ_1 (Fig. 1). The motion of the oscillator body M is governed by the equation

$$m\ddot{x} + cx + kx^3 = P\sin\gamma t, \tag{2.1}$$

where m is the body M mass; **x** its displacement from the neutral state; c the spring stiffness; P the parameter of the external force intensity; γ its frequency; t the time.

Assume that at the moment $\tilde{t}$ the impact occurs between the body and the obstacle. Changing to the new scales of time before the impact ($\tau=t/\omega_1$) and after it ($\tau=t/\omega_2$), the equation (2.1) can be reduced to the following forms:

$$\begin{aligned} &m\omega_1^{-2}\ddot{x} + cx + kx^3 = P\sin\gamma\omega_1\tau \ (0\le\tau\le1),\\ &m\omega_2^{-2}\ddot{x} + cx + kx^3 = P\sin\gamma(\omega_1 + \omega_2\tau) \ (0\le\tau\le1). \end{aligned} \tag{2.2}$$

The values ω_1 and ω_2 are correlated by the equalities

$$\omega_1 = \tilde{t}, \ \omega_2 = T - \omega_1 \ (T=2\pi/\gamma), \tag{2.3}$$

which issue from (1.24).

The T-periodical solution to the system (2.2) - (2.3) was constructed through the use of the above outlined approach for the following values of the parameters: m=1 kg, c=1 N/m, k=1 N/m^3, Δ_1=0.005 m. Fig. 2, a - e, shows the vibrations modes for the force parameter P values equal $2.5\cdot10^{-4}$ N, 10^{-2} N, $2\cdot10^{-2}$ N, $4\cdot10^{-2}$ N, $8\cdot10^{-2}$ N correspondingly. It can be seen that displacement in phase of the solution takes place with increasing parameter P owing to the body M impact with the obstacle. At large values of the parameter P the phase displacement between the functions $P\sin\gamma t$ and **x**(t) becomes significant with the resulting complication of the solution **x**(t) mode in the domain of its negative values.

Along with construction of the non-linear system solution, the determinant of the matrix of periodicity and impact conditions equations was investigate. The analysis testifies that in the considered diapason of the parameter P variation the critical states did not occur.

In investigation of motion of the body M with two unilateral constraints (Fig. 3) the vibration period T is divided into three scales. In parallel with the conditions of periodicity the solution is constructed with allowance made for conditions of interaction between the body M and the two constraints. The solution evolution is represented in Fig. 4, a - d for the cases P=0.06 N, 0.12 N, 0.16 N, 0.28 N correspondingly. One cans see that the vibro-impact motion mode undergoes tangible transformation with the parameter P increase, as this takes place in the time diapason between two impacts the additional wavelets arise whose amplitudes continue to increase. At $P \geq 0.303$ the body M motion becomes six-

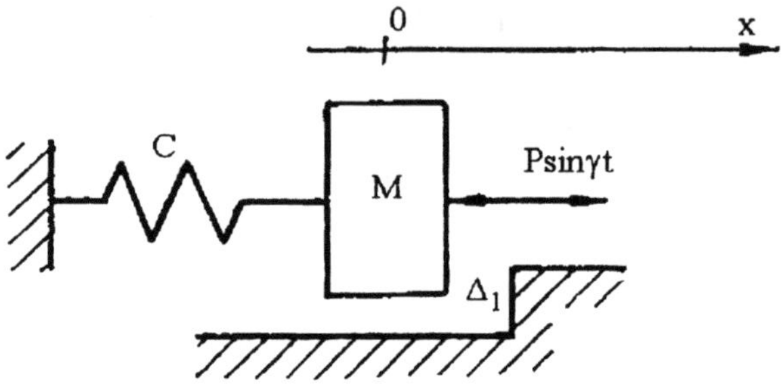

Fig. 1. The oscillator scheme

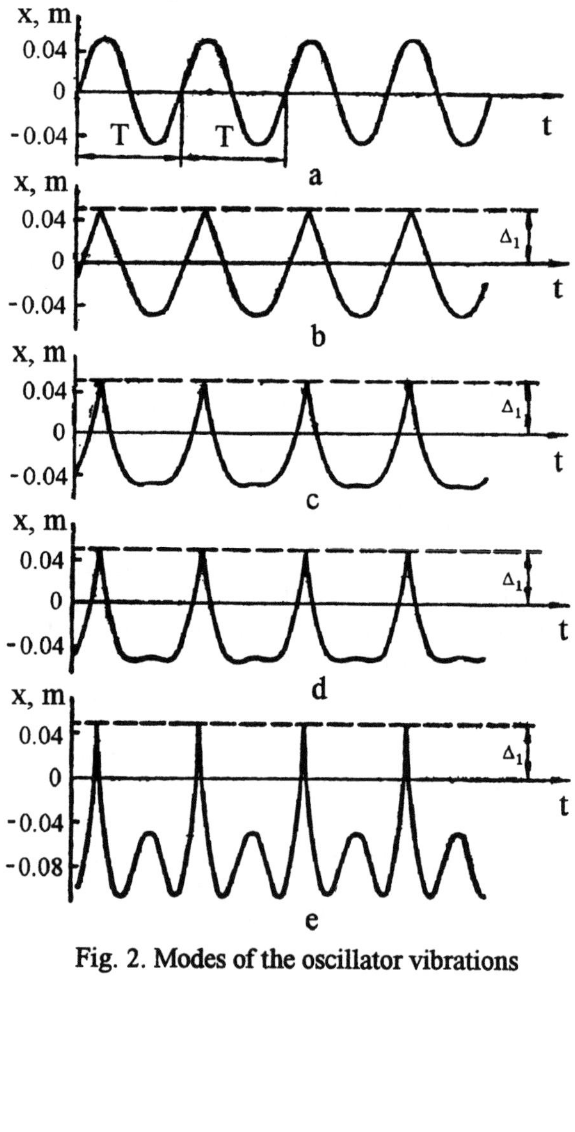

Fig. 2. Modes of the oscillator vibrations

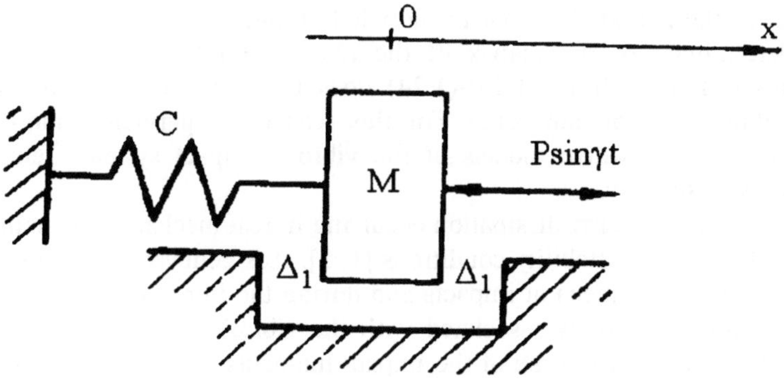

Fig. 3. The oscillator scheme

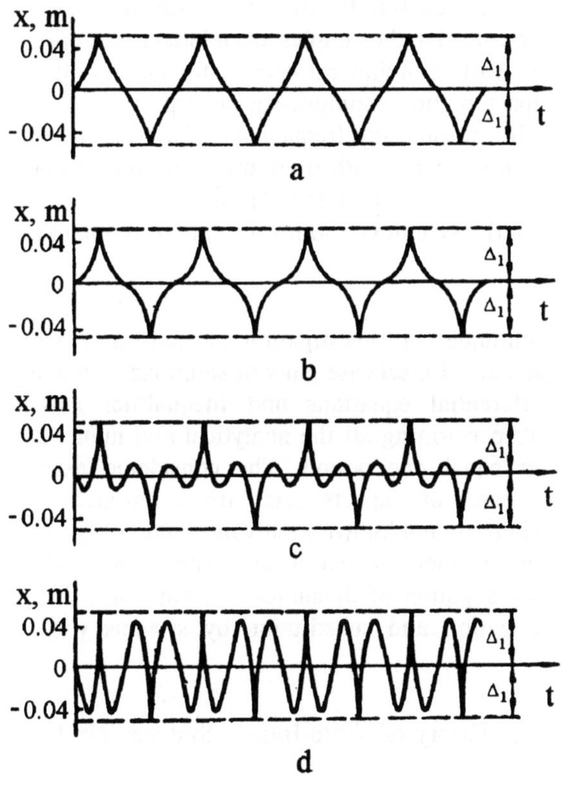

Fig. 4. Modes of the oscillator vibrations.

impact in the period.Note that in the whole range of the parameter P variation, the determinant of the matrix of the $(2n+1)(l+1)$ linear algebraic equations systems stemming from (1.20)-(1.24) does not transform to zero and so the bifurcation states do not come. For this reason the periodical motions of the considered conservative models of the vibro – impact systems remain in the boundary of stability.

Inasmuch as the energy dissipation occurring in real mechanical systems does not adversely affect the stability conditions [1, 3], it is believed that allowance made for damping phenomena at impacts and during the impactless motion will cause only quantity variations associated with the vibrations amplitude decrease and some displacements in time of the impact moments but it will not influence the dynamical process quality characteristics.

Other dynamical problems for several solid bodies, elastic beams, plates and shells with unilateral constraints are solved on the basis of the proposed methods[2, 4]. The considered examples are related to the cases when a thin-wall structure interacts with a unilateral elastic foundation. It is shown that the systems vibro-impact motions comprise a superposition of different periodical motion modes with low and high frequencies. The first ones dominate when the elastic object does not contact with the foundation, the second ones occur when the object vibrates together with the foundation.

It is discovered that the system vibrations may attain resonant character.

3 Conclusion

The proposed technique for investigation of non-linear vibro-impact devices permits to find periodical piecewise smooth solutions to the appropriate systems of non-linear differential equations and inequalities and to identify their bifurcational states performing all the analytical and numerical constructions in the time segment equal one period. The considered examples illustrate its efficiency for the cases of impacts many times repeated in period when the moments of collisions are not known in advance and change with variation of the external perturbation intensity parameter. The elaborated approach can be generalized for investigation of dynamical systems with distributed parameters after their discretization and substitution by systems with finite number of freedom degrees.

References

1 Babitsky V I, Theory of Vibro-Impact Systems, Nauka, Moscow, 1978. - 352 pp.

2 Gouliaev V I, et.al. Stability of Periodical Processes in Non-Linear Mechanical Systems, Vyshcha Shkola, Kiev, 1983. - 287 pp.

3 Kobrinsky A E, Kobrinsky A A, Vibro-Impact Systems, Nauka, Moscow, 1973. - 591 pp.

4 Lugovoy P Z, Mukoid V P, Meish V F, Dynamics of Shell Structures under the Explosive Loading, Nauk. Dumka, Kiev, 1991. - 280 p.

Subharmonic response of a single-barrier vibroimpact system to a narrow-band random excitation

*M.F.Dimentberg, D.V.Iourtchenko**

Sensitivity of subharmonics to imperfect periodicity of the excitation is studied for a vibroimpact system with a single barrier, slightly offset from the system's equilibrium position. The system is excited by a periodic force with random phase modulation, so that small random temporal variations of the excitation frequency are present, which make the force a narrow-band random process. Mean value of the excitation frequency is close to an even integer multiple of the system's natural frequency without a barrier. Analytical study is performed, based on the Zhuravlev transformation, which effectively removes velocity jumps from the equations of motion, combined with the Krylov-Bogoliubov averaging over the period. The reduced stochastic differential equations of motion are solved then by the method of moments for the mean square response amplitude. The solution is exact for the case of zero offset and approximate, using perturbation-based moment closure scheme for the general case. The results are verified by Monte-Carlo simulation, which was used also to demonstrate possibility for incorporating moderately large impact losses through the use of previously derived equivalent viscous damping factor. The results indicate possibility for significant reduction of peak resonant amplitudes of subharmonics due to imperfect periodicity of excitation, depending on the value of the excitation/system bandwidth ratio. Applications to dynamics of moored bodies under ocean waves' excitation are discussed.

1 Problem formulation

Fig. 1 illustrates a single-degree-of-freedom vibroimpact system with an one-sided barrier, excited by a force mz(t). The equation of motion between impacts, as divided by the system's mass m may be written as

$$\ddot{y}+2\alpha\dot{y}+\Omega^2 y=z(t), y>-h \qquad (1)$$

* Mechanical Engineering Department, Worcester Polytechnic Institute, Worcester, MA 01609, USA

where Ω is the system's undamped natural frequency of nonimpact motion and h is the offset between the barrier and the system's equilibrium position. This equation is supplemented by the impact/rebound condition

$$\dot{y}(t_* + 0) = -r\dot{y}(t_* - 0), y(t_*) = -h \tag{2}$$

with r being a restitution factor. The excitation z(t) is assumed to be sinusoidal with random phase modulation [1, 2], so that

$$z(t) = \lambda \sin\psi(t), \dot{\psi} = \nu + \sigma\xi(t),$$
$$\Phi_{zz}(\omega) = \frac{(\lambda^2\sigma^2/4\pi)(\omega^2 + \nu^2 + \sigma^4/4)}{(\nu^2 + \sigma^4/4 - \omega^2)^2 + \omega^2\sigma^4} \tag{3}$$

where $\xi(t)$ is a stationary zero-mean Gaussian white noise of unit intensity and $\Phi_{zz}(\omega)$ is a power spectral density (PSD) of z(t). Here σ is clearly seen to be a bandwidth of the PSD and also intensity parameter of "disorder", or random variations of the excitation frequency. Process z(t) is assumed to be narrow-band ($\sigma^2 << \nu$).

Solutions to this problem with perfectly periodic excitation ($\sigma = 0$) are reported in [3, 4] for resonant and nonresonant cases respectively. For the former case, where excitation frequency is assumed to be close to an arbitrary even integer multiple of Ω, an analytical solution has been obtained for the response amplitude [3] through the use of the Zhuravlev transformation [5] $y = |x| - h, \dot{y} = \dot{x}\,\mathrm{sgn}\,x$, which reduces the equation (1) to

$$\ddot{x} + \Omega^2 x = -2\alpha\dot{x} + h\Omega^2\,\mathrm{sgn}\,x + z(\mathrm{sgn}\,x) \tag{4}$$

The impact condition (2) is seen to be satisfied automatically by the transformed variable x(t), provided that the impact is elastic, i.e. r = 1. The case of nonresonant excitation has been studied numerically in [4], and possibility for chaos was demonstrated. Applications for moored vessels in harbours are described in [4]. It seems also, that the model (1), (2) may also be applicable for heaving motions of moored bodies with inextensible mooring lines in cases where slack of the line is possible. On the other hand, as long as the ocean waves are certainly not perfectly periodic, it seems worthwhile to study the effect of the imperfect periodicity on dynamics of moored bodies.

2 Solution for the system with elastic impacts

The transformed equation (4), with the excitation z(t) as defined by equation (3), will be studied in the same way as for the perfectly periodic case [3] - by Krylov-Bogoliubov (KB) averaging-over-the-period. Consider case of a subharmonic resonance of the n-th order, with magnitude of mean detuning $|\Delta_n|, \Delta_n = \nu - 2\Omega n$ being proportional to a small parameter, together with all coefficients α, h, λ in the RHS of the equation (4). (Condition for small h implies that the system is quasiisochronous, and actually h should be small compared with response amplitudes). Then the solution is represented in terms of slowly varying amplitude and phase A, θ as $x = A\sin\phi, \dot{x} = \Omega A\cos\phi, \theta = \psi - 2n\phi$. The RHSs of the resulting pair of first-order equations for amplitude and phase, with z(t) as defined according equations (3), are then averaged over the fast variable ϕ within their period. This leads to a pair of "shortened" equations, which can be further transformed by using state variables $x_c = A\cos\theta, x_s = A\sin\theta$. The transformed shortened equations are found to be

$$
\begin{aligned}
&\dot{x}_c = -\alpha x_c - \Delta_n x_s - h_n x_s/A + q_n - \sigma x_s \xi(t);\\
&\dot{x}_s = -\alpha x_s + \Delta_n x_c + h_n x_c/A + \sigma x_c \xi(t)\\
&A = \sqrt{x_c^2 + x_s^2},\ q_n = (\lambda/\pi\Omega)\left[\frac{4n}{4n^2-1}\right],\ h_n = 4n\Omega h/\pi
\end{aligned}
\qquad (5)
$$

Equations (5) can be solved exactly by the method of moments for mean square response amplitude if h = 0. For nonzero h the method can also be applied provided that some approximate closure scheme is used for the (nonlinear) terms with h. A simple perturbation-based scheme seems to be appropriate whereby random variations of the response amplitude are assumed to be small with respect to its mean value [6]. The latter, up to second-order small quantities, is equal to its deterministic value A_0 for the perfectly periodic case, and to the square root of mean square amplitude: $\langle A\rangle \cong A_0 \cong A_*, A_*^2 = \langle x_c^2 + x_s^2\rangle$, where angular brackets denote probabilistic averaging. Then the expectations of the nonlinear terms in the first and second stochastic differential equations (SDEs) (5) may be approximated by $-h_n m_s/A_*$ and $h_n m_c/A_*$ respectively, where $m_{c,s} = \langle x_{c,s}\rangle$. The "physical" SDEs (5) are then transformed to the Ito SDEs by adding Wong-Zakai corrections [2], which are found to be

$-\sigma^2 x_c/2$ and $-\sigma^2 x_s/2$ for the first and second SDEs respectively. Direct probabilistic averaging of the corresponding Ito SDEs (5) with linearized nonlinear terms yields then the following deterministic equations for the first-order response moments

$$\dot{m}_c = -\left(\alpha + \sigma^2/2\right)m_c - \left(\Delta_n + h_n/A_*\right)m_s + q_n;$$
$$\dot{m}_s = \left(\Delta_n + h_n/A_*\right)m_c - \left(\alpha + \sigma^2/2\right)m_s \tag{6}$$

Introducing a new state variable u(t), we obtain from SDEs (5)

$$u = x_c^2 + x_s^2, \dot{u} = 2\left(x_c\dot{x}_c + x_s\dot{x}_s\right) = -2\alpha u + 2q_n x_c$$

Direct probabilistic averaging of the latter (nonstochastic) equation yields a deterministic equation for the expected value of u(t), with the steady-state solution $\langle u\rangle = A_*^2 = q_n m_c/\alpha$. This formula is combined with steady-state solution to equations (6), which also relates A_* to m_c. These two relations after excluding m_c lead to a quadratic equation for the mean square response amplitude, with the following final solution [6]

$$A_* = \frac{-\Delta_n h_n \pm \sqrt{q_n^2\left(1+\sigma^2/2\alpha\right)\left(\alpha_e^2 + \Delta_n^2\right) - \alpha_e^2 h_n^2}}{\alpha_e^2 + \Delta_n^2};$$
$$\alpha_e = \alpha\left(1 + \sigma^2/2\alpha\right) \tag{7}$$

As long as any solution for amplitude should be real and positive, the domain for existence of a subharmonic of a given order may be established. Two cases may be identified here [6]. If, for given n and σ, $q_n > h_n$ (case of a "small" gap, or offset), then only positive root in (7) has physical meaning, and every amplitude frequency response curve has finite slope everywhere. In the special case of zero gap (h = 0) these (exact) curves are symmetric with respect to the corresponding resonant frequency.

If $q_n < h_n$ (case of a "large" gap), then the steady-state solution (7) exists only for negative detunings, specifically if $\Delta_n < -\alpha\sqrt{h_n^2/q_n^2 - 1}$. Two possible solutions may exist in this frequency range, and one with smaller value of amplitude should be unstable, at least in view of continuity reasoning, as long as the instability for the case of perfect periodicity has been demonstrated in [4]. Anyway, the perturbational reasoning behind the closure scheme used suggests that only the upper branch of the amplitude frequency response curve may be predicted, if at all, in the case of "large" gap. Indeed, the accuracy of the solution should increase with decreasing h and/or parameter of disorder σ, as long as the corresponding benchmark exact solutions would be approached. The convergence rate, or actual accuracy of the solution for finite level of disorder is established (later) by Monte-Carlo simulations as being very high. On the other hand, no claims for validity of the solution should be made for the frequency range where steady-state solution to the equation (7) does not exist in the case of a perfectly periodic excitation; indeed, as long as the closure scheme is based on perturbational reasoning, it should work only provided that $A_0 >> h$, and $A_* >> h$ is a necessary but not sufficient condition for accurate response prediction.

The solution obtained clearly shows, that random disorder, or deviations from perfect periodicity in the excitation, should lead to reduction of response amplitudes for sufficiently small magnitudes of detuning. The peaks in amplitude frequency response curves become "levelled", or less sharp with increasing σ - see these curves of A_*/h in Fig. 2. The reduction depends strongly on the excitation/system bandwidth ratio $\sigma^2/2\alpha$, and for moored systems may be quite large, as long as relative bandwidths σ^2/ν of the ocean waves spectra are typically of the order 0.1 - 0.2 or even higher. These results clearly correspond to the case of a "small" gap, with magnitudes of slope in amplitude frequency response curves being finite everywhere. It should be mentioned also, that when excitation/system bandwidth ratio $\sigma^2/2\alpha$ is much larger than unity, formula (7) reduces to one, obtainable by direct stochastic averaging [1, 2], the limiting value of mean square response amplitude being proportional to $\Phi_{zz}(2n\Omega)$ [6].

Solid symbols in Fig. 2 represent results of Monte-Carlo simulations, as obtained by numerical integration of the transformed differential equation (4) for n = 2 along N = 1274 cycles of the response (small-gap case). The mean square response amplitude was calculated as $A_*^2 = 2\left\langle (x - h\,\mathrm{sgn}\,x)^2 \right\rangle$, where angular brackets denote here common time averaging along the response sample. (Whilst amplitudes of the original and transformed responses, y(t) and x(t) respectively, are different for nonzero h, this difference is of the order of a small parameter, and thus cannot be resolved by the theoretical analysis). The results seem to be in a good agreement with predicted ones even for rather large disorder in periodicity - namely for large excitation/system bandwidth ratio - in spite of the fact, that the approximate closure scheme used relies on perturbational reasoning for small random variations of the steady-state response amplitude. Thus, the basic trend is predicted properly - introducing and increasing random "disorder", or frequency variations in a periodic excitation leads to reduction in peak amplitudes of subharmonic response.

Fig. 3 represents similar results for a large-gap case. Good agreement between theoretical predictions and simulation can be seen here for subresonant expected excitation frequencies. On the other hand, within a post-resonant range no agreement can be seen - even for the cases where increase of the disorder intensity lead to appearance of nonzero response amplitudes in this range. The reason had been explained already, and it may be added here, that the response was of a rather intermittent nature in this range, with instantaneous excitation frequency going in and out of a range with $A_0 = 0$.

3 Accounting for impact losses

It can be clearly seen from the impact/rebound condition (2) that the Zhuravlev transformation does not exclude velocity jump completely for the transformed variable x(t), but just reduces it whenever $r < 1$, i.e. when impact losses are present. For the case of small impact losses, i.e. when $1 - r$ is proportional to the small parameter, the remaining velocity jump can be transferred into the RHS of the transformed equation (4) using Dirac delta-function, as proposed in [5]. Subsequent KB-averaging-over-the-period of this term shows, that for small or zero h the result of this analysis is found to be equivalent to increasing available viscous damping coefficient by the amount $\alpha_{as} = (1-r)(\Omega/\pi)$ [1] (subscript "as" indicates that this formula is based on the asymptotic averaging method).

An improved version of this formula for the equivalent viscous damping, which reduces to one for α_{as} with r approaching unity, has been derived in [7] as

$$\alpha_{eq} = \left[\left(1-r^2\right)/\left(1+r^2\right)\right](\Omega/\pi) \tag{8}$$

The derivation was made for a model vibroimpact problem where z(t) is a zero-mean white noise with intensity D, no viscous damping is present and $h = 0$. Introducing then response energy H(t) as $2H = \dot{y}^2 + \Omega^2 y^2$, one can use stochastic differential equation calculus to derive the equation for expected energy between n-th and n+1-st impacts as $\dot{H} = D/2,\ H(t) = H_n + Dt/2$. Here subscript n is used for the value of energy at the n-th rebound, and no special symbols is used for the expected values. Applying above expression for H(t) to the n+1-st impact instant and imposing the impact condition (2), yields the following relation between expected rebound energies

$$H_{n+1} - H_n = -(1-r)H_n + r^2 DT_n/2 \tag{9}$$

Here $T_n \cong T = \pi/\Omega$ is an expected time between two impacts, whereas the approximate "universal" formula for this time is derived as a zero-order approximate solution to the PDE of the relevant first-passage problem [7]. For stationary response the expected rebound energy should be constant, and thus can be found from the condition for zero LHS in the equation (9). The mean energy may then be found as H(T/2), and equating it to the mean response energy of the viscously damped system without impact losses leads to formula (8). This formula should provide improved accuracy for not-too-small values of 1 - r, since it requires only negligibly small variations of the expected response period due to random excitation, whereas asymptotic averaging formula requires also that difference in magnitude between impact and rebound velocities should be small for every impact. Monte-Carlo simulation studies has shown this to be the case indeed when z(t) is a white noise [7].

Fig. 4 illustrates amplitude frequency response curves A_*/h vs. ν for the case of not-too-small impact losses, r = 0.8. Two pairs of curves are presented, one for the case of perfect periodicity, the other one - for the disordered excitation case with $\sigma = 0.4$. All these curves were obtained by Monte-Carlo simulation, but implemented differently for the solid and dashed curves. The former were obtained by numerical integration of the equation (1) with impact condition (2) (and excitation (3)), with "honest" tracing each impact. The dashed curves are based on numerical integration of the transformed equation (4) with an equivalent viscous damping factor as calculated by formula (8): $\alpha = \alpha_{eq} = 0.06987\Omega$. These results show, that the equivalent damping can be used indeed to account for impact losses in random vibration problems with narrow-band excitations.

4 Conclusions

Subharmonic response of a quasiisochronous vibroimpact system to periodic excitation with random phase modulation, or "disorder", has been studied both analytically and by Monte-Carlo simulation. Whilst analytical solution has been obtained originally for the case of elastic impacts, possibility for its use for systems with impact losses has been established, based on the use of the equivalent viscous damping. The results indicate possibility for drastic reduction of peak response amplitudes due to imperfect periodicity of the excitation.

Acknowledgement

This work was sponsored by the NSF, Grant CMS-9610363. This support is most highly appreciated.

References

1. Dimentberg, M.F. Statistical Dynamics of Nonlinear and Time-Varying Systems. Research Studies Press, Taunton, UK, 1988.

2. Lin, Y.K. & Cai, G.Q. Probabilistic Structural Dynamics. McGraw Hill, New York, 1995.

3. Zhuravlev, V.F. & Klimov, D.M. Applied Methods in Vibration Theory (in Russian). Nauka, Moscow, 1988.

4. Thompson, J.M.T. & Stewart, H.B. Nonlinear Dynamics and Chaos. Wiley, Chichester, UK, 1985.

5. Zhuravlev, V.F. A method for analyzing vibration-impact systems by means of special functions. Mechanics of Solids, vol. 11, 1976, pp.23 - 27 (English translation of the Russian journal "Mekhanika Tverdogo Tela").

6. Dimentberg, M.F., Iourtchenko, D.V. & van-Ewijk, O. Subharmonic response of a quasiisochronous vibroimpact system to a randomly disordered periodic excitation. To be published in the journal "Nonlinear Dynamics".

7. Dimentberg, M.F. & Iourtchenko, D.V. Towards incorporating impact losses into random vibration analyses: a model problem. To be published in the journal "Probabilistic Engineering Mechanics".

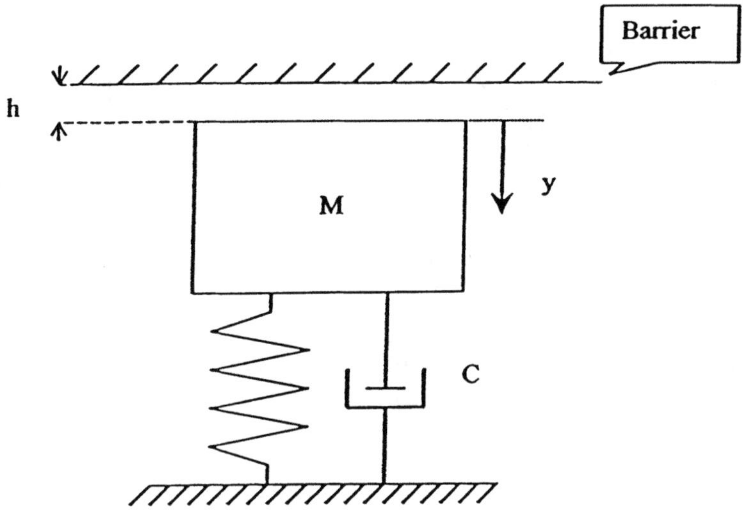

Figure 1. Model of SDOF system with impact.

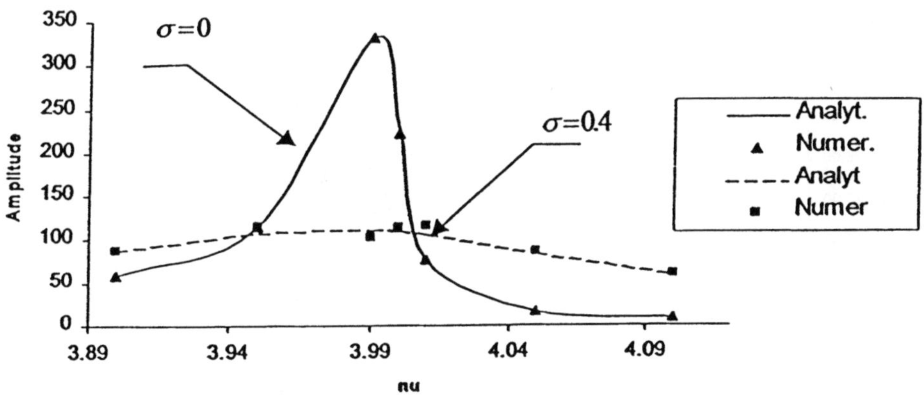

Figure 2. "Mean" amplitude frequency response curves for the case of small gap, with different values of σ , n=2 and $\alpha/\Omega = 0.01, h = 1, \lambda = 20$

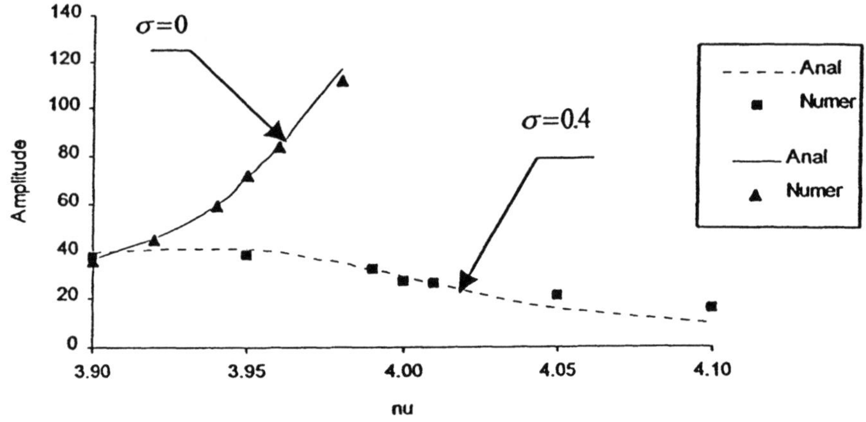

Figure 3. "Mean" amplitude frequency response curves for the case of large gap, with different values of σ, n=2 and $\alpha/\Omega = 0.01, h = 1, \lambda = 7$

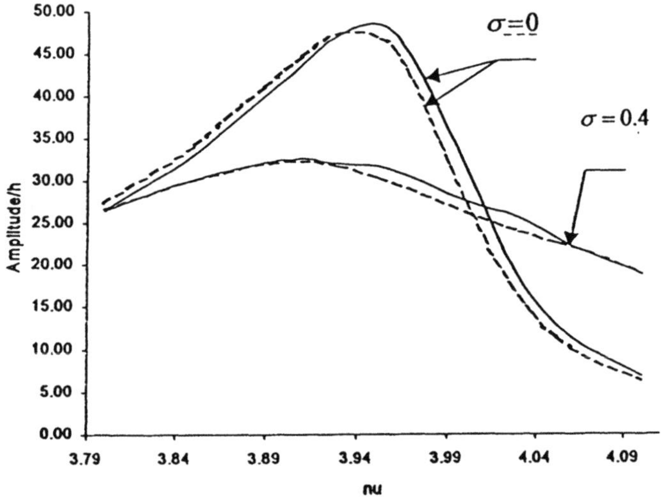

Figure 4. "Mean" amplitude frequency response curves, obtained numerically for the case of small gap, impact losses with r=0.8, λ=20,h=1;

Solid lines: from equations (1),(2)

Dashed lines: from equation (4)

Strain solitons in solids: physics, numerics and fracture

A.M. Samsonov, G.V. Dreiden, A.V. Porubov, I.V. Semenova [1]

Nonlinear elastic features of solids result in generation of new type of elastic waves - localized strain waves (solitons) even under short-run and weak reversible (elastic) loading. Based on continuum mechanics approach, the mathematical theory of long nonlinear strain waves in elastic wave guides was developed to provide the successful generation and observation of evolution of strain solitons in a rod. It was shown both in theory and experiments that the soliton propagates along the uniform waveguide without changing of its shape, while there is an amplification of the soliton in a tapered rod.

1 Introduction

We present a brief review of our recent results concerning the theoretical description and experimental observation of the propagation and amplification of a strain solitary wave (or, briefly, the soliton) in a cylindrical nonlinearly elastic rod with varying cross section.

Various solitons in fluids were observed and generated many times, and it was the surprising fact, however, that despite of almost similar description of stresses in fluids and solids, longitudinal strain solitons have not been observed in nonlinearly elastic wave guides till the last decade. The envelope solitary waves governed by the nonlinear Schrödinger equation were widely considered; another famous soliton found by Frenkel and Kontorova (1938) in a ball chain became useful for crystalline lattice models. However there must be a soliton in solids in the form of a nonlinear *long* quasistationary localized strain wave, propagating either along an interface (internal solitary wave) or inside a wave guide (a density soliton).

It is known that the soliton propagates without change of a shape in a uniform waveguide while its shape will vary in presence of inhomogeneities. Therefore focusing may occur, and the soliton amplitude will increase, while its width will decrease simultaneously. For this reason the localized area of plasticity and even fracture of a wave guide may happen to appear. Hence the study of the behaviour of such waves is of importance for an assessment of durability of elastic materials and structures, development of methods of nondestructive testing, determination of physical properties of both conventional (e.g., brass or steel) and new elastic materials, and, in particularly,

[1] The Ioffe Institute of the Russian Academy of Sciences, St.Petersburg, 194021 Russia. E-mail: samsonov@math.ioffe.rssi.ru

of polymers. Another possible application of strain nonlinear waves comes from their permanent shape property and from the dependence of their amplitude, phase velocity etc. upon the material properties of an elastic wave guide. In particular, it may help in development of a resolutive nondestructive test for pipelines because it was found that bulk waves better satisfy detection requirements than any surface strain waves.

The complete description of a three-dimensional nonlinear elasticity problem in a continuum is difficult, and, if possible, it is usually reduced to the one-dimensional form in order to clarify the simplest but qualitatively new solutions. On the contrary, the commonly used linearization of a problem seems to be unsatisfactory from the physical point of view, because the ratio of a finite deformation and its linear part is determined by a displacement gradient and its variation in time.

2 Theory of strain solitons in a rod

Introducing the cylindrical Langrangian coordinate system (x, r, φ), where x is the axis along the rod, $\varphi \,\epsilon[0, 2\pi]$, $0 \leq r \leq R(x) \leq R_0 - const$, one can write the displacement vector as $\vec{V} = (u, w, 0)$, when torsions are neglected. Basic equations, describing the nonlinear wave propagation in the initial configuration, are obtained from the Hamilton principle, requiring to be equal to zero the variation of the action $S = 2\pi \left(\int_{t_0}^{t_1} dt \left(\int_{-\infty}^{\infty} dx \int_0^R r\, \mathcal{L} dr \right)\right)$, where $\mathcal{L}$ is a Lagrangian. The Murnaghan approximation of the potential deformation energy is chosen, valid for a wide class of nonlinearly elastic materials. Details of derivation of the governing equation for longitudinal strain waves can be found in [1, 2], here we only note that the Cauchy-Green *finite* deformation tensor is used to describe the so-called geometrical nonlinearity, while the Murnaghan model with moduli (l, m, n) is taken into account for the so-called physical nonlinearity caused by elasticity of solids.

We have shown [1, 2] that in a long wave approximation the displacement u may be expanded in the power series of r,

$$u(x, r, t) = U(x, t) + r^2 C U_{xx} + ...,$$

where C is a constant defined by the material elastic features. Introducing the longitudinal strain: $v = U_x$, we obtained the governing nonlinear Double Dispersive Equation (DDE) for waves in a uniform rod, $r = R_0$:

$$v_{tt} - c v_{xx} = \left(\beta v^2 + a R_0^2 v_{tt} - b R_0^2 v_{xx}\right)_{xx}, \tag{1}$$

where a, b, c depend on the Lamé coefficients and the material density while the nonlinear parameter β depends also on the Murnaghan moduli. The terms in the l.h.s. of (1) constitute the linear wave operator, while terms in the r.h.s. govern nonlinearity and dispersive features of the rod. Eq. (1)

possesses an exact solution in the form of a localized bell-shaped travelling wave or the strain soliton,

$$v = \alpha \cosh^{-2}(k[\theta - \theta_0]). \tag{2}$$

where $\theta = x - Vt$, and α depends upon k and the equation coefficients.

More complicated equation has been obtained for the rod with slowly varying cross section: $R = R(\gamma x)$, and $\gamma << 1$ is a rate of nonuniformity [2]:

$$v_{tt} - c\,[(1/R^2)(R^2 v)_x]_x = \\ [(\beta R^2 v^2 + aR^4 v_{tt} - bR^4 v_{xx} - 4bR^3 R_x v_x)_x / R^2]_x, \tag{3}$$

Introduction of the phase variable θ as a function of a slow variable $X \equiv \gamma x$ by means of the coupled eqs. as follows: $\theta_t = -c, \quad \theta_x = A(X)$, lead to the asymptotic solution to Eq. (3) in new variables in the power series in $\gamma : v = v_0 + \gamma v_1 + \ldots$. In the leading order for v_0 we get from (3) the o.d.e. corresponding to the DDE, hence, v_0 is a strain soliton having slowly varying parameters,

$$v_0 = \alpha(X) \cosh^{-2}(k(X)[\theta - \theta_0(X)]). \tag{4}$$

while A, k are expressed in terms of α and R, see [2]. Then the type of the strain wave (4) (compression or tensile one) is defined by the sign of the nonlinear coefficient β.

Studying a distortion of the solitary strain wave due to the nonuniformity along the axis, we solved the nonlinear 1st order equation for α, see [2], that has an exact solution:

$$\frac{R^6 \alpha^3 \left[\nu + \alpha(2\nu - 6/5) - \alpha^2(1-\nu)\right]^2}{(1-\nu)[\nu - \alpha(1-\nu)]} = \text{const}, \tag{5}$$

that can be written in a simplest form for small α, together with relationships for other soliton parameters:

$$\frac{\alpha}{\alpha_0} = \left(\frac{R_0}{R}\right)^2, \quad A = 1, \; k^2 = \frac{\alpha}{4R^2(a-b)}. \tag{6}$$

The routine analysis showed that when the soliton (4) propagates along the tapered rod, its amplitude increases while its width decreases, and simultaneously, the pulse front becomes steeper while the back - smoother. Vice versa, the amplitude of the solitary wave, moving along the widening rod, decreases, the pulse width increases and the front becomes smoother, while the back -steeper.

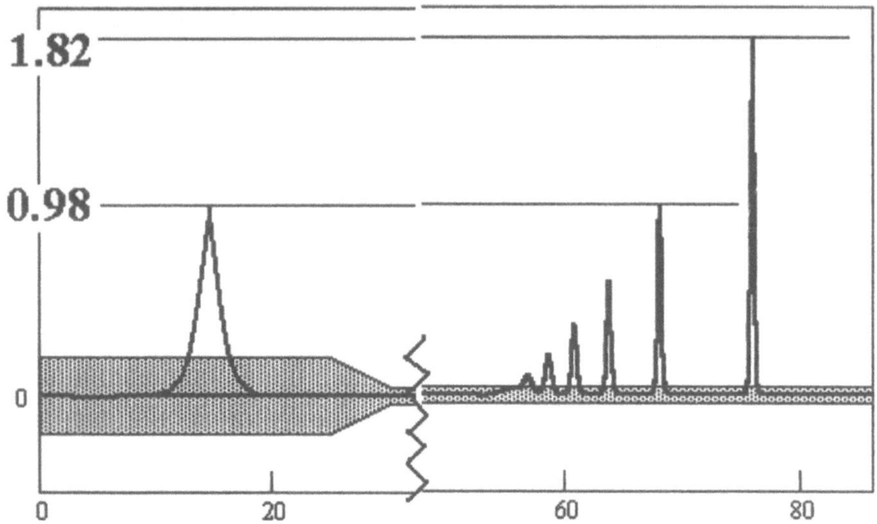

Figure 1: Recurrence of solitons and the train formation due to tapering of a rod.

3 Numerical simulations of strain solitons

The equations (1),(3) obtained are, in general, nonitegrable, therefore a solution to nonstationary problem of the soliton formation can be done by means of numerical simulations. In particular, the generation of strain solitons in a uniform rod from the initial rectangular pulse was studied, and it was found that depending of elasticity of the rod material the initial compression or tensile pulses are transformed into a train of solitons, and each of them is governed by (2).

These solitons are stable with respect to small perturbations and keep its shape when propagate, and even after interaction with other solitons. We studied the soliton evolution in a tapered rod [2], and it was shown that the strain soliton, being a reversible elastic perturbation (a bulk density wave), can provide not only transfer but also transformation of a wave energy. In a tapered (conical) rod there are two parameters for the soliton shape control: a length on which soliton is focusing (measured in its own length units) and the ratio of cross section areas. In dependence of them soliton can be transformed either into a train of sharp pulses (weak shocks), or, of 2÷8 secondary solitons (smooth deformation waves). The evolution of soliton in a rod with varying cross section was studied numerically in [3], and, being in the good agreement with the theory, the change of the soliton parameters

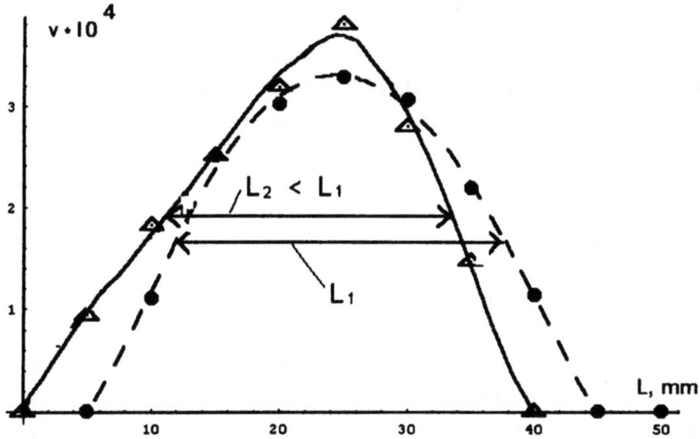

Figure 2: Steepening of the soliton front and simultaneous smoothing of the back. Dashed line used for interpolated data of soliton in the uniform rod, the solid one - of amplified soliton in the tapered rod.

was observed in either tapered or widening part of a rod. Moreover, being passed through a part with variable cross section, the soliton is recovered and again keeps its features in another uniform part of the rod, see Fig.1, while the first soliton in a train has almost doubled amplitude.

4 Experiments on strain solitons observation

The first generation and observation of the solitary nonlinear longitudinal strain wave (a soliton) were made in elastic rod by means of laser-produced impact generation and detection of solitons by means of holographic interferometry. The detailed description of experimental setup can be found in [4].

We have observed that initial weak shock wave entering the tip of the rod quickly transforms into the bell-shaped localized wave, that keeps the pulse width 6-7 times more then the cross section radius (5 mm in our experiments) and propagates with constant amplitude and velocity along the rod. Measurements of the wave parameters was shown to be in a good agreement with the predictions given in (2) for the strain solitary wave. It has to be noted that the rod made of polystyrene was used, i.e., of material, highly absorbing almost any elastic pulse, and widely used for protection against impacts, and then we shown the transparency of PS for solitary strain waves.

The amplification (focusing) of the soliton was observed in experiments in a perfect agreement with the theory, briefly described above, including the steepening of the front of the strain soliton and simultaneous smoothing of its back, as it shown in Fig.2.

5 Conclusions

The possibility was shown to transfer elastic energy at a long distance without considerable losses even in materials, having remarkable absorption (dissipation) features with respect to elastic shocks, nothing to say about harmonic waves. We proved in experiments that the elastic strain soliton does not dissipate even at the distances much greater than the typical linear dissipation length for polystyrene. Presumably, it means that the nonlinear absorption is much less than the linear one, which does not affect the strain soliton, however this problem requires further analysis.

The recurrence of such 'an energy cluster' as the strain soliton and focusing of them, as it was shown in Fig.1, provides an opportunity to develop a new striker without considerable energy absorption and to fracture another material at the end of a tapered wave guide, using nonlinear localized or periodical (cnoidal) waves generation.

Due to lack of space the problem on waves in the rod, having varying cross section and free lateral surface, was considered only. However, the strain solitons were observed also in a nonlinearly elastic plate, see [5]. Another interesting problem is to study tentative influence of an external surrounding medium on the soliton, propagating in an embedded wave guide, that is quite common problem for seismology. First attempt to study it was already done, when the sliding contact on the lateral surface of the rod is assumed, see [6].

The support by the Russian Fund for Basic Research (grants 98-01-01109 and 98-01-10906) is gratefully acknowledged.

References

1 Samsonov, A.M. Nonlinear strain waves in elastic waveguides. In: "Nonlinear waves in solids"; Jeffrey, Engelbrecht eds., Springer, 1994, 349- 382.
2 Samsonov, A.M. et al., Phys. Rev. B**57**, 5778-5787, (1998).
3 Bukhanovsky, A.V., Samsonov, A.M. Numerical simulation of strain solitons in elastic wave guides using vector and massive parallel supercomputers. In: Lect. Notes in Comp. Science, **1401,** 869-871 (Springer, 1998).
4 Dreiden, G.V. et al, Tech. Phys. Lett. **21,** 415 (1995).
5 Samsonov, A.M et al, Tech. Phys. Lett. **22,** 891(1996).
6 Porubov, A.V. et al, Phys. Rev. E **58**, 3854-3865 (1998).

Multiple impacts of a bar with external dry friction

*L.V.Nikitin**

Vibration due to impact of an elastic bar interacting with surrounding in accordance to the Coulomb dry friction law is considered. Qualitative analysis of the non-linear equations governing the problem is given. Multiple impacts of a finite and semi-infinite bars are studied.

1 Introduction

The simplest widely known idealisation of the vibro-impact system is a mass elastically joint by a spring with a rigid wall. Taking into account dry friction approaches this idealisation to reality but brings to rather nontrivial mathematical problem for discontinuous vibrations. Here extension of the idealisation on one dimensional continuum, i.e. on an elastic bar interacting with environment according to the Coulomb dry friction law is considered. It turned out that this more advanced idealisation permits to obtain a number of exact solutions for important for engineering application problems[1-5]. Impact with constant stress in the frame of this model firstly was considered in [1], impact by a moving mass in [2] and impact with constant velocity in [3]. In all cases exact solutions were obtained. Experience of solving of these problems showed that in contrast to gas dynamics fronts of shock of waves propagate with the sound velocity while fronts of weak waves may propagate with any velocity. Here we give qualitative analysis of governing equations and prove the stated facts. Using these findings exact solutions for two problems of multiple impact are obtained:

1) multiple step-like resonance vibro-impact of a finite bar and

2) vibro-impact of a semi-infinite bar performed by the piecewise linear periodical load. Analysis of results of these problems permits to conclude regarding influence of dry friction on resonance and dissipation.

2 Basic equations and their qualitative analysis

Equation of longitudinal motion of a bar interacting with environment in accordance with the Coulomb dry friction law is

$$\frac{\partial \sigma}{\partial x} = \rho \frac{\partial v}{\partial t} + \kappa\tau \tag{1}$$

Here σ and v are the normal stress and velocity of a cross section located at position x at time t, ρ is density, τ is the reduced friction force at the bar

---

* United Institute of Physics of the Earth, Russian Academy of Sciences, Moscow, Russia

surface. The dry friction switch κ in the case of motion is the sign of velocity

$$\kappa = \operatorname{sgn} v, \qquad v \neq 0 \tag{2}$$

When $v = 0$ the quantity κ is unknown and is to be found from the equation of static equilibrium

$$\frac{d\sigma}{dx} = \kappa\tau \tag{3}$$

The second equation for unknowns σ and v is derived from the Hooke's law after differentiation it in respect of time

$$\frac{\partial\sigma}{\partial t} = E\frac{\partial v}{\partial x} \tag{4}$$

where E is the Young module.

In terms of the displacement u instead of the system of the equations (1) we have a single equation

$$a^2\frac{\partial^2 u}{\partial x^2} = \frac{\partial^2 u}{\partial t^2} + \frac{\kappa\tau}{\rho} \tag{5}$$

where $a = (E/\rho)^{1/2}$ is the bar sound velocity. The quantity κ being a function of velocity is unknown, sought for function. Sometimes the sign of velocity is evident from the formulation of the problem. However even in these cases the problem is nontrivial since it is necessary to find an unknown boundary between regions of motion and rest.

Characteristics of the system (1), (4) or, what is the same, of the equation (5) and the relationships along them are

$$dx = \pm a dt, \qquad \pm\, d\sigma + a\rho dv + \kappa\tau a dt = 0 \tag{6}$$

From the system (1), (4) it follows that at fronts of strong discontinuity, i.e. discontinuity of σ and v propagating with the velocity c, jumps of stress $\langle\sigma\rangle$ and velocity $\langle v\rangle$ must satisfy the equations

$$\langle\sigma\rangle + \rho c\langle v\rangle = 0, \qquad c\langle\sigma\rangle + E\langle v\rangle = 0 \tag{7}$$

Here $\langle\sigma\rangle = \sigma^+ - \sigma^-$, where superscripts "+" and "-" refer a quantity to the side behind and beyond a shock front, respectively.

The system (7) has nontrivial solution when $c = a$ only. Thus strong discontinuities propagate along the characteristics of the system (6). According to (7) along characteristics the next relations are hold

$$\langle\sigma\rangle \pm a\rho\langle v\rangle = 0, \qquad x = \pm at + const \tag{8}$$

However wave fronts of weak discontinuities i.e. derivatives of σ and v may not coincide with the characteristics. Let as discuss this problem in more details.

Let $x = x_*(t)$ is a front separating domains of rest and motion. Inside a domain of rest including the front we have

$$v = 0, \qquad \sigma = \sigma_*(x) \tag{9}$$

Stress $\sigma_*(x)$ in the domain of rest must be in equilibrium with the friction force

$$\frac{d\sigma_*(x)}{dx} = \kappa_* \tau$$

Differentiation of velocity and stress in respect of x along the front $x = x_*(t)$ gives

$$\frac{\partial v}{\partial x} + \frac{\partial v}{\partial t}\frac{dt}{dx_*} = 0, \qquad \frac{\partial \sigma}{\partial x} + \frac{\partial \sigma}{\partial t}\frac{dt}{dx_*} = \frac{d\sigma_*}{dx} \tag{10}$$

The ifferences of these expressions across the front $x = x_*(t)$ are

$$a_*\left\langle \frac{\partial v}{\partial x} \right\rangle + \left\langle \frac{\partial v}{\partial t} \right\rangle = 0, \qquad a_*\left\langle \frac{\partial \sigma}{\partial x} \right\rangle + \left\langle \frac{\partial \sigma}{\partial t} \right\rangle = 0 \tag{11}$$

where $a_* \equiv dx_*/dt$ is the velocity of the front. The ifferences of equations (1) and (4) across the front with account of (3) give two more equations

$$\left\langle \frac{\partial \sigma}{\partial x} \right\rangle = \rho\left\langle \frac{\partial v}{\partial t} \right\rangle + (\kappa - \kappa_*)\tau, \qquad \left\langle \frac{\partial \sigma}{\partial t} \right\rangle = E\left\langle \frac{\partial v}{\partial x} \right\rangle \tag{12}$$

The set of the equations (11) and (12) constitutes the complete non homogeneous system of linear equations for determination of the jumps of derivatives along the front $x = x_*(t)$. If $a_* \neq a$, then the determinant of the system is nonzero and it has an unique solution. Resolving the system in respect of the jump of acceleration we obtain

$$\left\langle \frac{\partial v}{\partial t} \right\rangle = \frac{a_*^2 \tau(\kappa - \kappa_*)}{\rho\left(a^2 - a_*^2\right)} \tag{13}$$

From (13) the very important for the subsequent analysis expression for the velocity of the wave front is obtained

$$a_* = \pm a\left(1 + \frac{\tau(\kappa - \kappa_*)}{\rho\langle \partial v/\partial t \rangle}\right)^{-1/2} \tag{14}$$

First of all from (14) it follows that in general when, $\kappa \neq \kappa_*$, fronts of weak discontinuity, along which by definition acceleration is bounded, propagate with velocity different from a. When $\langle \partial v/\partial t \rangle \to \infty$ the previous conclusion that strong discontinuities propagate with the sound velocity a is confirmed.

If the front $x = x_*(t)$ separates a region of rest from that of motion, i.e. is the leading front of motion then the sign of κ in the vicinity of the front coincides with the sign of acceleration and hence with the jump of acceleration since in the region of rest acceleration is zero. Then

$$\left(\kappa - \kappa_*\right)/\langle \partial v/t \rangle > 0 \quad \text{and} \quad |a_*| \leq a \tag{15}$$

Thus the leading front of motion propagates with velocity the module of which does not exceed the sound velocity a.

In the contrary if the front $x = x_*(t)$ separates a region of motion from that of rest i.e. is the back front or the front of the motion arrest then the signs of velocity and acceleration in the vicinity of the front are opposite. Hence

$$\left(\kappa - \kappa_*\right)/\langle \partial v/\partial t \rangle < 0 \quad \text{and} \quad |a_*| \geq a \tag{16}$$

Thus the front of motion arrest propagates with velocity the module of which exceeds the sound velocity a.

In particular velocity of fronts of the motion arrest may be infinite. In the case of the limiting preliminary stress $|\kappa| = 1$ and both the leading and the back fronts propagate with the sound velocity a.

From (14) it is plane that deviation of fronts of weak discontinuity is due to discontinuity of dry the friction force or κ along the boundary between the regions of motion and rest.

In terms of the characteristic variables

$$\xi = at - x, \qquad \eta = at + x$$

the equation (5) takes the form

$$4a^2 \frac{\partial^2 u}{\partial \xi \partial \eta} = \frac{\kappa \tau}{\rho} \tag{17}$$

Integrating of (17) brings to extension of the d'Alambert solution on taking into account friction

$$\sigma = \frac{1}{2}\kappa\tau x + f(at + x) - \varphi(at - x)$$

$$a\rho v = -\frac{1}{2}\kappa\tau a t + f(at + x) + \varphi(at - x) \tag{18}$$

$$Eu = -\frac{1}{4}\kappa\tau\left(a^2 t^2 - x^2\right) + \int_0^{at+x} f(\eta)d\eta + \int_0^{at-x} \varphi(\xi)d\xi$$

where f and φ are arbitrary functions. In some cases direction of velocity may be guessed hitherto. Then with the help of the above proved statements solutions (18) may be found in the closed form.

3 Multiple stepwise resonant type impacts of a finite bar

Consider firstly a multiple impact of a bar with length l one end of which, $x = l$, is clamped while at the other end, $x = 0$, the applied stress changes sign stepwisely with the period of the free vibrations $4l/a$

$$\sigma = \sigma_0 H(t) + 2\sigma_0 \sum_{k=1}^{\infty} (-1)^k H\big(at - 2(k-1)l\big) \qquad \text{for } x = 0 \tag{19}$$

where $H(t)$ is the Heavyside unit step function. At the initial moment, $t = 0$, a bar is assumed to be at rest and free of stress

$$v = 0, \quad \sigma = 0 \qquad \text{for} \quad t = 0 \tag{20}$$

The difficulties of solving this non-linear problem consist of the dependence of κ on the unknown velocity. However in the case under consideration κ may be established beforehand. This is based on the fact that since the friction force is passive by its physical nature, it can not change the sign of velocity Therefore it should be expected in problems with stepwise in time loading in which wave fronts propagate along characteristics, that the distribution of velocity signs is the same as in the corresponding problems without friction. This guess was checked in [3]. Omitting details here we write down only the final result for κ in our case

$$\kappa = -\sum_{k=1}^{\infty} (-1)^k H\big(at - x - 2(k-1)l\big) + \sum_{k=1}^{\infty} (-1)^k H\big(at + x + 2kl\big) \tag{21}$$

Substitution of κ from (21) into the equation (1) reduces the initially non-linear problem to solving of linear equations. It is then possible to use the Laplace transform. The Laplace transforms of equations (1) and (4) with κ according to (21) have the form

$$\frac{d\bar{\sigma}}{dx} = p\rho\bar{v} + \tau\frac{\sinh p(l-x)/a}{\cosh pl/a}, \qquad E\frac{d\bar{v}}{dx} = p\bar{\sigma} \tag{22}$$

where a bar over a letter denotes the Laplace transform. The boundary condition (21) reduces to

$$\bar{\sigma} = -\sigma_0 \tanh\frac{pl}{a} \qquad \text{for} \quad x = 0 \tag{23}$$

the solution of the equation (22) obeying to the initial condition (20) and boundary condition (23) is the next

$$\begin{aligned} \bar{\sigma} &= -\sigma_0\frac{\sinh pl/a \cosh p(l-x)/a}{\cosh^2 pl/a} + \frac{\tau l}{2}\frac{\sinh px/a}{\cosh^2 pl/a} - \\ &\quad -\frac{\tau x}{2}\frac{\sinh p(l-x)/a}{\cosh pl/a} \\ a\rho\bar{v} &= \sigma_0\frac{\sinh pl/a \sinh p(l-x)/a}{\cosh^2 pl/a} + \frac{\tau l}{2}\frac{\cosh px/a}{\cosh^2 pl/a} - \\ &\quad -\frac{\tau a}{p}\frac{\sinh p(l-x)/a}{\cosh pl/a} \end{aligned} \tag{24}$$

The inverse transforms of these rather cumbersome expressions are very simple in the concrete regions. Consider firstly the regions of the type

$$x + 2nl < at < -x + 2nl, \qquad 0 < x < l$$

where n =1,2... is the number of the wave fronts. Numeration of direct fronts starts from the front $x = at$ while inverse fronts from the front $x + at = 2l$. After calculation of contributions of direct and inverse waves taking part in generation of motion and some rearrangement we obtain

$$\sigma = (-1)^n\left(\sigma_0 - \frac{\tau x}{2}\right), \qquad a\rho v = (-1)^{n-1}\left((2n-1)\sigma_0 - \frac{\tau at}{2}\right) \qquad (25)$$

For the regions of the type

$$-x + 2nl < at < x + 2nl; \qquad 0 < x < l$$

solution is

$$\sigma = (-1)^n n(2\sigma_0 - \tau l), \qquad v = 0 \qquad (26)$$

The considered regions completely cover the domain where solution exists. Fig.1 shows the.displacement of the rod end in the presence and absence of friction. The periodical loading (19) produces resonance in the absence of friction, Fig.1. It is seen that the dry friction does not prevent resonance but rather retards appearance of large amplitudes.

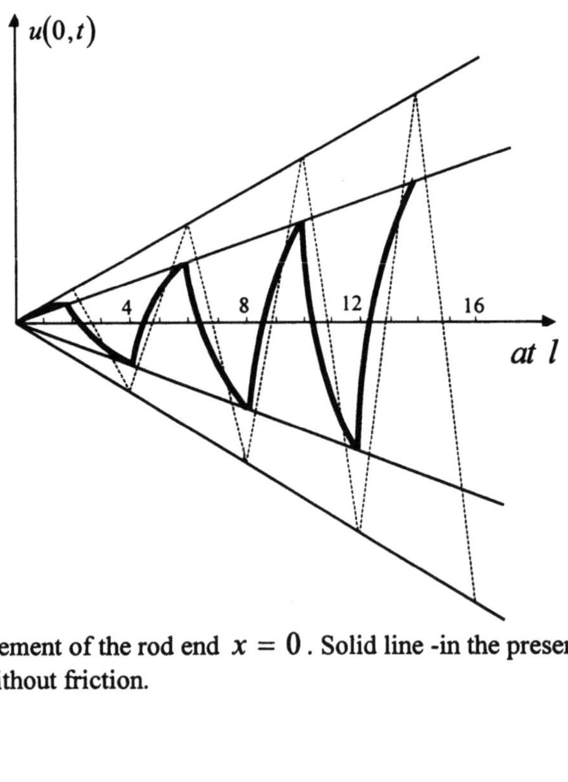

Fig.1. Displacement of the rod end $x = 0$. Solid line -in the presence of friction, dashed line-without friction.

4 Vibro-impact of a bar produced by a periodical saw-tooth stress impulse

Consider now stationary vibrations of a semi-infinite bar produced by the sign alternating saw-tooth stress impulse $\sigma_0(t)$ with the period T and amplitude σ_m. The length of the bar is taken long enough so that disturbances would cease before they reach the remote end. Due to piece linearity of the exciting stress wave fronts are expected to be rectilinear and solutions σ and v piece linear in terms of x and t [4]. We suggest the wave fronts and regions of motion and stagnation are as shown in Fig.2

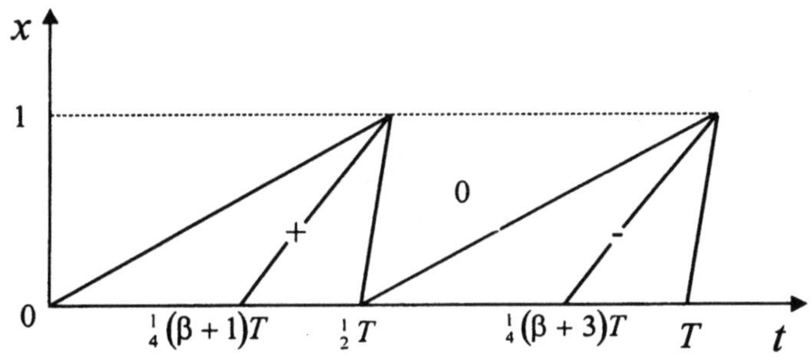

Fig.2. Wave fronts, regions of motion and stagnation in the case of saw-tooth excitation. The signs "+" and "-" mark regions with positive and negative velocities, respectively, the sign "0" marks the region of stagnation.

In the region "+" velocity is positive, in the region "-" it is negative. Between the regions of motion with different signs of velocity the region of stagnation is situated. According to the above proved velocity of the wave front separating stagnation and motion is $a_1 \leq 1$, separating motion and stagnation is $a_2 \geq 1$. Due to the skew-symmetry of the stress impulse motion in the regions "+" and "-" differs by sign only. Therefore it is enough to consider only a half of the period, Fig.2. Stress in the region of stagnation can be presented in the form

$$\sigma_* = \beta\sigma_m - \kappa_0 \tau x \tag{27}$$

where β is the phase difference between stress and velocity and κ_0 determines the friction stress in the regions of stagnation. Then κ take the form

$$\kappa = H(x - a_1 t) - (1 - \kappa_0) H\left(x - a_2\left(t - \frac{1}{2}T\right)\right) \tag{28}$$

In virtue of the periodicity at the boundaries of the region of motion the next conditions are hold

$$\sigma = \sigma_*(x), \quad v = 0 \qquad \text{along } x = a_1 t$$
$$\sigma = -\sigma_*(x), \quad v = 0 \qquad \text{along } x = a_2\left(t - \frac{1}{2}T\right) \tag{29}$$

Fig.3 shows the change of stress and velocity during a period. The boundary condition at the end of the bar for stationary motion has the form

$$\sigma^0(t) = \sigma_m\left(\beta - \frac{4t}{T}\right)H(t) + 2\sigma_m\left\{\frac{4t}{T} - \beta - 1\right\} \tag{30}$$

where here and in what follows figure brackets mean $\{z\} = zH(z)$.

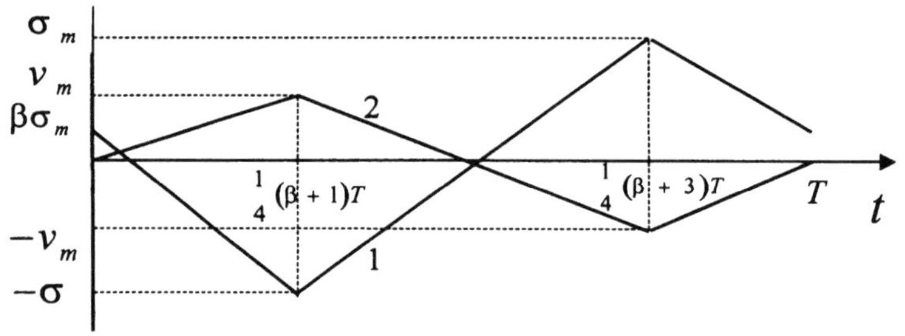

Fig.3. Change of stress (1) and velocity (2) during a period of vibrations.

Solution of the equations (1), (4) with κ taken from (27) and satisfying (28-30) in the region "+" left to the characteristic $x = a(t - (\beta + 1)T/4)$ is

$$\sigma = \beta\sigma_m - \kappa_0 \tau x - \frac{4\sigma_m}{T}\left(t - \frac{x}{a_1}\right), \qquad v = \frac{4\sigma_m}{ET}(a_1 t - x) \tag{31}$$

In the region "+" right to the characteristic $x = a(t - (\beta + 1)T/4)$ solution is

$$\sigma = \frac{4\sigma_m}{T}\left(t - \frac{2 + \beta}{4}T\right) + \left(1 - \frac{a_2\omega}{a}\right)\tau x, \tag{32}$$
$$v = \frac{4\sigma_m}{ET}\left(a_2\left(\frac{1}{2}T - t\right) + x\right)$$

Here dimensionless frequency $\omega = 4\sigma_m / aT\tau$ is introduced while parameters β, a_1 and a_2 are found to be

$$\beta = (2\omega + \kappa_0)^{-1}, \quad a_1 = (1 - \beta)a, \quad a_2 = (1 + \beta)a \tag{33}$$

where κ_0 is a root of the next eqvation satisfying the inequality $0 < \kappa_0 < 1$.

$$\kappa_0^3 + 4\omega\kappa_0^2 + \left(4\omega^2 - 1\right)\kappa_0 - \omega = 0 \tag{34}$$

The depth of disturbances penetration l and the maximum of the velocity amplitude v_m, respectively, are

$$l = \frac{a_1 a_2 T}{4\beta a}, \qquad v_m = \frac{a_1 a_2 \sigma_m}{aE} \tag{35}$$

The displacement of the bar end $u^0(t)$ has the form

$$u^0(t) = -\frac{a_1 a_2 \sigma_m T}{8Ea} + \frac{2a_1\sigma_m}{ET}t^2 - \frac{4a\sigma_m}{ET}\left\{t - \frac{a_2 T}{4a}\right\}^2 + \tag{36}$$
$$+\frac{4\beta a\sigma_m}{ET}\left(t - \frac{1}{2}T\right)^2 H\left(t - \frac{1}{2}T\right) + \frac{4a\sigma_m}{ET}\left\{t - \frac{3+\beta}{4}T\right\}^2$$

The maximum displacement amplitude is

$$u_m = \frac{a_1 a_2 T_m}{8aE} \tag{37}$$

Fig.4 shows the hysteresis loop plotted according to (31), (32) and (37).

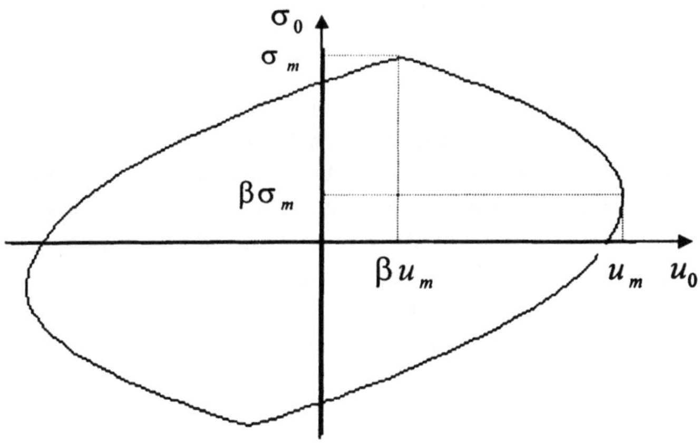

Fig.4

The energy dissipation W per cycle of vibration is

$$W = \int_0^T \sigma^0(t) v^0(t) dt = \frac{a_1 a_2 \left(2 - \beta^2\right)\sigma_m^2 T}{6aE} \tag{38}$$

In the case of the low frequency $\omega << 1$ the pertinent parameters of the process are

$$l \cong \frac{\sigma_m}{\tau}, \quad u_m \cong \frac{\sigma_m^2}{2\tau E}, \quad v_m \cong \frac{\omega a \sigma_m}{E}, \quad W \cong \frac{2}{3}\frac{\sigma_m^3}{E\tau} \tag{39}$$

In this approximation dissipation does not depend on the frequency and is the same as for quasistatics. In the case of the high frequency $\omega >> 1$

$$l \cong \frac{2\sigma_m}{\tau}, \quad v_m \cong \frac{a\sigma_m}{E}, \quad u_m = \frac{a\sigma_m T}{8E}, \quad W \cong \frac{4}{3}\frac{\sigma_m^3}{\omega E\tau} \tag{40}$$

In this case dissipation is inversely proportional to the frequency while the depth of penetration of disturbances is twice of the quasistatic one. For the fixed period T the asymptotic solutions for $\omega << 1$ and $\omega >> 1$ may be interpreted as cases of strong and weak damping respectively.

Acknowledgement

Support of the Russian Foundation for Basic Researches under the grant 96-05-64347 is gratefully acknowledged.

References

1. Nikitin L.V. Wave propagation in an elastic rod in the presence of dry friction. Engineering Journal, USSR Acad., Sc.**1**, 3, pp.126-130, 1963 (in Russian).
2. Nikitin L.V. Impact by a rigid body of an elastic rod with external dry friction. Engineering Journal. Solid Mechanics, **2**, pp.166-170, 1967 (in Russian).
3. Nikitin L.V., Tyurekhodgaev A.N. Wave propagation and vibration of elastic rods with interfacial frictional slip. Wave motion, **12**, pp.513-526, 1990.
4. Mogilevsky R.I., Nikitin L.V. Dynamics of rods with interfacial dry friction. Journal of Mechanical Behaviour of Materials, **5**, 1, pp.85-93, 1993.
5. Nikitin L.V. Statics and dynamics of solids with external dry friction. Moskovsky Licey, Moscow, 272p., 1998 . (in Russian).

Solving wave propagation problems symbolically using computer algebra

*B. Hu**, *P. Eberhard** *and W. Schiehlen**

Wave propagation analysis of longitudinal impacts of rods has fascinated many famous scientists over centuries like Newton, Cauchy, Poisson and St. Venant. In this paper, it is shown that such complicated problems can be solved much more easily using computer algebra systems. As an example, the program MAPLE is used to analyze the wave propagation in a thin rod struck by a moving rigid body. Using only a few of MAPLE's powerful commands, we can get analytical results about time responses of physical quantities such as displacements and stresses which has taken years of research in the past. It is found that some data appearing in a number of textbooks are not correct and fairly restricted. Moreover, amplitude bounds for the maximum displacements of the rod with one fixed end are presented.

1 Introduction

Digital computers are a fundamental basis for numerical computations. Besides, in recent decades it has been realized more and more that computers can also be of great importance for symbolical computations. Nowadays there exist many computer algebra systems, e.g. MAPLE, MATHEMATICA, MACSYMA, DERIVE or REDUCE. The main advantages of computer algebra systems are their ability to handle complicated algebraic computations and the fact that results of symbolic computations are exact and not subject to approximation errors. Often it is advantageous in terms of computation time to simplify expressions algebraically before evaluating them numerically. This way, the expressions become simpler and less prone to numerical errors. In addition to their symbolic capabilities, computer algebra systems usually also provide numerical capabilities for dealing with numbers and graphical features to visualize computation results. This makes computer algebra systems powerful tools for scientific computation. During the last decades, formula manipulation systems have been widely applied in many fields. In applied dynamics, e.g. symbolical equations of the motion for modeling of multibody systems can be obtained using the software package NEWEUL, see []. In this paper we investigate the impact dynamics of a fairly simple vibro-impact system using computer algebra methods. Here the program system MAPLE [1] is used to analyze the wave propagation due to longitudinal impacts on rods.

This problem has fascinated many famous scientists, see Szabó [2]. Vibrations resulting from longitudinal impacts were first examined by Bernoulli,

*Institute B of Mechanics, University of Stuttgart, Pfaffenwaldring 9, 70550 Stuttgart, Germany. Email: [hbi,pe,wos]@mechb.uni-stuttgart.de

Navier and Poisson. A complete treatment was provided by St. Venant [3] and later by Boussinesq [4]. Many results for longitudinal waves in rods produced by impacts were reviewed by Goldsmith [5] and Timoshenko and Goodier [6]. In the form of Laguerre polynomials, some solutions were given in Nikitin [7]. In this paper we analyze in detail the wave propagation in a thin rod struck by a moving rigid body and show how to obtain symbolical results for time responses of physical quantities such as displacements, velocities, stresses and strains at an arbitrary position along the rod. Moreover, we also use MAPLE to compute numerical data for the duration of contact as well as the coefficient of restitution. The wave motion in the rod is visualized and animated. It is shown that even such complicated impact problems, which famous scientists in the past kept busy for years of research, can be solved with relatively few effort using computer algebra. It is found that some data appearing in standard textbooks such as Goldsmith [5] and Timoshenko and Goodier [6] are not correct. Moreover, amplitude bounds for the maximum displacements of the rod with one end fixed are presented.

2 Mathematical modeling of longitudinal impacts

Consider a slender homogeneous rod struck at contact point $x = 0$ by a moving rigid body with mass m_2 and initial velocity v_2. At the point $x = L$ the rod may be free or fixed as shown in Fig. 1. The rod is initially at rest. It has Young's modulus E, density ρ, cross-sectional area A, length L and mass m_1, respectively. It is assumed for simplicity that transverse waves of the rod and vibrations of the striking body can be neglected and the contact surface is considered as perfectly plane, see [5].

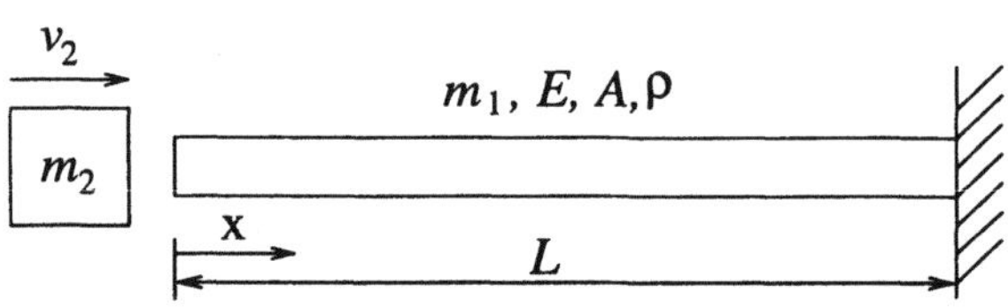

Figure 1: Longitudinal impact of a mass on a rod

Important quantities are the axial displacement $u(x,t)$ and the stress $\sigma(x,t)$ in the rod at position x and time t. The following relations are used to compute the velocity

$$v(x,t) = \dot{u}(x,t) = \frac{\partial u(x,t)}{\partial t} \tag{1}$$

and the strain

$$\epsilon(x,t) = \frac{\partial u(x,t)}{\partial x} \; . \tag{2}$$

The material is described by Hooke's law for linear elasticity

$$\sigma(x,t) = E\epsilon(x,t) \; . \tag{3}$$

The governing partial differential equation for the longitudinal wave is

$$\frac{\partial^2 u(x,t)}{\partial t^2} = c^2 \frac{\partial^2 u(x,t)}{\partial x^2} \ , \tag{4}$$

where $c = \sqrt{E/\rho}$ is the wave propagation velocity which can be calculated from material constants. The general solution of the wave equation first presented by D'Alembert reads as

$$u(x,t) = f(ct - x) + g(ct + x) \ , \tag{5}$$

where f and g are real functions. Prior to the impact with the rigid body, the rod is at rest, i.e. for $0 \le x \le L$, we have

$$u(x,0) = f(-x) + g(x) = 0 \ , \tag{6}$$
$$v(x,0) = cf'(-x) + cg'(x) = 0 \ , \tag{7}$$

where $f'(\bullet)$ and $g'(\bullet)$ denote the derivative of $f(\bullet)$ and $g(\bullet)$ with respect to $\bullet$. Integrating Eq. (7) yields

$$-f(-x) + g(x) = C \ , \tag{8}$$

where C is a constant. From the sum and the difference of Eqs. (6) and (8), we obtain for $0 \le x \le L$

$$f(-x) = -C/2 \quad \text{and} \quad g(x) = C/2 \ . \tag{9}$$

The constant C can be set to any value since it has no influence on the displacement $u(x,t)$. For simplicity, here C is set to zero. Therefore, we get

$$g(x) = f(-x) = 0 \quad \text{and} \quad g'(x) = f'(-x) = 0 \quad \text{for} \quad 0 \le x \le L \ . \tag{10}$$

At the instant of the impact, the velocity of particles at the contact end of the rod ($x = 0$) is assumed to become v_2 immediately and then a compressive wave is produced. It travels forward along the rod and is reflected at the other end $x = L$. As long as the striking body and the rod have persistent contact, a resistance force $A\sigma(0,t)$ from the rod is acting on the striking body. Its displacement $u_2(t)$ and velocity $\dot{u}_2(t)$ are the same as those of the rod at the contact end $x = 0$, i.e. for $t > 0$ we have

$$u_2(t) = u(0,t) \quad \text{and} \quad \dot{u}_2(t) = \dot{u}(0,t) \ . \tag{11}$$

The motion of the striking body is therefore determined by

$$m_2 \frac{d\dot{u}_2(t)}{dt} = A\sigma(0,t) \tag{12}$$

with initial conditions $u_2(0) = 0$ and $\dot{u}_2(0) = v_2$. Defining a new parameter

$$\alpha := \frac{EA}{m_2 c}\frac{L}{c} = \frac{m_1}{m_2} \tag{13}$$

representing the mass ratio of the striking body to the rod, and using Eqs. (2), (3) and (11), the equation of the motion (12) for the striking body turns into

$$\frac{d}{dt}[f'(ct) + g(ct)] + \frac{\alpha c}{L}[f'(ct) + g'(ct)] = 2\frac{\alpha c}{L}g'(ct) \ . \tag{14}$$

Substituting $y = ct$ into Eq. (14) and solving this differential equation, we obtain

$$f'(y) + g'(y) = [f'(y_0) + g'(y_0)]e^{-\frac{\alpha}{L}(y-y_0)} + 2\frac{\alpha}{L}ce^{-\frac{\alpha}{L}y} \int_{y_0}^{y} e^{\frac{\alpha}{L}y}g'(y)dy \ , \tag{15}$$

where y_0 denotes the arbitrary initial value of y. Setting $y_0 = 0$ and using Eqs. (10) and (15) yields

$$f'(y) = \frac{v_2}{c}\,\mathrm{e}^{-\frac{\alpha}{L}y} \quad \text{and} \quad f(y) = \frac{v_2 L}{\alpha c}\left(1 - \mathrm{e}^{-\frac{\alpha}{L}y}\right) \quad \text{for} \quad 0 < y < L \ . \tag{16}$$

The stress at the struck end $\sigma(0,t)$ is compressive. When it becomes equal to zero, the impact terminates. Denoting this instant by t_c, we certainly have

$$\sigma(0,t) < 0 \quad \text{for} \quad 0 < t < t_c \quad \text{and} \quad \sigma(0,t_c) = 0 \ . \tag{17}$$

If the time t_c, i.e. the duration of contact, is available, the coefficient of restitution

$$e := -\dot{u}(0,t_c)/v_2 \tag{18}$$

can also be computed.

After the impact, the struck end of the rod is free and the stress at this end must vanish. The boundary condition for the struck end then becomes

$$\sigma(0,t) = 0 \quad \text{for} \quad t \geq t_c \tag{19}$$

and results in

$$f'(ct) = g'(ct) \quad \text{for} \quad t \geq t_c \ . \tag{20}$$

The boundary condition at the struck end is given by Eqs. (17) and (19). However, the solution of the wave equation also depends on the boundary condition of the rod at the other end. In the following sections, we first discuss the impact responses of the rod with one end free and then with one end fixed at $x = L$.

3 Impact response of the rod with one end free

For the rod with one end free, the stress at the end $x = L$ is always equal to zero. The boundary condition then reads as

$$\sigma(L,t) = 0 \ , \quad \forall t \ . \tag{21}$$

Using Eq. (21) yields

$$-f'(ct-L)+g'(ct+L)=0\,. \tag{22}$$

Integrating Eq. (22) and using Eq. (10), we have then

$$f(ct-L)=g(ct+L)+f(-L)-g(L)=g(ct+L)\,. \tag{23}$$

Hence, the function g has the following relation with the function f

$$g(y)=f(y-2L)\quad\text{for}\quad y\geq L\,. \tag{24}$$

The stress at the contact end now reads as

$$\sigma(0,t)=E[-f'(ct)+f'(ct-2L)]\,. \tag{25}$$

When the reflected wave reaches the contact end, the stress $\sigma(0,t)$ turns into zero instantly and the contact terminates. Hence, for a rod with one end free, we can easily obtain the duration of contact $t_c=2L/c=:T$ and the coefficient of restitution $e=\mathrm{e}^{-2\alpha}$. After the impact, i.e. for $y>2L$, it follows from Eqs. (19) and (25)

$$f'(y)=f'(y-2L)\ \text{and}\ f(y)=f(y-2L)+\frac{v_2L}{\alpha c}(1-e)\,. \tag{26}$$

Therefore, the stress $\sigma(x,t)$ and the velocity $v(x,t)$ are periodic with period T. The displacement $u(x,t)$ is composed of a wave motion and a rigid body motion with average velocity $v_2(1-e)/\alpha$. The more energy the striking body looses during the impact, the greater is the average velocity of the rod. Using MAPLE, we can easily visualize the wave motions in the rod and even animate them. For $\alpha=1$, Fig. 2 shows the dimensionless displacements and stresses, respectively. The motion at the two ends of the rod is shown in Fig. 3. It can also be seen that the free rod has a rigid motion.

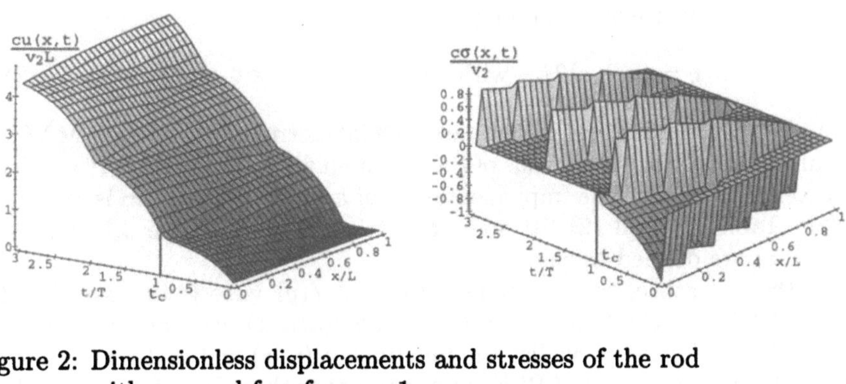

Figure 2: Dimensionless displacements and stresses of the rod with one end free for $\alpha=1$

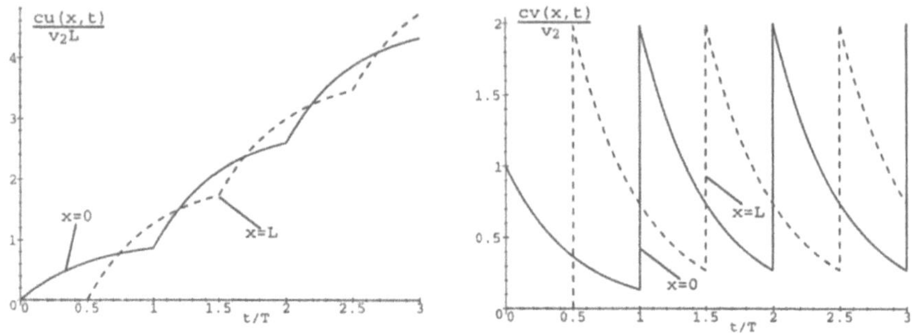

Figure 3: Dimensionless displacements and velocities at the two ends of the rod with one end free for $\alpha = 1$

4 Impact response of the rod with one end fixed

For the rod with one end fixed, the displacement at the end $x = L$ is equal to zero. The boundary condition reads as

$$u(L,t) = 0 \ , \quad \forall t \tag{27}$$

and results in

$$g(y) = -f(y-2L) \quad \text{for} \quad y \geq L \ . \tag{28}$$

It was proven that the function $f'(y)$ and $f(y)$ can be written as

$$f'(y) = \frac{v_2}{c} \mathrm{e}^{-2\alpha z} p_s(n,z) \ , \tag{29}$$

$$f(y) = \frac{Lv_2}{\alpha c} \left(\mathrm{e}^{-2\alpha z} p_{ss}(n,z) + C_{ss}(n)\right) \ , \tag{30}$$

where $p_s(n,z)$ and $p_{ss}(n,z)$ are two polynomials of order n and $C_{ss}(n)$ are constants, see [8]. The variable y and the couple (n,z) are related with each other by the one–to–one mapping

$$y = (n+z)2L \quad \text{with} \quad n \in \mathbf{Z}^+ \quad \text{and} \quad z \in [0,1) \ , \tag{31}$$

where $\mathbf{Z}^+$ denotes the set of non-negative integer numbers. Using MAPLE, we can compute the expressions of the polynomials $p_s(n,z)$ and $p_{ss}(n,z)$ with relatively few effort. The implementation of a MAPLE program is described in [8]. Table (1) and (2) list some analytical results, where $p_s(-1,z)$ and $p_{ss}(-1,z)$ are defined as zero.

In the literature, the functions $f'(y)$ and $f(y)$ were given explicitly for $0 \leq y \leq 8L$, i.e. the functions $p_s(n,z)$ and $p_{ss}(n,z)$ were given for $n \leq 3$, see [5]. In the following we list the functions $p_s(n,z)$ and $p_{ss}(n,z)$ explicitly for $n \leq 6$. In fact, using MAPLE, we can compute them for arbitrary numbers of n restricted only by the available memory of the used computer.

Table 1: Analytical expressions for the polynomial $p_s(n,z)$

n	$p_s(n,z) - \mathrm{e}^{-2\alpha}\, p_s(n-1, 1+z)$
0	1
1	$1 - 4\alpha z$
2	$1 - 8\,\alpha\, z + 8\,\alpha^2 z^2$
3	$1 - 12\,\alpha\, z + 24\,\alpha^2 z^2 - \frac{32\,\alpha^3 z^3}{3}$
4	$1 - 16\,\alpha\, z + 48\,\alpha^2 z^2 - \frac{128\,\alpha^3 z^3}{3} + \frac{32\,\alpha^4 z^4}{3}$
5	$1 - 20\,\alpha\, z + 80\,\alpha^2 z^2 - \frac{320\,\alpha^3 z^3}{3} + \frac{160\,\alpha^4 z^4}{3} - \frac{128\,\alpha^5 z^5}{15}$
6	$1 - 24\,\alpha\, z + 120\,\alpha^2 z^2 - \frac{640\,\alpha^3 z^3}{3} + 160\,\alpha^4 z^4 - \frac{256\,\alpha^5 z^5}{5} + \frac{256\,\alpha^6 z^6}{45}$

Table 2: Expressions for the polynomials $p_{ss}(n,z)$ and constants $C_{ss}(n)$

n	$p_{ss}(n,z) - \mathrm{e}^{-2\alpha}\, p_{ss}(n-1, 1+z)$	$C_{ss}(n)$
0	-1	1
1	$1 + 4\alpha z$	0
2	$-1 - 8\,\alpha^2 z^2$	1
3	$1 + 4\,\alpha\, z + \frac{32\,\alpha^3 z^3}{3} - 8\,\alpha^2 z^2$	0
4	$-1 - 16\,\alpha^2 z^2 + \frac{64\,\alpha^3 z^3}{3} - \frac{32\,\alpha^4 z^4}{3}$	1
5	$1 + 4\,\alpha\, z - 16\,\alpha^2 z^2 + \frac{128\,\alpha^3 z^3}{3} - 32\,\alpha^4 z^4 + \frac{128\,\alpha^5 z^5}{15}$	0
6	$-1 - 24\,\alpha^2 z^2 + 64\,\alpha^3 z^3 - \frac{224\,\alpha^4 z^4}{3} + \frac{512\,\alpha^5 z^5}{15} - \frac{256\,\alpha^6 z^6}{45}$	1

In contrast to the rod with one end free, the duration of contact of the rod with one end fixed depends on the mass ratio α and is always greater than T. The limiting mass ratios α are listed in Table (3). In the literature, only the first three limiting ratios were given, where the third one is not correct, see [5]. Moreover, for $\alpha = 1/6$, St. Venant [3] computed the duration of contact $t_c = 7.419T/2$ whereas the coefficient 7.419 correctly must be 7.408 instead. This error also appeared in the textbook by Timoshenko and Goodier [6]. The visualization of the duration of contact, the coefficient of restitution and the maximum stress as well as the wave motions of the rod can be found in [8].

The maximum displacement of the rod

$$u_{\max} := \max_{t} \max_{0 \le x \le L} |u(x,t)| \tag{32}$$

is difficult to compute. From our mechanical intuition, we guess this maximum should appear at the contact end, i.e. $u_{\max} = max_t|u(0,t)|$, however,

Table 3: The limiting mass ratios α for the duration of contact t_c

t_c	$2T$	$3T$	$4T$	$5T$	$6T$
α	0.57859	0.24090	0.13566	0.087680	0.061527
$1/\alpha$	1.7283	4.1512	7.3713	11.4051	16.2529

this still needs a strict mathematical proof. Here, we present amplitude bounds for the displacements $u(x,t)$ utilizing the finite element method. The rod is discretized into n bodies, see Fig. 4.

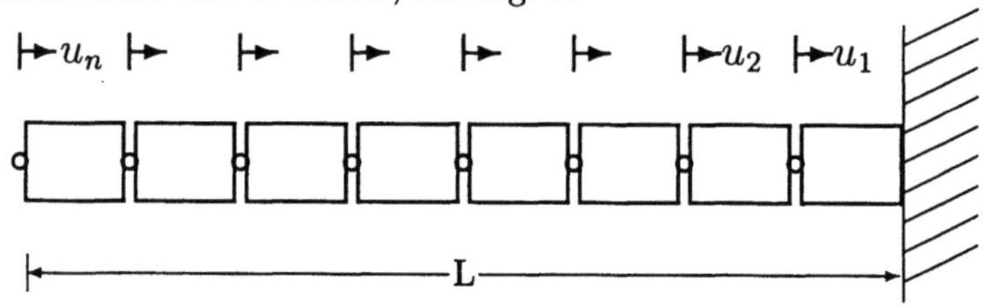

Figure 4: Discretization of the rod with the finite element method

The corresponding stiffness matrix then reads as

$$\mathbf{K} = \frac{nAE}{L}\begin{bmatrix} 2 & -1 & 0 & \cdots & 0 & 0 \\ -1 & 2 & -1 & \cdots & 0 & 0 \\ \cdot & \cdot & \cdot & \cdots & \cdot & \cdot \\ 0 & 0 & 0 & \cdots & 2 & -1 \\ 0 & 0 & 0 & \cdots & -1 & 1 \end{bmatrix}, \tag{33}$$

see [], and its inverse matrix is

$$\mathbf{K}^{-1} = \frac{L}{nAE}\begin{bmatrix} 1 & 1 & 1 & \cdots & 1 & 1 \\ 1 & 2 & 2 & \cdots & 2 & 2 \\ \cdot & \cdot & \cdot & \cdots & \cdot & \cdot \\ 1 & 2 & 3 & \cdots & n-1 & n-1 \\ 1 & 2 & 3 & \cdots & n-1 & n \end{bmatrix}, \tag{34}$$

which was also obtained using MAPLE. Denoting the mass matrix of the rod by $\mathbf{M}$, the energy of the rod follows as

$$E = \frac{1}{2}\dot{\mathbf{u}}^T\mathbf{M}\dot{\mathbf{u}} + \frac{1}{2}\mathbf{u}^T\mathbf{K}\mathbf{u} \tag{35}$$

with $\mathbf{u} = [u_1, u_2, \ldots, u_{n-1}, u_n]^T$. Before the impact, the energy of the striking and the resting rod together is equal to $\frac{1}{2}m_2v_2^2$. Hence, the inequality

$$\frac{1}{2}\mathbf{u}^T\mathbf{K}\mathbf{u} \le E \le \frac{1}{2}m_2v_2^2 \tag{36}$$

holds true. Based on a theorem presented in [9], it follows from Eq. (36)

$$|u_i(t)|^2 \leq K_{ii}^{-1} \mathbf{u}^T \mathbf{K} \mathbf{u} \quad \text{for} \quad i = 1, 2, \ldots, n \, , \tag{37}$$

where K_{ii}^{-1} is the i–th diagonal element of the matrix $\mathbf{K}^{-1}$. Hence, we have

$$\max_t |u_i(t)| \leq \sqrt{K_{ii}^{-1} m_2} \, v_2 = \sqrt{\frac{iLm_2}{nAE}} \, v_2 = \frac{v_2 L}{c\sqrt{\alpha}} \sqrt{\frac{i}{n}} \, . \tag{38}$$

Since for $n \to \infty$, the discrete model converges to the continuous model, it follows for $n \to \infty$ and $i/n \to (L - x)/L$ that

$$\max_t |u(x,t)| \leq \frac{v_2 L}{c\sqrt{\alpha}} \sqrt{1 - \frac{x}{L}} \leq \frac{v_2 L}{c\sqrt{\alpha}} \, . \tag{39}$$

For $x = 0$, the maximum $max_t|u(0,t)|$ is smaller than the quantity $v_2 L/(c\sqrt{\alpha})$ which can also be obtained by replacing the rod with a massless linear spring, see [5]. There it was shown that for different mass ratios this quantity is close to the exact maximum displacement. The discussion above shows that an upper limit on the maximum displacement for a given position can be computed. Equation (39) can also be used to obtain a bound for the maximum displacement $u_{\max}$ according to (32) along the rod.

5 Conclusions

In this paper, it has been shown that the complicated impact problem of a moving rigid body hitting a rod can be solved using computer algebra systems. Except for the duration of contact t_c, all other important variables, such as the stresses $\sigma(x,t)$, the displacements $u(x,t)$ and the maximum stress $\sigma_{\max}$, can be computed symbolically. Computer algebra systems not only allow to avoid the tedious numerical integration of the wave motion and the equation of the motion of the striking body manually, but also to avoid human formula manipulation errors. In this paper, some analytical expressions for the wave motion in the rod are supplemented and some errors appearing in the standard literature are corrected. Moreover, amplitude bounds for the maximum displacements of the rod with one end fixed are given.

Computer algebra systems, which are used nowadays, are permanently improved and further developed, but it should be noted, that computer algebra systems also have their limitations. For a very small mass ratio α, i.e. a large striking mass, the computation of the wave propagation is very time consuming and requires a large computer memory. In the literature, a model using the multibody system method is mentioned, where the rod is replaced by a massless linear spring and Boussinesq [4] already proposed an empirical formula for the maximum stress $\sigma_{\max}$, see [6]. However, it is difficult to validate this formula for a very small mass ratio α. A satisfactory solution for this problem is of great importance both in theory and in practice.

Finally, it is of great fascination to be able to perform analytical computations with relatively few effort using computer algebra systems. The MAPLE code used for computing results presented in this paper is available at `http://www.uni-stuttgart.de/mechb/people/Eberhard/wave_maple.txt` as MAPLE–ASCII file (.txt) and MAPLE worksheet (.ms).

Acknowledgment

The first author acknowledges the support of the German Research Foundation (DFG), SFB 404, Project B1.

References

1 Schiehlen, W.: *Technische Dynamik.* Stuttgart: Teuber, 1986.

2 Char, B.W., Geddes, K.O., Gonnet, G.H., Monagan, M.B. and Watt, B.M.: *First Leaves, A Tutorial Introduction to Maple.* Waterloo: Watcom Publications, 1989.

3 Szabó, I.: *Geschichte der mechanischen Prinzipien und ihrer wichtigsten Anwendungen.* Basel: Birkhäuser, 1977.

4 St. Venant, B.D. and Flamant, M.: Résistance vive ou dynamique des solids. Représentation graphique des lois du choc longitudinal. *C.r. hebd. Scéanc. Acad. Sci., Paris,* **97**, pp. 127, 214, 281 and 353, 1883.

5 Boussinesq, J.: *Applications des potentiels à l'étude de l'équilibre et du mouvement des solides élastiques.* Paris: Gauthier–Villars, 1885.

6 Goldsmith, W.: *Impact: The Theory and Physical Behaviour of Colliding Solids.* London: Edward Arnold Ltd, 1960.

7 Timoshenko, S.P. and Goodier, J.N.: *Theory of Elasticity.* New York: McGraw Hill, 1970.

8 Nikitin, L.: *Statics and dynamics of solids with external dry friction* (in Russian). Moscow: Moscow Liceum, 1998.

9 Hu, B., Eberhard, P. and Schiehlen, W.: Symbolic Wave Propagation Analysis of Longitudinal Impacts of Rods. Submitted to the journal *MapleTech*, 1998.

10 Hu, B. and Schiehlen, W. : Eigenvalue, frequency responses and variance bound of linear damped systems. *Eur. J. Mech. A/Solids*, **15**, pp. 617 – 646, 1996.

Stiction modeling: first order differential equation approximation for nonholonomic inequality constraint

*A. P. Kovacs**

Current trends in computational multibody dynamics address stiction between rigid bodies by using the complementarity conditions of optimization theory to develop the unilateral constraint. This paper presents an alternative Newtonian approach employing a nonhomogenous first order differential equation to approximate joint kinematics of a nonholonomic inequality constraint. A stiction algorithm is developed as a function of joint displacement, velocity and a kinematic state variable defined by the first order dynamic equation. System topology remains constant during stiction as the formulation computes proper stick-slip reaction loads. The technique is demonstrated for contacting bodies in lateral motion and offers a smooth approximation for stiction in mechanical systems analyzed by standard ODE and DAE stiff integrators.

1 Introduction

The unilateral constraint with time variant topology (e.g. impact, stiction) is a current research topic in multibody dynamics [1]. True joints exhibit some level of variant structure but multibody formalisms focus on the bilateral form in one of three ways: Lagrange multipliers and appended constraints, coordinate reduced minimal dynamic equation set and projected dynamics orthogonal to the constraints. Newtonian mechanics simulates stick-slip by the Amontons-Coulomb law [2]. This method works well for macroscopic rigid body motions but fails to capture small elastic interaction and micro-slip [3]. Finite element modeling [2,4] can analyze small local contact but generates too many dynamic equations for a multibody solution. This paper presents revised kinematics of the Amontons-Coulomb law for stiction's small scale motion within the scope of computational multibody dynamics.

To begin, non-Eulerian or Lagrangian equations of motion avail systems appended by equality constraints. This formalism is divided into type I, dynamic equations plus algebraic constraints, and type II, dynamic equations only [5]. Type I Lagrange equations of motion will be considered in this paper and evolve from

$$\frac{d}{dt}\left(\frac{\partial L}{\partial \dot{q}_i}\right) + \frac{\partial L}{\partial q_i} = Q_i + \lambda_m \frac{\partial \Phi_m}{\partial q_i} \tag{1}$$

* Senior Project Engineer, Meritor Automotive, 2135 West Maple Road, Troy, MI 48084-7186, email: kovacsap@meritorauto.com

where q_i are generalized coordinates, L is the Lagrangian L=T-V, T is system kinetic energy, V is potential energy, Q_i are generalized applied forces and $\lambda_m\,(\partial\Phi_m / \partial q_i)$ are force reactions from constraints Φ_m, defined as,

$$\sum_{n=1}^{N} A(q_n,t)_{mn}\,\dot{q}_n + A(q_n,t)_m = 0 = \Phi_m \cdot \tag{2}$$

Each of the m constraint equations Φ_m can be classified as holonomic, non-holonomic, catastatic and acatastatic with definitions readily found in the literature. This paper will only consider mechanical joint constraints.

Mechanical joints characterize attachments between system bodies derivable by a variety of geometric methods. Examples include spherical, revolute, translation and planar joints. Until recently computational type I Lagrangian codes exclusively applied equality or bilateral constraints forcing state variables to algebraic equation agreement within DAE numerical error. Constraint forces are computed from Lagrange multipliers but tangential motion between contacting bodies can be affected by unilateral stiction and will be the focus of this paper.

Some multibody codes [6,7] directly manage unsteady topology structures typified by unilateral inequality constraints,

$$\sum_{j=1}^{L} A(q_j,t)_{ij}\,\dot{q}_j + A(q_j,t)_i \leq 0 \text{ or } \geq 0 = \Phi_i \cdot \tag{3}$$

These conditions change system degrees of freedom with the onset of motion by applied loads exceeding static friction limits. An extensive reference overview is given by Ferris and Pang [8]. Implementing the transition between stick and slip is a formidable computational task and arbitrary occurrences require a combinatorial procedure. Klarbring and associates [9-16] programmed numerous friction problems using techniques of linear complementarity, sequential quadratic programming algorithms, Newton methods for generalized equations and B-differentiable equations. Much of this work was performed in the continuum mechanics regime discretized by finite element methods. In the multibody dynamic area Pfeiffer [6] has produced exceptional unilateral results with optimization and linear complementarity theory. Further progress in time integration techniques were developed by Meijaard [17] that applied test functions of the final states in the variant structure, using them to restart the integration with altered multibody topology.

This paper explores a kinetic modeling option of the unilateral friction constraint. First, constraint micromechanics are presented and manipulated into a first order, nonhomogenous, ordinary differential equation tied to a force relationship. Next, this technique is augmented and applied to a computational example (ADAMS®) simulation) of a nonholonomic stick-slip illustration.

2 Micromechanics constraint modeling

Kinematic constraints in multibody formalisms avoid explicit definition of joint deformations and force-displacement algorithms. But revisiting Lanczos' treatise [18] on variational mechanics, real physical connections between bodies under applied load are maintained by strong forces that yield joint deformation. He begins with the sum of constraint and applied force potential energy in the Lagragian ($L = T - \nabla$) as

$$\nabla = V + \lambda_1 \Phi_1 + ... + \lambda_m \Phi_m \tag{4}$$

where ∇ is total system potential energy, V is applied force potential energy, Φ_i are algebraic constraints and λ_i are corresponding Lagrange multipliers. The ideal bilateral constraint

$$\Phi(q_i) = 0 \tag{5}$$

contributes zero potential energy to equation (4), but elastic structures are not infinitely rigid and kinematic pairs possess potential energy as a function of the constraint formulation, or

$$V_c = \Psi(\Phi)\ . \tag{6}$$

Only small values of Φ contribute to V_c and a Taylor series expansion of $\Psi(\Phi)$ around $\Phi(q_i) = 0$ (with constant set to zero) gives

$$V_c = \Psi'(0)\,\Phi + \frac{1}{2}\,\Psi''(0)\,\Phi^2 + \text{h.o.t.} \tag{7}$$

Since V_c must have a minimum at $\Phi = 0$, the first derivative is $\Psi'(0) = 0$ and the second derivative must be positive, $\Psi''(0) = \frac{1}{\varepsilon}$, ε is a small positive constant. The constraint potential energy now takes the form

$$V_c = \frac{1}{2\varepsilon}\,\Phi^2\ . \tag{8}$$

The smallness of ε gives a "strong" reaction force and exact fulfillment of Φ, $\varepsilon = 0$, requires an infinite force. The real system allows joint deformation and Lanczos treats this by introducing an error term to the RHS (right hand side) of equation (5),

$$\Phi(q_i) = \varepsilon\,\rho(t) \tag{9}$$

where $\rho(t)$ is time dependent and the RHS is the "microscopic error" of Φ. Numerically, the RHS can be zero or non-zero which suggests a nonholonomic inequality constraint. The nature of $\rho(t)$ is found from equation (8), where

$$F_i = -\frac{\partial V_c}{\partial q_i} = -\frac{1}{\varepsilon}\Phi\frac{\partial \Phi}{\partial q_i} \tag{10}$$

The Lagrangian λ - method alternatively gives this force as

$$F_i = -\lambda\frac{\partial \Phi}{\partial q_i} \tag{11}$$

concluding, the Lagrange multiplier $\lambda = \rho(t)$ and the small positive constant ε is a compliance. This paper will compute $\rho(t)$ from multibody and micromechanic variables related by a dynamic equation.

Consider two contacting bodies with static friction large enough to resist lateral motion under external loads, figure (1a). This condition is shown in two ways: (1) a multibody model, figure (1b) and (2) a micromechanics interface model, figure(1c). Noting body coordinates x_1, x_2 and displacement variable α, comparison between the force balance ($\Sigma F_x = 0$) for each model give

$$-k(x_1 - x_2)_{\text{multibody}} = F_{\text{applied}} \rightarrow -(c\,\dot{\alpha} + k\,\alpha)_{\text{micromechanics}} = F_{\text{applied}} \tag{12}$$

where k and c are tangential stiffness and damping respectively. The multibody joint ignores damping (imposed by equation (9) "stiffness" kinematics) which is present in the micromechanics interface, modeled as a linear spring and damper. Because the applied force F_{applied} is the same in each model, equation (12) becomes an equality and dividing both sides by the stiffness, k, gives,

$$(x_1 - x_2) = \frac{c}{k}\dot{\alpha} + \alpha = \tau\,\dot{\alpha} + \alpha \tag{13}$$

a relation between multibody coordinates, micromechanic displacement and velocity plus a very small time constant τ. The state space form of equation (13) is,

$$\tau\,\dot{\alpha} = (x_1 - x_2) - \alpha \tag{14}$$

where the RHS is the difference between the multibody and micromechanics joint gap and the LHS (left hand side) is a micro-kinematic component overlooked in the multibody formalism. Multiplying the RHS by the stiffness $-k_J$ gives the restoring force between body 1 and body 2 to approximate $(x_1 - x_2) \approx 0$,

$$F_x = -k_J \cdot [(x_1 - x_2) - \alpha] \tag{15}$$

Hence, the strong force required to hold the joint has been derived by a "vectorial mechanics" approach in a multibody context where a micro-kinematic error is used to compute restoring force, much like a linear spring. Modeling a bilateral constraint is a trivial example of this method and the next section expands equation (14) for the kinematics in unilateral stiction.

3 A kinetic stiction formulation

The rectilinear motion between two contacting bodies, figure (1), involves tangential micromechanic friction forces influenced by juncture weldments, surface oxide layer and various impurities [2,3]. The kinetics are best described by partial differential equations of motion reduced to finite elements. Alternatively, multibody methods view friction between rigid bodies as a function of states and time. This paper assumes smooth transition of the stiction motion (no instantaneous jumps) that must address three kinematic zones:

- zone 1: absolute stick, $(\dot{x}_1 - \dot{x}_2) = 0$
- zone 2: stick-slip transition or micro slip, $0 \le (\dot{x}_1 - \dot{x}_2) \le \delta(\dot{x}_1 - \dot{x}_2)$
- zone 3: macro slip, $(\dot{x}_1 - \dot{x}_2) >> 0$

Most Newtonian methods model zone 1 and/or zone 3 while zone 2 is a difficult region of micro-slip to be described kinematically. Modification of equation (14) to accommodate the three zones leads to the following equation,

$$\tau \cdot \dot{\alpha} = Z_{12}(\dot{x}_1 - \dot{x}_2) \cdot \left[(x_1 - x_2) - \alpha - Z_3(\dot{x}_1 - \dot{x}_2)\right] \tag{16}$$

where $(x_1 - x_2)$ is the multibody joint displacement, α is a micro-kinematic variable and τ is a time constant that controls gap convergence and the equation's numerical "stiffness" [19]. $Z_{12}(\dot{x}_1 - \dot{x}_2)$ is a cubic Heaviside unit step function [7] of joint velocity ($Z_{12} = \text{step}(\dot{x}_1 - \dot{x}_2, 0 \rightarrow 1.0)$ and figure (2)) and admits the micro-slip kinematics for zone 1 and zone 2. $Z_3(\dot{x}_1 - \dot{x}_2)$ is also a cubic Heaviside step function ($Z_3 = \text{step}(\dot{x}_1 - \dot{x}_2, 0 \rightarrow x_{saturated})$) for saturating displacement when the joint velocity is large, giving zone 3. Bilateral constraints are obtained for the special case, $Z_{12}(\dot{x}_1 - \dot{x}_2) = 1$ and $Z_3(\dot{x}_1 - \dot{x}_2) = 0$. The RHS terms of equation (16) provides the excitation to a first order lag feedback dynamic and α converges to the $(x_1 - x_2)$ value with small error. During stiction (zone 2) the small displacement error, $(x_1 - x_2)$ - α, interpolates the reaction load to restrict joint motion. Figure (3) shows the force-displacement relationship, implicitly using a static and dynamic friction force function that is not shown. Applied loads in excess of the static limit cause the joint to slip (zone 3) triggering dynamic friction interpolation by the saturated displacement, $Z_3(\dot{x}_1 - \dot{x}_2)$. In zone 3 $Z_{12}(\dot{x}_1 - \dot{x}_2)$ and $Z_3(\dot{x}_1 - \dot{x}_2)$ take on their maximum values and $(x_1 - x_2)$ - α is very small. Also, equation (16) provides multiple stick-slip locations depending on the applied load time history which is demonstrated in the next section.

4 Computational results

An ADAMS simulation for the two bodies in contact, figure (1), is presented using the stiction equation (16). A time varying lateral force is applied to body 1 and slowly increased from 0 to 220 force units, exceeding the static friction force of 200 and causing the body to move laterally under dynamic friction, note the drop in reaction load in figure (4). The friction force function is given as

$$F_{friction} = \mu_{static} \cdot F_n \cdot \text{step}(v_z, 1 \rightarrow 0.7) \tag{17}$$

where a Heaviside cubic step function proportions the static friction value from 1.0 to 0.7 when the relative joint velocity exceeds a specified small value. The applied force is maintained for a short period and abruptly removed at time = 2. The body remains in free motion until the friction force, 140, stops it and remains at rest from time = 2.75 to 4. The applied force is again applied at time = 4 from 0 to 220. Figure (4) gives the applied force time history (dotted line) and the corresponding friction reaction force (solid line). Figure (5) shows the displacement time history of body 1, note the rest period between time = 0 and 1 and time = 2.75 and 4. Figure (6) displays the corresponding graphical initial condition, (6a), and time sequence superimposed over 5 units, (6b).

The small velocity limits used in the Heaviside step functions must be determined empirically for accurate simulations. The limit values and friction coefficients were arbitrarily selected for demonstration purposes and must be appropriately selected for dimensional consistency.

5 Conclusions

A single ordinary differential equation evolves from a force balance comparison between multibody and micro-mechanic models exhibiting translation friction contact without slip. This kinematic relationship determines the displacement difference between the two methods and is used in computing the corresponding interface friction load. The differential equation is then modified with relative joint velocity to capture stiction's three zones of absolute stick, micro slip and macro slip. Since the differential equation involves a small time constant and small velocities, the method is numerically stiff and frequent stiction applications would reduce the time step size, increasing computation CPU through the mico-slip zone and possibly causing integration failure. Yet, the process does not change system topology and avoids special purpose code needed for DAE equations of motion and inequality constraint solution. This procedure has succeeded in unpublished but completely validated multibody full vehicle dynamic simulations [20]. An ADAMS vehicle drift sensitivity study, dependent upon ball joint stiction and tire conicity, corroborated test DOE (design of experiment) coefficients evaluated for a linear response method analysis. The analysis required equation (16) to accurately predict the DOE parameters.

References
1 Schiehlen, W. 'Multibody system dynamics: roots and perspectives', *Multibody System Dynamics*,**1**(2), 1997, 149-188
2 Kikuchi N. and Oden, J.T., *Contact Problems in Elasticity: A Study of Variational Inequalities and Finite Element Methods*, SIAM, Philadelphia, 1988
3 Ibrahim, R. A., 'Friction-induced vibration, chatter, squeal, and chaos part I: mechanics of contact and friction', *Applied Mechanics Reviews*, **47**(7), 1994, 209-226
4 Amirouche, F. M. L., Xie, M., Shareef, N. H. and Valco, M., 'Finite element modeling of contact conditions in multibody system dynamics', *Nonlinear Dynamics*, **4**, 1993, 83-102
5 Roberson, R. E. and Schwertassek, R. *Dynamics of Multibody Systems*, Springer-Verlag, New York, 1988
6 Pfeiffer, F. and Glocker, C., *Multibody Dynamics with Unilateral Contacts*, John Wiley & Sons, New York, 1996
7 ADAMS Reference Manual, Mechanical Dynamics, Inc., Ann Arbor, 1997
8 Ferris, M. C. and Pang, J. S., 'Engineering and economic applications of complementarity problems', *SIAM Review*, **39**(4), 1997, 669-713
9 Klarbring, A., 'A mathematical programming approach to three dimensional contact problems with friction', *Comput. Methods Appl. Mech. Engrg.*, **58**, 1986, 175-200
10 Klarbring, A., 'Mathematical programming and augmented Lagrangian methods for frictional contact problems', *Proceedings Contact Mechanics International Symposium*, A. Curnier, ed., Presse Polytechniques et Universitaires Romandes, Lausanne, 1992, 409-422
11 Klarbring, A., Bjorkman, G., 'Solution of large displacement contact problems with friction using Newton's method for generalized equations, *Internat. J. Numer. Methods Engrg.*, **34**, 1992, 249-269
12 Kwak, B. M., 'Complementarity problem formulation of three-dimensional frictional contact', *J. Appl. Mech.*, **58**, 1991, 134-140
13 Lee, S. S., 'A computational method for frictional contact problems using finite element method', *Internat. J. Numer. Methods Engrg.*, **37**, 1994, 217-228
14 Lötstedt, P., 'Mechanical systems of rigid bodies subject to unilateral constraints', *SIAM J. Appl. Math.*, **42**, 1982, 281-296
15 Stavroulakis, G. E., Panagiotopoulos, P. D., and Al-Fahed, A. M., 'On the rigid body displacements and rotations in unilateral contact problems and applications', *Comput. & Structures*, **40**, 1991, 599-614
16 Sun, S. M., Natori, M. C., and Park, K. C., 'A computational procedure for flexible beams with frictional contact constraints', *Internat. J. Numer. Methods Engrg.*, **36**, 1993, 3781-3800
17 Meijaard, J. P., 'Efficient numerical integration of the equations of motion of non-smooth mechanical systems', *ZAMM*, **77**, 1997, 419-427
18 Lanczos, C., The Variational Principles of Mechanics, Dover Publication, New York, 1986

19 Gear, C. W., Hsu, H. H., and Petzold, L., 'Differential-algebraic equations revisited', *Proc. Numerical Methods for Solving Stiff Initial Value Problems,* Obersolfach, W. Germany, 1981

20 Kovacs, A. P., Knapp, R., and Sigman, D., department memorandums and reports at Ford Motor Company, Vehicle Dynamics Department, Chassis Engineering, Dearborn, Michigan, 1992-1993

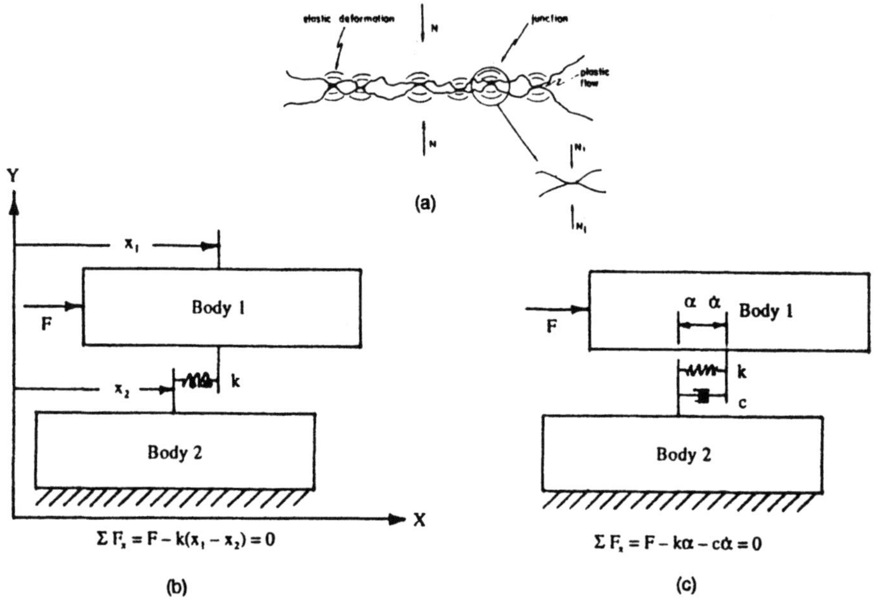

Figure 1, (a) two bodies in contact under lateral motion, from Kikuchi and Oden [2], (b) multibody model, (c) micromechanics interface model

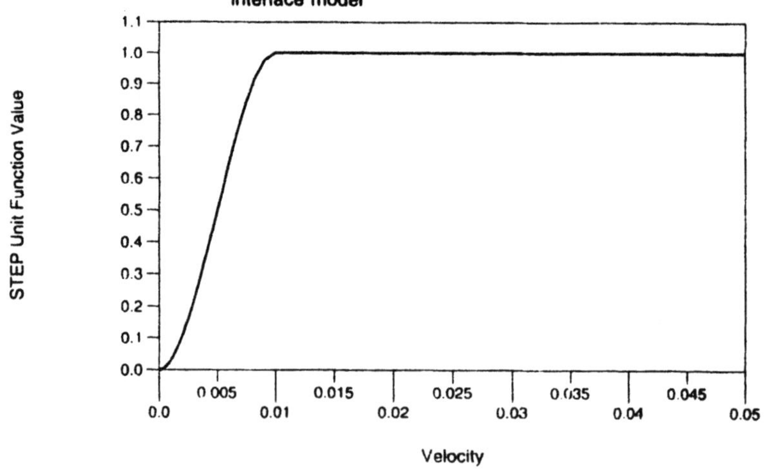

Figure 2, Heaviside unit step function

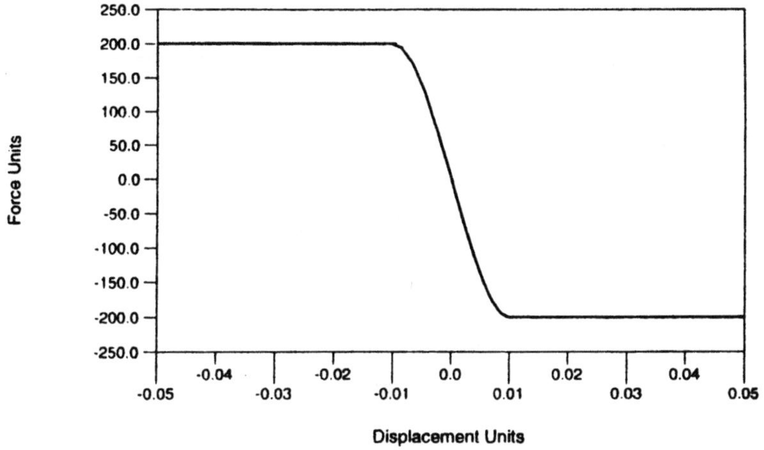

Figure 3, Heaviside stiction force-displacement function

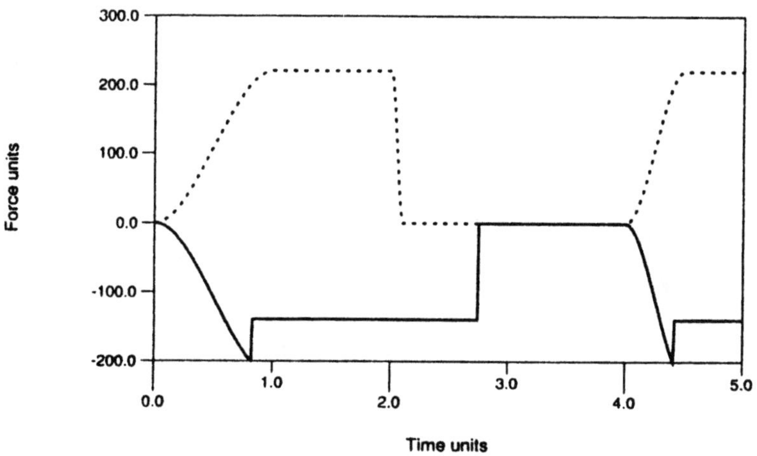

Figure 4, Applied load (dotted line) and stiction reaction (solid line) time histories

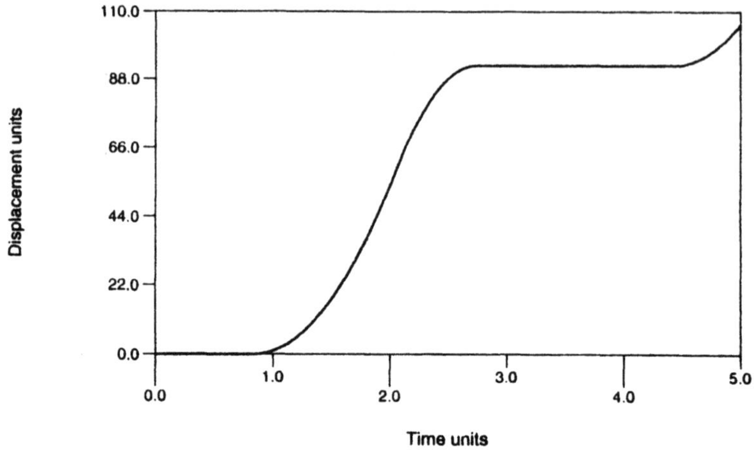

Figure 5, Body displacement time history

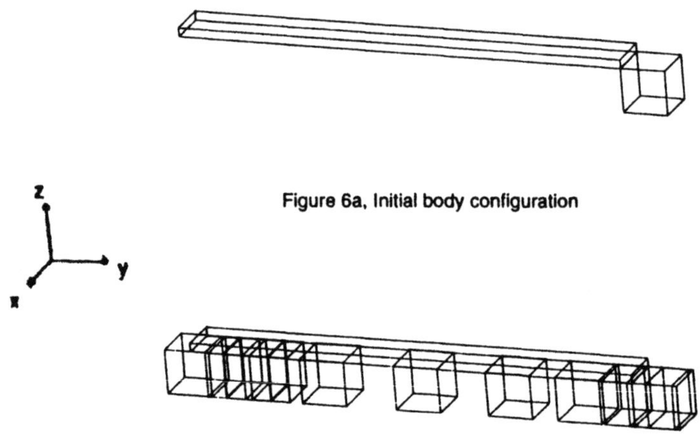

Figure 6a, Initial body configuration

Figure 6b, Body motion time sequence

Vibro-impacts induced by non-linear resonances in hertzian contacts

J Perret-Liaudet and J Sabot**

A single-degree-of-freedom non-linear oscillator modelling a loaded sphere-plane Hertzian contact is studied. Dynamic responses are induced by an external harmonic normal force. Non-linear resonances which lead to vibro-impact responses are investigated. Effects of the non-linear Hertzian contact stiffness and effects of some non-linear damping laws are also considered.

1 Introduction

Many mechanical systems exhibit dynamic behaviour characterized by vibro-impacts which leads to large vibratory level, wear and excessive noise. A simple way commonly used to model these impact systems is to introduce standard coefficient of restitution [1-4]. This approach is based on the assumption of short impact time. In several applications, this assumption cannot be justified and it needs to introduce contact stiffness and contact damping. For this end, piecewise linear systems are generally considered [5-9] but few studies are based on a Hertzian contact law and on a non-linear damping law.

In the present paper, we propose to consider the fundamental case of a loaded sphere-plane Hertzian contact. In our analysis, we examine the principal, subharmonic and superharmonic resonances which can lead to vibro-impact responses. Numerical results allow to conclude about some effects of the non-linear Hertzian contact stiffness and the non-linear damping. Finally, some experimental results are reported.

2 Dynamic model

Consider the single-degree-of-freedom impact oscillator shown in Figure 1. A movable mass, in contact with a flat surface through an hemispherical pin, is loaded by a static normal load. Assuming elastic deformation, the non-linear contact stiffness is deduced from Hertz's theory [10]. When the system is excited by a harmonic normal force, the equation of motion may be written as :

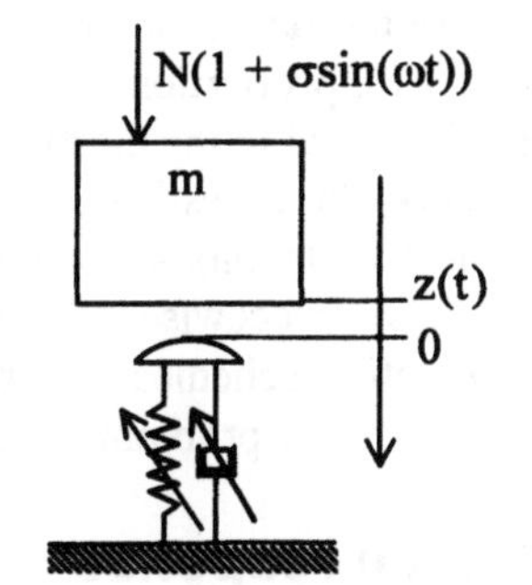

Fig. 1 : Dynamic model

$$\begin{cases} m\ddot{z} + f_c(z,\dot{z}) + kz^{3/2} = N(1+\sigma \sin \omega t) & z \geq 0 \\ m\ddot{z} = N(1+\sigma \sin \omega t) & z < 0 \end{cases} \quad (1)$$

* Laboratoire de Tribologie et Dynamique des Systèmes, UMR 5513,
Ecole Centrale de Lyon, BP 163, F-69131 Ecully Cedex, France.

where z is the normal displacement of the mass m (z < 0 corresponds to contact loss), k is a constant deduced from the Hertzian theory, N is the static load and f_C is the damping force. Amplitude and circular frequency of the external harmonic force are controlled by parameters σ and ω. Considering the static contact compression $z_S=(N/k)^{2/3}$ and the linearized contact natural frequency Ω defined by $\Omega^2=(3k/2m)z_S^{1/2}$, the equation of motion can be written in dimensionless form as :

$$\text{NL model :} \begin{cases} \ddot{q} + f_C(q,\dot{q}) + (1+2q/3)^{3/2} = 1+\sigma\sin\varpi\tau & q \geq -3/2 \\ \ddot{q} = 1+\sigma\sin\varpi\tau & q < -3/2 \end{cases} \tag{2}$$

where $q = 3\,(z/z_S - 1)/2$, $\tau = \omega t$ and $\varpi = \omega/\Omega$. Overdot indicates now differentiation with respect to the non-dimensional time τ.
In order to employ some commonly used methods, one can expand the non-linear restoring force in a one and a third order Taylor series yielding to two simplified models governed by the following equations :

$$\text{NL1 model :} \begin{cases} \ddot{q} + f_C(q,\dot{q}) + q = \sigma\sin\varpi\tau & q \geq -1 \\ \ddot{q} = 1+\sigma\sin\varpi\tau & q < -1 \end{cases} \tag{3}$$

$$\text{NL3 model :} \begin{cases} \ddot{q} + f_C(q,\dot{q}) + q + q^2/6 - q^3/54 = \sigma\sin\varpi\tau & q \geq -3/2 \\ \ddot{q} = 1+\sigma\sin\varpi\tau & q < -3/2 \end{cases} \tag{4}$$

To describe the contact damping force, one may assume different viscous damping laws which can be expressed for the NL model case as:

$$f_C(q,\dot{q}) = 2\zeta(1+2q/3)^r\dot{q} \quad q \geq -3/2 \tag{5}$$

where the damping force can be linear (r = 0), proportional to the contact radius (r = 1/2), proportional to the elastic deformation and to the contact area (r = 1), or proportional to the elastic restoring force (r = 3/2) (for the last case see [11]).
Finally, to investigate dynamic responses, we have used both analytical and numerical methods, more precisely multiple scales method (for the NL1 and NL3 models), piecewise analytical resolution (for the NL1 model), numerical time integration schemes (central-difference) and specific computing methods devoted to non-linear problems like shooting method with continuation technique.

3 Free response

For the undamped NL1 model case, analytical value of the natural frequency of oscillation as a function of amplitude can be easily obtained. As shown in Figure 2, it is 1 up to contact loss and then begins to quickly decrease, approaching null value. So, one can confirm that contact loss implies softening behaviour. One can now consider numerical results obtained from the NL model to analyse the

Hertzian contact stiffness influence on this behaviour. As shown in Figure 2, one can clearly conclude that clearance masters the non-linear behaviour of the loaded system. However, one can note that Hertzian contact law implies softening behaviour even without contact loss [12]. Also, one can retain that, for a same response level, the natural frequency is higher. It can be explained by the hardening character of the contact stiffness when it is compressed.

In the NL damping model case, we have considered different laws of damping (see equation 5) for the free response of the system. After some numerical experiments, negligible effect of an eventual non-linear damping law has been observed when there was no contact loss. Thus, the logarithmic decrement technique can be yet used to estimate damping ratio in real systems. We propose to calculate the mean value of the logarithmic decrement obtained from maxima and minima of the response. To give an idea, one obtains a 1% precision on the damping ratio when the non dimensional acceleration has a minimum amplitude 40% less than the non dimensional mean static load. Considering these numerical results, we were able to identify damping ratio in a similar experimental system [13-14], revealing low values for dry contacts (around 0.5%) and a damping ratio which increases linearly with the dynamic viscosity for lubricated contacts. Figure 3 illustrates this experimental result. In contrast, when there are impacts, the non-linearity of the damping law becomes deciding on the temporal response decrease. In particular, this decrease can be almost linear (for $r = 0$ in equation 5) to almost exponential (for $r = 3/2$). We have performed some experiments but no significant results have been obtained, due to the small value of damping. Finally, other experimental studies are necessary to conclude.

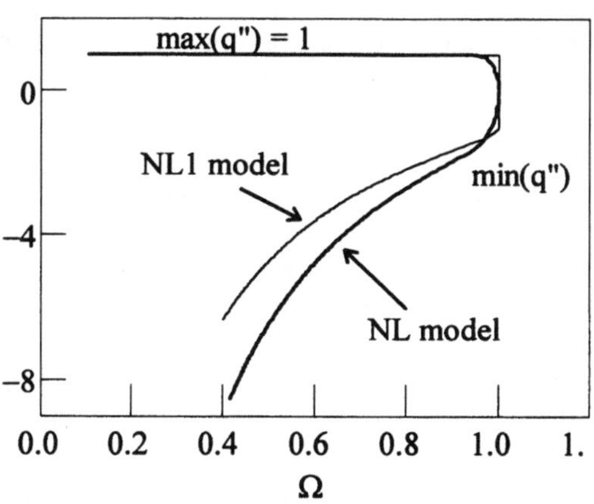

Fig. 2 : Natural frequency *vs.* acceleration amplitude.

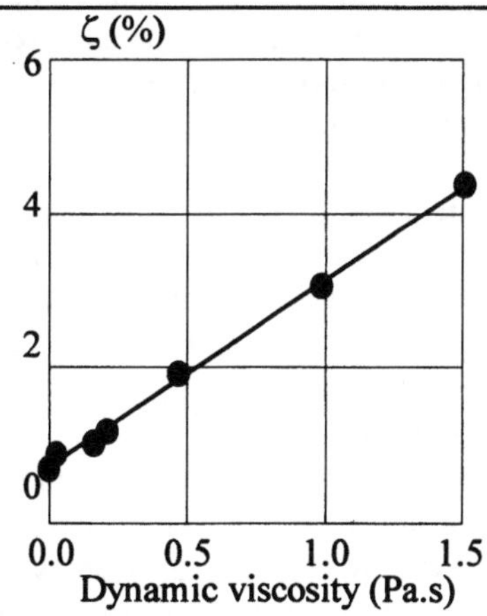

Fig. 3 : Damping ratio *vs.* dynamic viscosity of the lubricant (● experimental results).

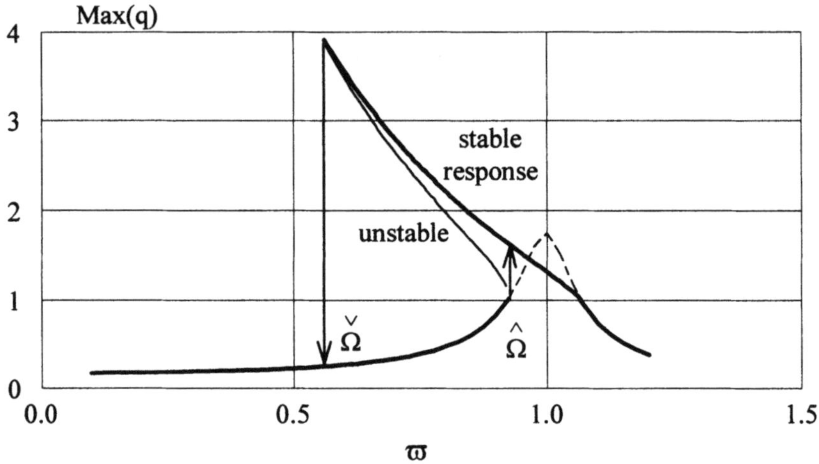

Fig. 4 : Primary resonance curve Max(q) *vs.* ϖ for the NL1 model. $\hat{\Omega}$ and $\check{\Omega}$ correspond respectivaly to upwards and downwards jump frequencies. Parameter values are $\zeta = 0.01$ and $\sigma = 0.035$.

4 Primary resonance

In order to describe the principal resonance, we have considered firstly the NL1 model with a linear viscous damping. Thus, the only non-linear effect associated to the loss of contact can be analysed.

Figure 4 shows one example of the evolution of the maximum displacement amplitude versus the external frequency, exhibiting the primary resonance close to $\varpi = 1$. According to what happened for the unforced and undamped case, the system has a typical softening behaviour. Actually, the effect of the geometrical non-linearity is to bend the amplitude frequency curve to lower frequencies. One can notice that this effect is strong as a relatively low excitation level induces a high response amplitude over a large frequency range. Moreover, quick growing of the displacement amplitude versus excitation level shown in Figure 5 illustrates this result.

Now, effect of the Hertzian contact law is investigated considering the NL model with linear viscous damping (equation 5 with $r = 0$). All the numerical simulations allow us to conclude as follows :

- without contact loss, softening behaviour can be observed as the frequency response curve is bent to lower frequencies, but over a short range [12-15].
- with loss of contact, closed frequency response curves have been obtained between models which include or not the hertzian contact non-linearity.

These results are illustrated in Figure 6, in which downwards jumps obtained from the both models NL and NL 1 are shown in a plane $(\sigma/2\zeta)$.

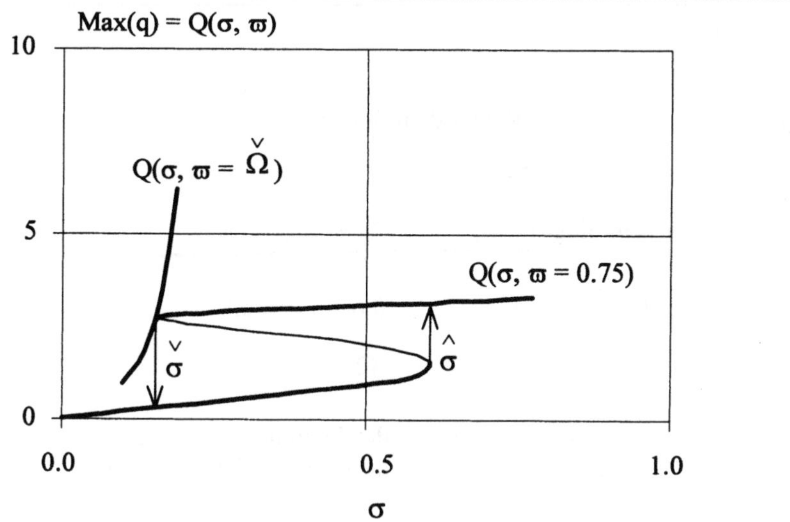

Fig. 5 : Load response curve Max(q) *vs.* σ for $\varpi = 0.75$, and for $\varpi = \check{\Omega}$. The last case correspond to the peak resonance value of Max(q); ζ=0.01 - NL1 model.

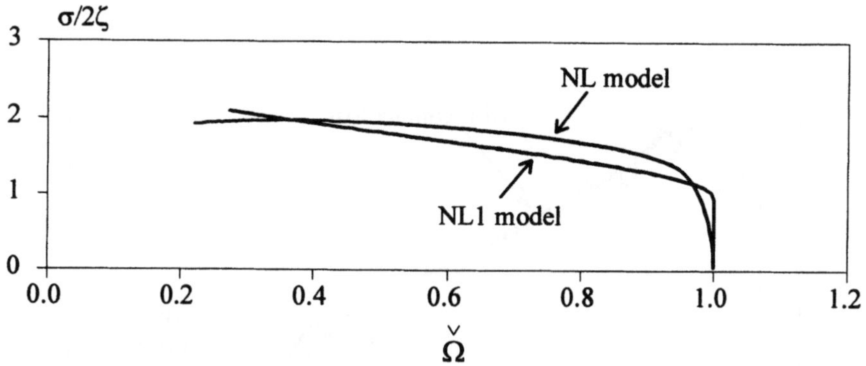

Fig. 6 : Downwards jump frequencies for the NL and the NL1 model, with linear damping (r = 0 - eq.5).

For analysing damping law effect on the NL model, we have compared results obtained by introducing the four laws described in equation 5. We have conclude that the kind of the damping law has a negligible effect on the resonance for oscillations smaller than those required for loss of contact. On the contrary, strong influence has been observed considering the impact case. In particular, with same excitation level, one can note large modifications of the peak resonance amplitude as shown in Figure 7.

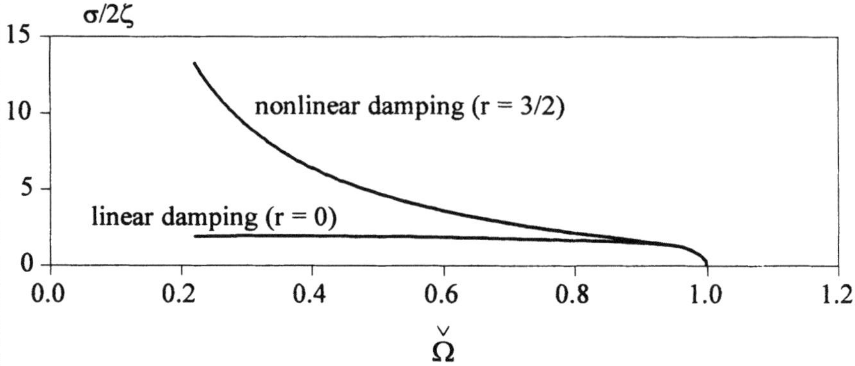

Fig. 7 : Effect of the damping law on the downwards jump frequencies for the NL model

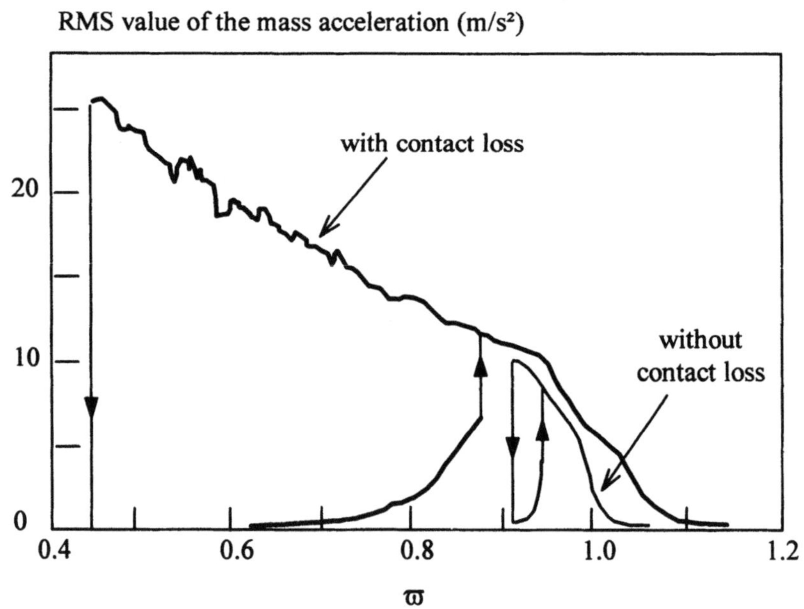

Fig. 8 : Typical measured response curves $\ddot{z}$ *vs*. forcing frequency.

In conclusion, clearance non-linearity governs the primary resonance but precise knowledge of damping sources is essential to describe it correctly. As shown in Figure 8, good agreement has been observed between experimental and numerical studies, but more refined experiments are necessary to analyse the damping sources.

5 Secondary resonances

In order to describe some secondary resonances, we have considered both the 2-superharmonic and the 1/2-subharmonic resonances, and we have used multiple scales method and shooting method applying to the NL3 and the NL models. For these two resonances, excitation levels are much higher than that required to induce vibro-impacts around the primary resonance. In fact, they remain sufficiently low to consider these problems. To give an idea, for a damping ratio of about 1%, excitation levels needed to induce contact loss for the both secondary resonances are about 35% of the mean static load.

Figure 9 shows a typical 2-superharmonic resonance which reveals the softening behaviour induced by the Hertzian contact non-linearity. At this excitation level, no contact loss occurs and the downwards jump remains closed to $\varpi = 1/2$. Increasing this excitation level by a small value initiates vibro-impacts and leads to a sudden widening of the resonance down to $\varpi = 0.25$. So, this resonance governed by the Hertzian contact law constitutes a precursor of the vibro-impact conditions.

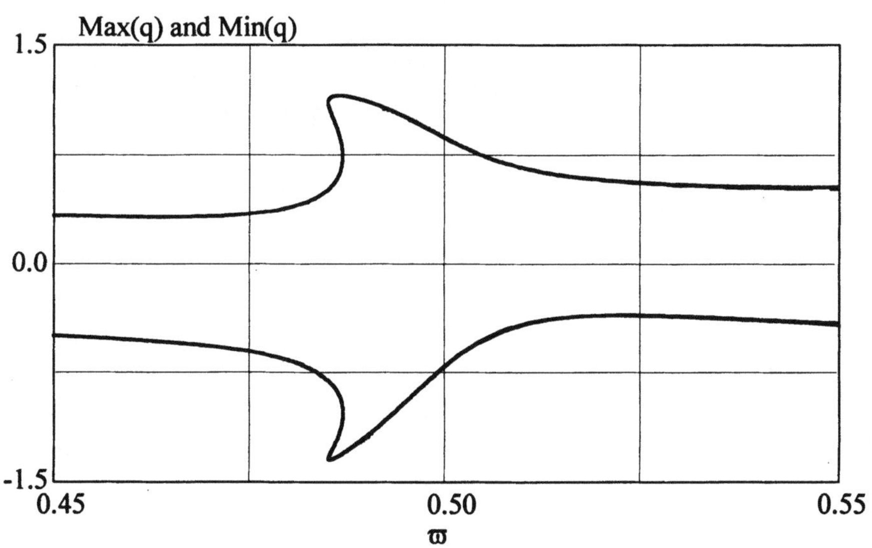

Fig. 9 : Typical 2-superharmonic resonance for the NL model with r = 0 (eq.5). Parameter values are $\sigma = 0.32$ and $\zeta = 0.01$.

The 1/2-subharmonic resonance is a typical non-linear behaviour. It merges from a pair of two flip bifurcations around $\varpi = 2$ (or a saddle node bifurcation when the period 1 motion remains stable). Typical frequency response curve is shown in Figure 10. It again reveals the softening behaviour of the system. As one can see, it leads to vibro-impacts (period 2 response) when the external frequency is slowly decreased over a large range. Finally, it has also been found that the appearance of this resonance was almost sudden for increasing excitation level with a critical level given approximately by $\sigma^* \approx 36\zeta$ [16].
In conclusion, the non-linearity induced by the loss of contact mainly controls the superharmonic and the subharmonic resonances but the non-linearity related to the Hertzian stiffness constitutes a necessary precursor of these secondary resonances. Actually, they cannot be described considering simplified models such as the piecewise NL1 model.

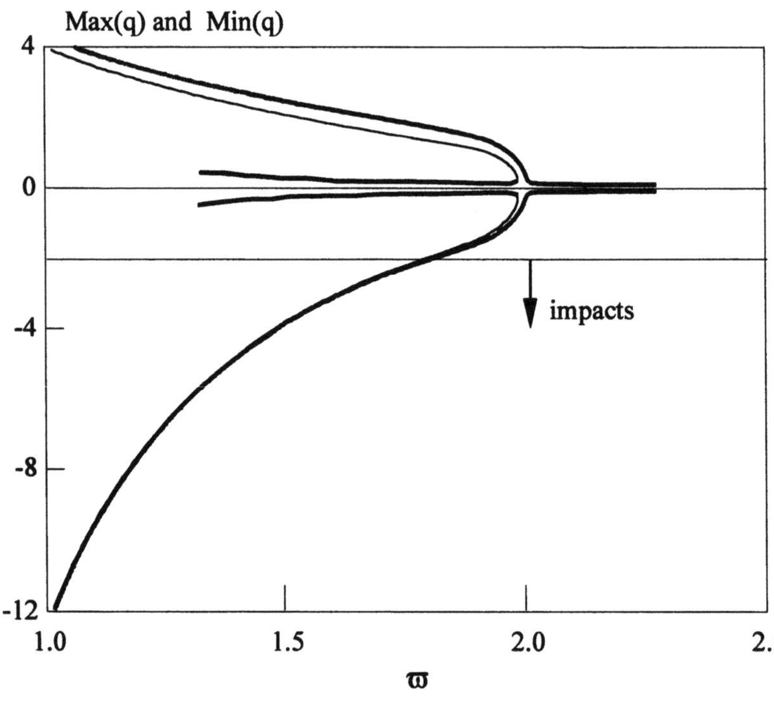

Fig. 10 : Typical 1/2-subharmonic resonance for the NL model with r = 0 (eq.5). Parameter values are $\sigma = 0.35$ and $\zeta = 0.01$.

6 Conclusion

For a single-degree-of-freedom impact oscillator modelling a loaded sphere-plane Hertzian contact, we have performed numerous numerical simulations and some experimental investigations. Our results lead to the main following conclusions :

a) contact loss implies a softening behaviour,

b) the damping non-linearity has a negligible effect on the free response when no contact loss occurs but becomes very important when there are impacts,

c) clearance non-linearity mainly governs the primary resonance but precise knowledge of damping is needed to correctly describe it,

d) clearance non-linearity controls the superharmonic and the subharmonic resonances but the non-linear Hertzian contact stiffness constitutes a precursor of these secondary resonances.

References

1. BAPAT C N AND POPPLEWELL N, Several similar vibroimpact systems. *Journal of Sound and Vibration*, **113**(1), 17-28, 1987.

2. WHISTON G S, The vibro-impact response of a harmonically excited and premloaded one dimensional linear oscillator. *Journal of Sound and Vibration*, **115**(2), 303-319, 1987.

3. WHISTON G S, Global dynamics of a vibro-impacting linear oscillator. *Journal of Sound and Vibration*, **118**(3), 395-429, 1987.

4. PETERKA F AND VACIK J, Transition to chaotic motion in mechanical system with impacts. *Journal of Sound and Vibration*, **154**(1), 95-115, 1992.

5. SHAW S W AND HOLMES P J, A periodically forced piecwise linear oscillator. *Journal of Sound and Vibration*, **90**(1), 129-155, 1983.

6. COMPARIN R J AND SINGH R, Non linear frequency response characteristics of an impact pair. *Journal of Sound and Vibration*, **134**(2), 259-290, 1989.

7. NATSIAVAS S, Stability and bifurcation analysis for oscillators with motion limiting constraints. *Journal of Sound and Vibration*, **141**(1), 97-102, 1990.

8. NATSIAVAS S, On the dynamics of oscillators with bi-linear damping and stiffness. *Int. J. Non-Linear Mechanics*, **25**(5), 535-554, 1990.

9. MAHFOUZ I A AND BADRAKHAN F, Chaotic behaviour of some piecewise-linear systems. Part II: systems with clearance. *Journal of Sound and Vibration*, **143**(2), 289-328, 1990.

10. JOHNSON K L, *Contact mechanics*, Cambridge Univ. Press, 1979.

11. LANKARANI H M AND NIKRAVESH P E, A contact force model with hysteresis damping for impact analysis of multibody systems. *ASME, Journal of Mechanical Design*, **112**, 369-376, 1990.

12. RANGANATH NAYAK P, Contact vibrations. *Journal of Sound and Vibration*, **22** (3), 297-322, 1972.

13. SABOT J, KREMPF P and JANOLIN C, A device for laboratory courses in non-linear contact vibrations. *Proceedings of the 2nd European Nonlinear Oscillations Conference, Euromech,* Prague, 9-13 Septembre, 1996.

14. SABOT J, KREMPF P and JANOLIN C, Non-linear vibrations of a sphere-plane contact excited by a normal load. *Journal of Sound and Vibration*, **214** (2), 359-375, 1998.

15. HESS H P, SOOM A AND KIM C H, Normal vibrations and friction under harmonic loads: Part I - Hertzian contacts. *ASME, Journal of Tribology*, **113**, 80-86, 1991.

16. PERRET-LIAUDET J, Résonance sous harmonique d'ordre deux dans un contact sphère plan, *C. R. Acad. Sci.*, **325** (II *b*), 443-448, 1997

Effect of prestressing on durability at repeated impacts

I. Dumitru, T.Babeu*, S. Babeu*, L. Marşavina**

In present time, there is considerable interest in the durability of some steels and alloys loaded by repeated impacts, phenomenon known under the name of "impact fatigue". So, we can mention a lot of researches regarding the influence upon durability at impact fatigue of the dimensional factor, the shape and dimensions of stress concentrators, the impact speed, the temperature, thermal treatments. But very few studies approach the influence of prestressing on durability at impact fatigue in the light of initiation and propagation of a fatigue crack.

In this way the authors made a study regarded of effect of prestressing on durability at repeated impacts, using Charpy V - notch specimens in the same bearing conditions as at a Charpy test - the impacts being applied on the opposite side of the notch.

The prestressing of Charpy V notch specimen has been performed by repeated compression tests corresponding to some levels that have represented certain percentages of static compression strength.

For certain values of the compression forces as well as for certain number of cycles applied there has been studied the influence of prestressing upon the durability at impact fatigue, having in view both the periods of fracture initiation and the period of propagation the crack. There has been pointed out that prestress level and the number of cycles applied have significant influence upon the periods of initiation and propagation the crack.

1 Introduction

At present, a special attention is conferred upon some particularities regarding the durability of certain steel and alloys to stresses by repeated impacts, phenomenon known under the name of impact fatigue. Repeated impacts durability of a structure depends on many factors by which the state of prestress as a result of applying of some working stresses.

* *Department Strength of Materials, University POLITEHNICA*
Bv.M.Viteazul, Nr.1, Timişoara, 1900 ROMANIA
E-Mail: msvina@linux1.mec.utt.ro

In literature then are a series of researches regarding the influence of some prestresses upon the impact strength. T. P. Gosh has performed an experimental study upon the influence of variable stress (rotative bending) upon the impact strength of a mild steed, using circular section specimens with a V notch bar at 45^0 [1]. These researches have shown that the impact strength decreases considerably when the prestress level exceeds the limit of fatigue. S.V. Kulkarm has also studied the influence of cyclical stresses upon the impact strength using Charpy V notch [2]. In this case also the researches have pointed out a decrease of the impact strength depending on the stress level and the number of applied cycles. But very few studies approach the influence of prestresses upon the impact fatigue durability by the point of view of initiation phases and fatigue fracture propagation.

Starting from this aspect we performed a study regarding the influence of some repeated compressions (prestresses) upon the durability at repeated impacts of a steel in the cases of there different thermal treatments.

2 Experimental Methods

A series of fundamental researches in the field of impact fatigue base point out the special role upon durability of various factors such us: shape and size of specimens, shape and size of stress concentrations the specimens supporting conditions, mod of loading, etc, [3] [4] [5]. All these factors make difficult a comparison of a experimental data obtained by various authors at impact fatigue tests. Starting from this aspect the authors have proposed a standard specimen for repeated impacts tests. The classic Charpy V notch bar for impact tensile testes used also for dynamic tenacity testing K_{id}. There has been used the usual Charpy specimen (10 x 10 x 55) mm having a V notch bar with $\alpha = 45\,^0$ and $r = 0.25$ mm.

3 Mechanical Characteristics of the Analyzed Steel

The mechanical characteristics of the analyzed steel for the variants of thermal treatment have been determined on Charpy specimens and have referred to: notch ultimate strength and Charpy impact strength. The notch ultimate strength has been defined on the basis of max. force of compression and on the area at notch bar section. The tests have been performed on a universal machine with max. force of 200 KN.

The Charpy impact strength has determined by tests on a Charpy impact machine provided also with the possibility of forces - time recording. In Table 1. there are indicated these characteristics for three variants of the analyzed steel treatment.

Table 1

Treatment	Ultimate Strength in Compression [MPa]	Ultimate Strength in Tension [MPa]	Charpy impact Strength [J]
I	1110	880	70
II	1310	1050	44
III	1170	940	64

For comparation in Table 1 has been put also the ultimate strength at static tension obtained by testes on standard specimens.

4. Testing Machines

The specimens prestresing has been performed without repeated tests of compression on a hydraulic universal machine with a max. capacity of 200 kN. The repeated impacts bests have been performed on an installation concerned by authors which provides the same supporting to Charpy V specimen as at single impact tensile test (fig. 1). The repeated impacts have been applied on the apposite side of the notch.

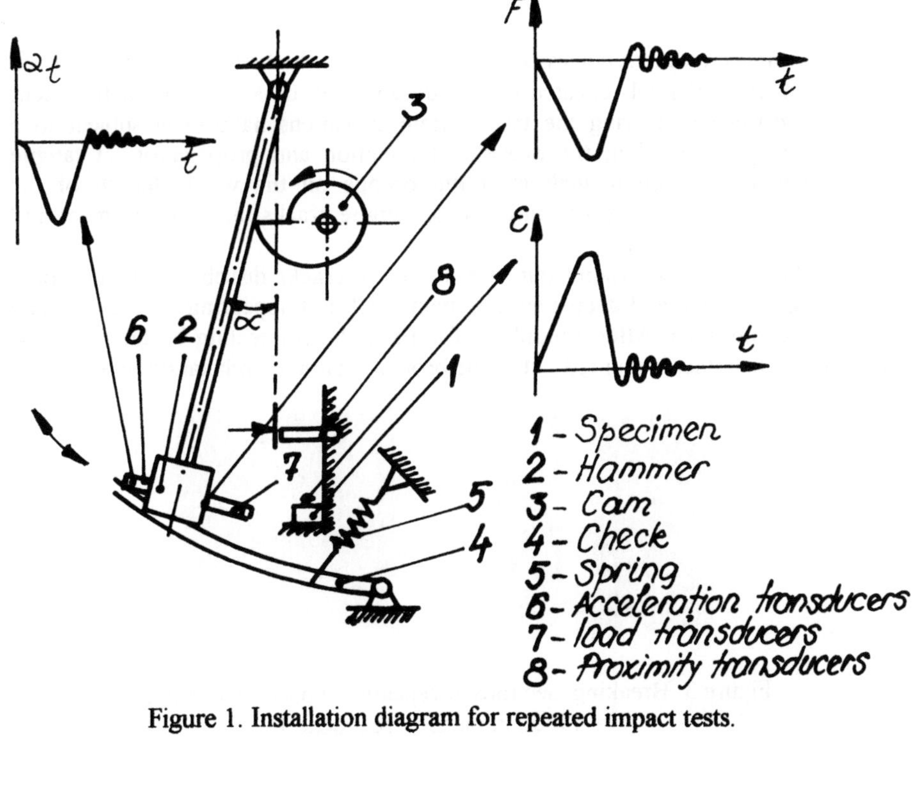

Figure 1. Installation diagram for repeated impact tests.

The installation is provided with four simultaneous tests stands the weights of pendulum hammers differing from one test to another.

In fig. 2 there is presented the variation of impact force corresponding to test 1 at an impact energy $W_0 = 2$ J.

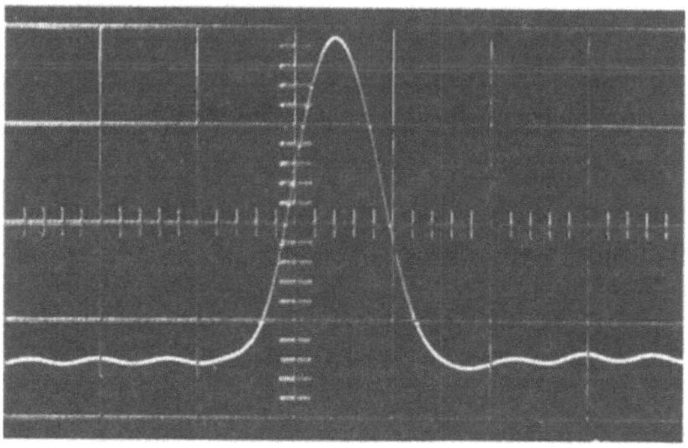

Figure 2. Oscilograme F-t test 1

5 Experimental Procedures

The experimental research consisted of two stages. In the first stage there has been made an analysis of steel repeated impact durability in three variants of treatment without prestressing. During the tests certain specimens have been subject to a rigorous analysis regarding the process of initiation and propagation of fatigue cracks. The notch section analysis at microscope on the whole length of the specimen has permitted the pointing out of fatigue cracks initiation moment and forming of the main crack.

There has been found out that the micro-cracks distributed at the notch section are elongated and deepened, the main crack not appearing on the two side faces of the specimen. After a number of impacts the connecting bridges between there are broken, fact is pointed out in the broken section by relaxation lines (fig.3).

Figure 3. Breaking sections at repeated impacts $W_0 = 0.8$ J;
N= 43200 impacts ; treatment 3

In fig. 4 there are seen these lines of relaxation located at the notch section of which number and orientation depends on the level of intensity of the impact applied. After a certain number of impacts the main crack becomes also visible on the lateral faces of the specimen (fig.4).

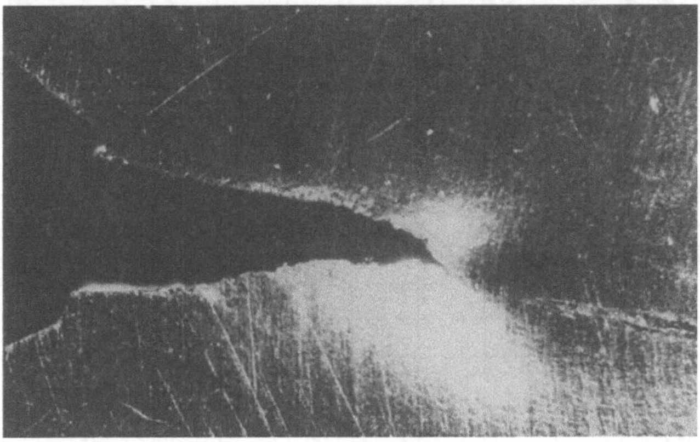

Figure 4. Crack extension for $W_0 = 1.2$ J ; N= 8900 impacts ; treatment 3

Because the crack length measuring on side faces is difficult for valuation of its evolution under the action of repeated impacts there has been applied the oxidation method. So after the application of a certain number of impacts the specimens have been introduced in a furnace and kept at 300-400^{0}C, for one hour in which time the area broken by fatigue is colored by oxidation. In order to limit as exact as possible the oxidated area the specimens have been subject further some fatigue tests at repeated impacts according to the most reduced energetic level $W_0 =$ 0.8 J where they have been kept until the crack had been extend with 0,5-1 mm. The specimens with the area broken by oxidated fatigue and where the crack extension has been configured with another 0.5-1 mm have been broken then by a single impact on a Charpy impact machine. In this conditions the breaking section with clear limiting of the crack length

Within the second stage there has been worked on prestressed specimens by repeated compression for 5 cycles of stressing according to a load equal with 50% of maximum breaking force. The analysis at microscope of these specimens after prestressing hasn't pointed out the presence of micro-cracks.

With these specimens have been repeated the program presented earlier.

6 Results

The curves of durability for both the unstressed and prestressed specimen (impact energy W_0 against number of applied impacts) are shown in fig.5.

In the same diagram there are indicated the lines of regression obtained by statistic processing of experimental data.

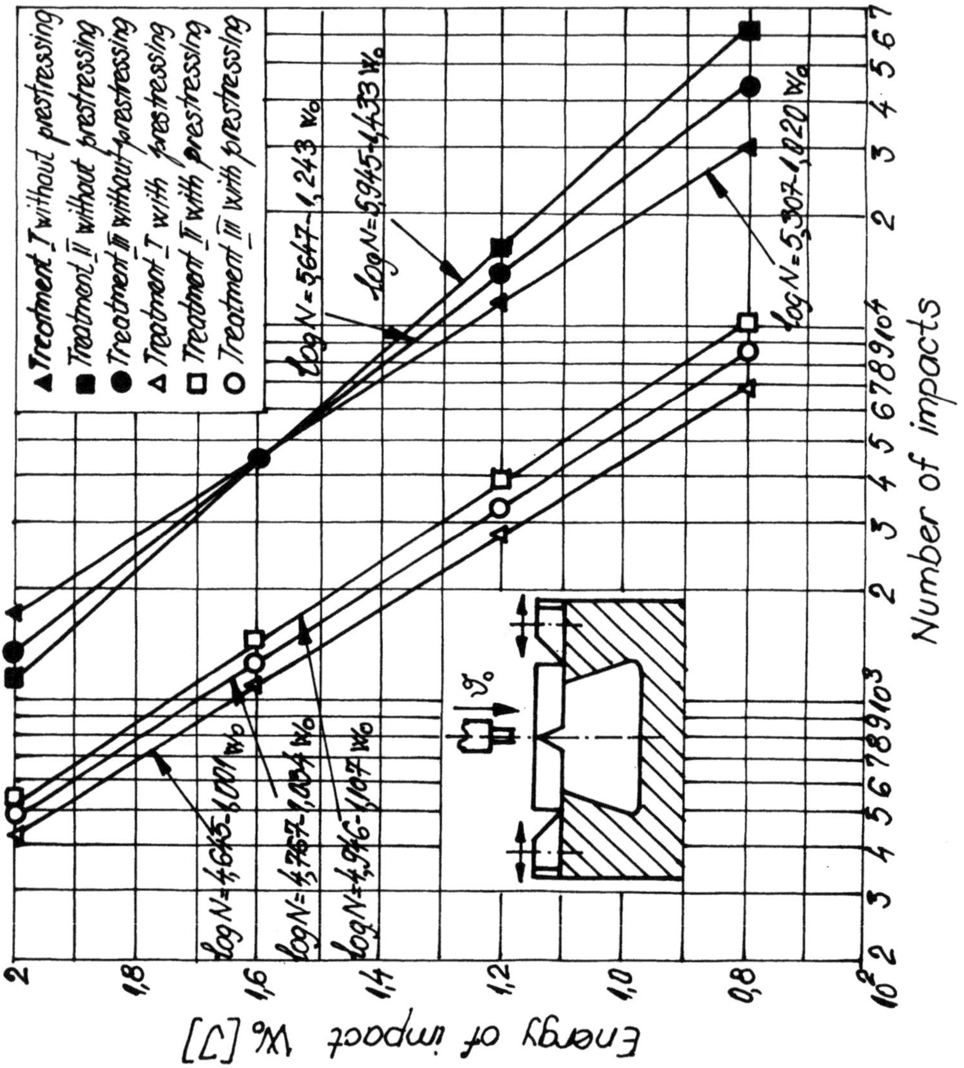

Figure 5. Durability curves W_0 - N.

There is noted the special influence of prestressing upon the durability at repeated impacts. The sting concentration of plastic deformations on the notch sections has conducted to decrease of durability for the three variants of treatment. Applying the oxidation method there have been studied for each stress level the propagation of crack having an initial length $\lambda = 0.1$ mm and which has represented the length from which its extension becomes uniform on the specimen width.

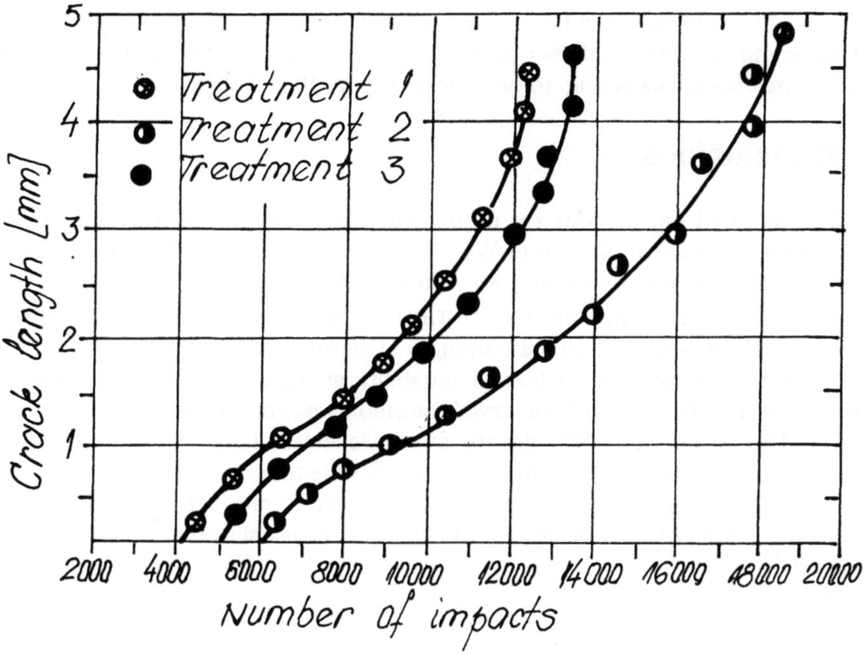

Figure 6. Variation of crack length function of numbers of impacts applied for the unprestressed specimens

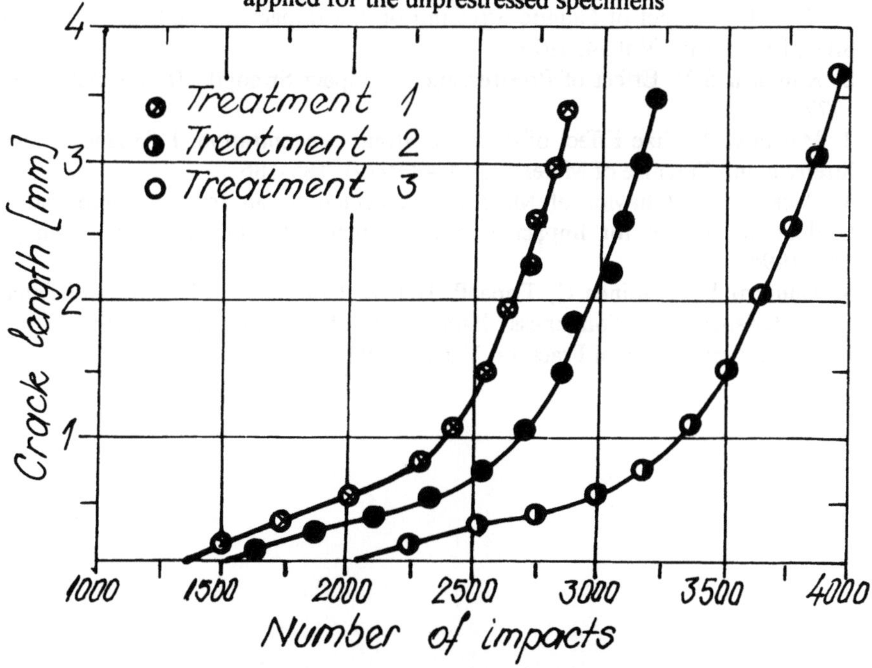

Figure 7. Variation of crack length function of number of impacts applied for the prestressed specimens

For comparison in fig. 6 and 7 are indicated the variation diagrams of the crack length function of number of impacts for the unstressed specimens and those earlier prestressed according. to one impact energy $W_0 = 1.2$ J.

7 Conclusions

* The repeated impact tests on unstressed specimens according to those three treatments have shown that up to $W_0 = 1.55$ J, the highest durability have had the steel corresponding to treatment I for which Charpy impact strength is higher.
Under this energy the highest durability has the steel corresponding to treatment II for which the ultimate strength in tension was the highest.
* * Prestress caused a reduction of durability at repeated impacts for the three treatments and the previously observed transition was not present.
* * * The analysis of crack distribution at repeated impacts in the absence of prestress has pointed out three distinct stages for the three treatments.
* * * * Prestressing decreased the number of cycles member required for crack initiation. The two stages of crack formation that are usually present in fatigue were observed.

References

1 Gosh T P, Effect of Fatigue Stressing on the Impact Strength of Notched mild Steel, I.E.Journal, Vol 54, 1973.
2 Kulkurni S V, Effect of Prestressing on Impact Strength, IE Journal, Vol. 58, 1977.
3 Maekawa I, Size Effect of dynamic Strain concentration, J. Japanese Soc. for Strength and Fracture of Materials, 14-4 (1979), 185-186.
4 Maekawa I, Behavior of Materials Subjected to an Impact Force and New Evaluation Method for Impact Strength, Nuclear Engineering and Design, vol. 150, 1994.
5 Dumitru I, Cristuinea C, Tripa P, Blaj L, Correlation of Static and Dynamic Characteristics for Toughness Estimation of Thermoresistant Steels, Ninth European Conference of Fracture, Varna, 1992.

Identification of damage in bars using PZT sensors and regression techniques

*K.T. Feroz * and S.O. Oyadiji **

A methodology is developed for the identification of defects in solid bars using longitudinal stress wave propagation data in conjunction with regression techniques. For the experimental study, a series of notches of different depths are cut at a fixed axial location of bars. Longitudinal waves are generated within the bars by means of the indirect collinear impact of a surface hardened spherical steel ball on one of the plane ends of the bars. In order to prevent local plastic deformation of the bar ends and to ensure that the waves induced in the bars are plane longitudinal waves, anvils are attached to the impacted ends of the bars. PZT (lead zirconium titanate) tiles of dimensions 5 $\times$ 3 mm, which are cut from standard PZT patches of dimensions 30 $\times$ 30 mm, are bonded on the surface of the bars and used as strain sensors. It is shown that the use of PZT sensors has several advantages over the use of conventional resistance strain gauges. A regression analysis technique is used to relate the strain histories from the notched bars to the strain history from the defect free bar. It is shown that the technique readily indicates the presence and size of a structural defect. In particular, it is shown that the area enclosed by the regression curve increases as the defect size increases.

1 Introduction

Non-destructive damage detection of structures or structural component is an important area, since early detection of the defect avoids the catastrophic failure. These defects can be in the form of cracks, voids and inclusion which can change the characteristics of the structures adversely. This paper demonstrate a non-destructive testing (NDT) method of characterising defects in a square bar-section using wave propagation techniques. The numerical predictions were carried out using the finite element (FE) code ABAQUS [1] while the experimental measurements were made using PZT sensors. Longitudinal waves

*Division of Mechanical Engineering, Manchester School of Engineering University of Manchester, Manchester M13 9PL, U.K.

were generated within the bars by means of the indirect collinear impact of a surface hardened spherical steel ball on one of the plane ends of the bars.

A recent paper by Kalinke and Weber [2] presented a similar method of NDT to detect defect in a rectangular section bar. The bar was modelled with a defect in the form of a notch of depth equal to half the depth of the bar. The maximum frequency of interest was 25 kHz which gave a shortest wavelength of 0.2 m which was found to be much larger than the longest lateral dimension of the bar. A further investigation was proposed to increase the sensitivity of the method so that cracks much smaller than half the diameter of the bar will be detectable. Another approach of characterising the structural waves in notched beams using neural network was demonstrated by Wendal et al [3].

Ultrasonic stress pulse propagation in structures is commonly monitored by means of electrical resistance strain gauges. These gauges consist basically of a very thin foil or fine wire of a metallic material. The filament materials commonly used in these conventional strain gauges are constantan, karma and isoelastic whose gauge factors are 2.0, 2.1 and 3.5 respectively. These gauges are usually used in a Wheatstone bridge arrangement which supplies an excitation voltage to the gauges. The voltage V_{out} equation for a Wheatstone bridge for strain gauge applications is given by [4]

$$V_{out} = \frac{K\epsilon N V_{in}}{4} \tag{1}$$

where K=gauge factor, V_{in} = bridge volts, ϵ = strain and N = number of active arms of the bridge. For an electrical resistance strain gauge, the values commonly used are K=2, $V_{in} = 1.2$ Volts and N = 1 (active arm). For a value of strain $\epsilon = 1\ \mu\epsilon$, Eqn. 1 gives an output voltage of 0.6 μvolts. Because the outputs of these conventional gauges via the Wheatstone bridge arrangement are very small, as illustrated by this example, and because the frequency content of the stress waves measured extend into the ultrasonic range, it is always necessary to employ amplifiers of very high amplification and a very wide frequency bandwidth.

On the other hand, PZT (lead zirconium titanate) tiles of the same physical dimensions as the conventional gauges, when used as strain sensors, give voltage outputs of about $1/12\ V/\mu\epsilon$ which is a factor of more than 10,000 greater than the unamplified output voltages from electrical resistance strain gauges. Consequently, PZT tiles do not require any amplification system but produce signals which can be directly sampled.

Also, a time domain FE simulation of wave propagation in mild steel bars of length 2.44 m was performed. The bars were analysed with and without defects. For the experimental measurements the surface of the bars were bonded with PZT tiles of dimensions 5 x 3 mm, which were cut from standard PZT patches of dimensions 30 x 30 mm using a diamond wheel. These sensors were used for monitoring the propagation of stress pulses induced in the bars by the collinear impact of spherical balls. The defects were introduced in the bar in the form of slots. Furthermore, analyses of the predicted and measured pulses were carried out in order to magnify the defect and facilitate its identification. These analyses were performed using regression techniques.

2 Finite element analysis of the bar

The FE analysis was performed on mild steel bars free of defects and with defects. The mild steel bars were of dimensions 16x16x2440 mm long and the defects were in the form of slots of size 2x2 mm and 2x8 mm. The defects were located at a distance of 915 mm from the impacted end as shown in Figure 1(a). The strain pulses were predicted using the FE tool ABAQUS at one element position. The bars were modelled with 3D solid elements of type C3D20. The model had free boundary conditions. Discretization of the model is shown in Figure 1(b). Longitudinal impact was applied at the central node (node 7448 in Figure 1(b)) located at the impacted end. The impact force - time curve was calculated using Hertz theory of impact for an impact velocity of 4.5 m/sec [5].

3 Experimental measurements of wave propagation in bars

The longitudinally impacted mild steel bar of 2.44 m length was held at three points, one point in the middle and two points at a quarter of the length from each end as shown in Figure 1(c) using piano wires.

For maintaining the same end conditions for each test, an anvil was attached to the impacted end of the long bar using an extremely fine layer of 'petroleum jelly'. The anvil consisted of a 7 mm thick cylindrical disc of the same diameter and material as the long bar. After a single impact, the anvil was replaced by a new one.

The piezoceramic PZT sensors were cut out of 30 by 30 mm square pieces to an approximate size of 5mm x 3mm which were used for capturing the signals and were cemented to the surface of the bar.They were located at the position shown in Figure 1(a). The signals were sampled and stored in a four channel 50 MHz transient recorder card which was installed inside a computer which was also used to display and plot the signals.

A rigid pendulum mechanism was used to impact the bars in the longitudinal direction. The mechanism consisted of a 1m long rod with a spherical head at the lower end and a main cylinder fixed to a rigid base, which contained another small cylinder and a plunger as shown in Figure 1(c). The pendulum was raised to the required height and released. The lower spherical head of the pendulum impacted the plunger which then transferred the energy to the ball to strike the bar with a certain velocity which caused the wave to propagate through the bar.

4 Predicted and measured strain pulses

The predicted and measured strain pulses obtained at the first element and gauge positions, located at a distance of 610 mm are shown in Figures 2(a), (b)& (c). The first trough 'a' in Figures 2(a), (b) & (c) represents the primary compressive pulse. This is the point at which the primary compressive pulse reaches the element or gauge position 1 along the bars as shown in Figure 1(a). For the defect-free bar, the pulse travels to the other end of the bar where it is reflected as a primary tensile pulse (peak 'd') as shown in Figure 2(a). This reflected pulse propagates to the impacted end of the bar where it is reflected as a primary compressive pulse (trough 'e').

For the bars with the 2x2 mm and 2x8 mm slots, the pulse travels to the slot where it is reflected and returned as a secondary tensile pulse, which is small in amplitude and represented by the first peak 'b' in

Figures 2 (b)&(c). This secondary pulse travels to the impacted end where it is reflected as a secondary compressive pulse which is represented by the trough 'c' in Figures 2 (b)&(c). Meanwhile, the original primary compressive pulse travels to the other end of the bar where it is reflected as a primary tensile pulse represented by peak 'd'. This primary tensile pulse then travels to the impacted end and is reflected as a primary compressive pulse (trough 'e'). The pulses 'b' and 'c' indicate the presence of the slot. The procedure for estimating the location of the slot from this data is detailed in previous work [6], [7]. These figures also show a good comparison between the measured and predicted strain pulses.

5. Regression analysis of pulses

Comparing the measured longitudinal strain pulses for the bar without any slot (Figure 2(a)) and for the bar with the 2x2 mm slot (Figure 2(b)), it is seen that there is not much difference between their profiles. Since the amplitudes of the reflected secondary pulses were very small for slots of small sizes, it was necessary to magnify these pulses using signal analysis techniques.

Regression analysis was carried out by plotting the amplitude of the strain pulse obtained for the bar with a defect against the strain pulses for the bar free of defect. The results are shown in Figure 3 (a) & (b). Oval-shaped loops formed due to the presence of the slot would otherwise have been a straight line in the absence of the slot. It can be seen that the size and area enclosed by these loops increase with increase in the depth of the slot. Figure 4 represents an example chart obtained by plotting the measured strains of bars with 2x2 and 2x8 mm slots against the predicted strains of a defect free bar. As expected loops are formed which indicate the presence of the slot; in the absence of any defect, these loops would have been straight lines. A more detailed description of the technique can be seen in elsewhere [7].

6. Conclusions

A procedure has been developed to detect defect in a bar using finite element and regression techniques. The experimental results show that PZT sensors can be used successfully for detecting the defect in a bar by capturing propagated pulses. The regression technique causes formation of oval-shaped loops and the area enclosed by the regression curve increases as the defect size increase. These oval-shaped loops indicate the presence of the defect which otherwise would have been a straight line in the absence of any defect. This techniques enable detection of slots whose depths are only 10 % of the cross-sectional dimensions, which are generally not satisfactorily detected using other techniques.

References

1 ABAQUS User's Manual, HKS(UK) Ltd., Vol. 1,2 and 3, 1992.

2 Kalinke P and Weber T, A non destructive testing method to detect cracks in structures, International Symposium Non Destructive Testing in Civil Engineering (NDT-CE), 1995, pp1275-1281.

3 Wendel R and Dual J, Application of neural networks to quantitative non-destructive evaluation, Proceedings of the 16th ultrasonics international conference, Intelligent Signal Processing, UK, Ultrasonics 34, 1996, pp 461-465.

4 Window A L and Holister G S, Strain Gauge Technology, Applied science publishers, 1982.

5 C Barton S, Volterra E G and Citron S J, On elastic impacts of spheres on long rods, Proc 3rd US Natl Cong Appl Mech, 1958, pp89-94.

6 Feroz K T and Oyadiji S O, Defect Detection in rods using PZT sensors, Sensors and Actuators Conference, Manchester Business School, July 1996.

7 Feroz K T, Impact dynamics of Rods, Beams and Panels , PhD Thesis, Manchester School of Engineering, University of Manchester, 1997.

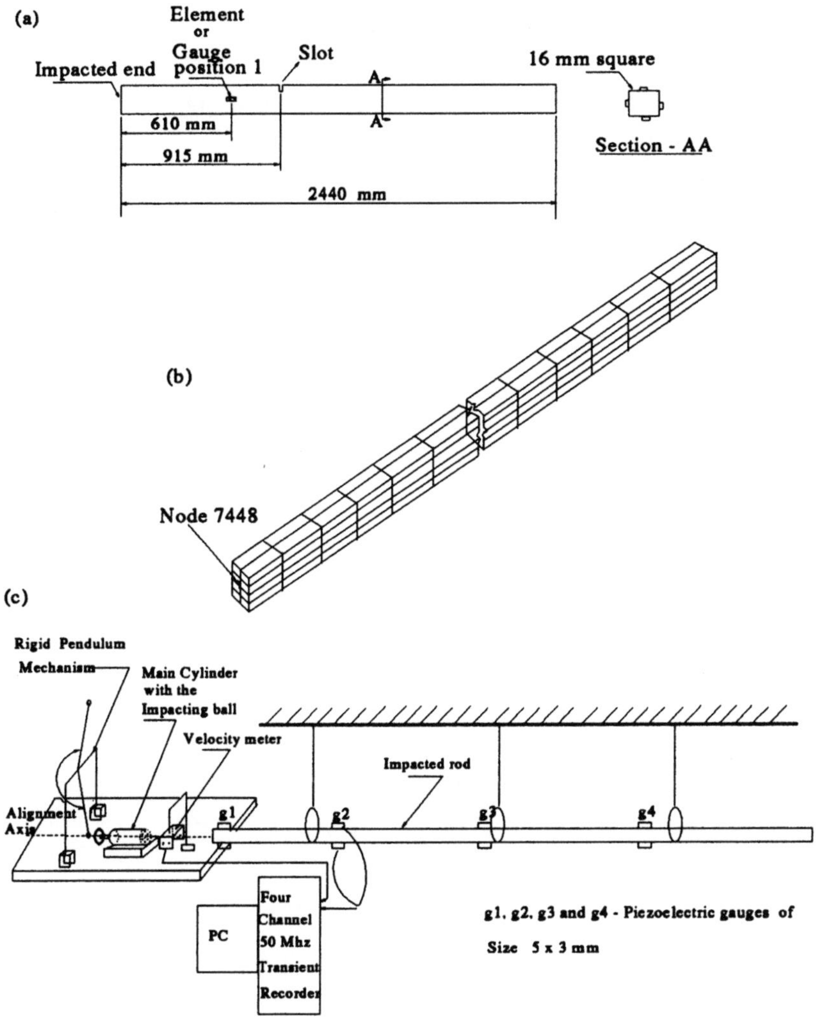

Figure 1: (a) Slot and element or gauge locations on the bar, (b) FE mesh generation using 3D solid elements C3D20, and (c) schematic representation of the experimental set-up for the propagation of pulses in bars.

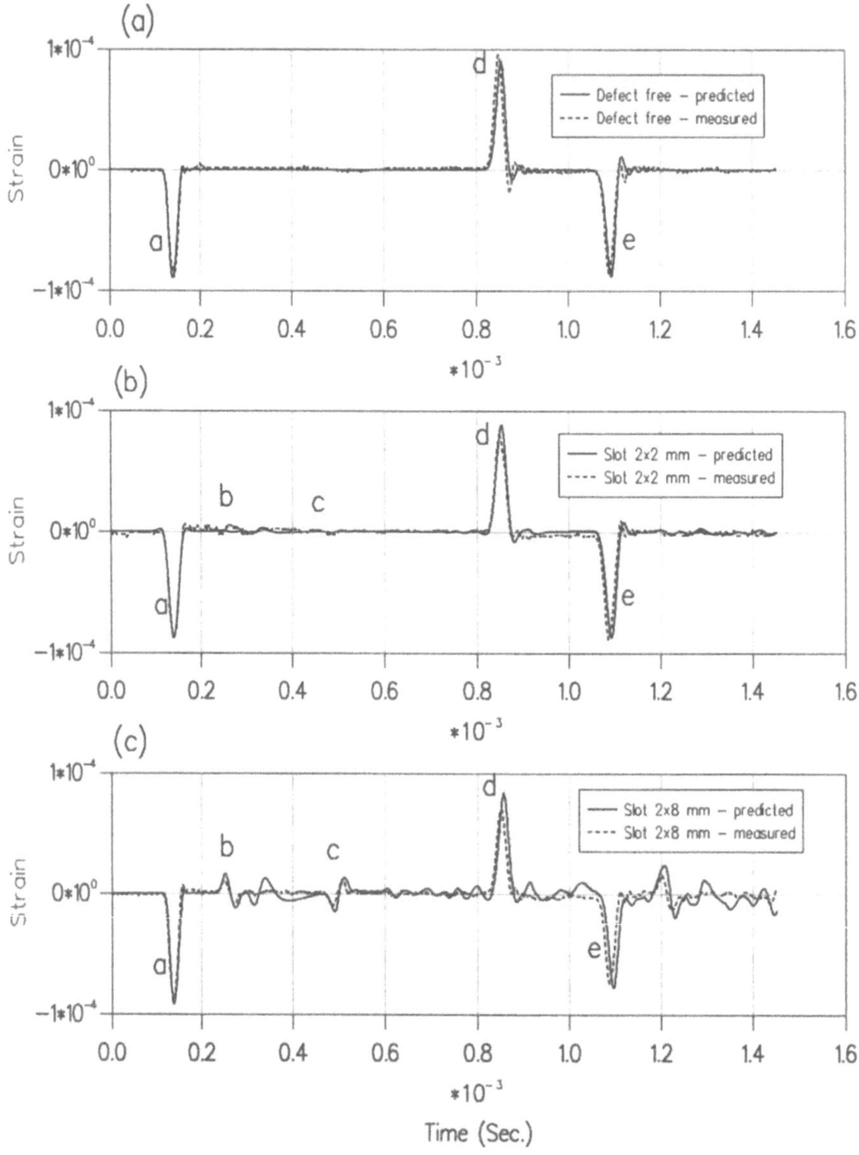

Figure 2: Comparison of measured (dashed) and predicted (solid) pulses at a distance 610 mm from the impacted end (i.e at the first gauge or element positions), (a) for the bar free of defects, (b) for the bar with a slot size of 2x2 mm, and (c) for the bar with a slot size of 2x8 mm.

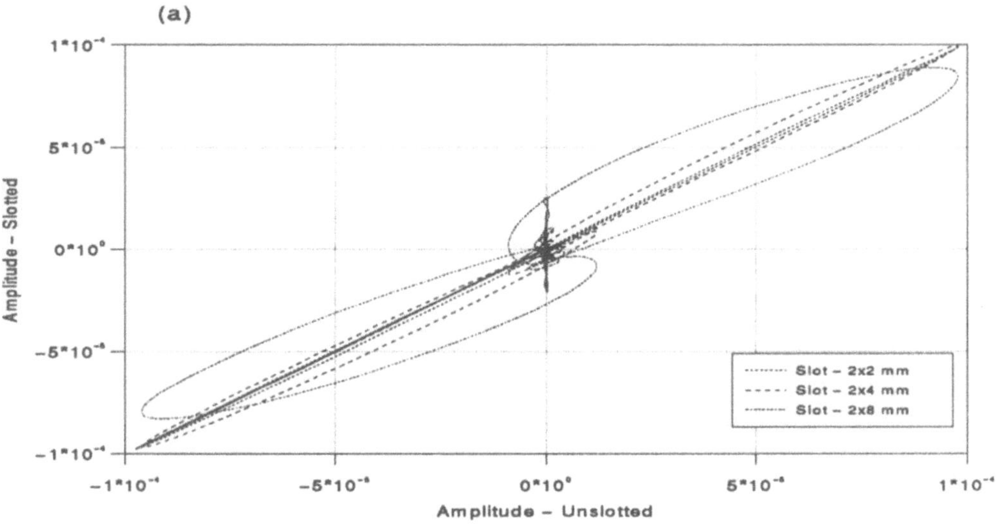

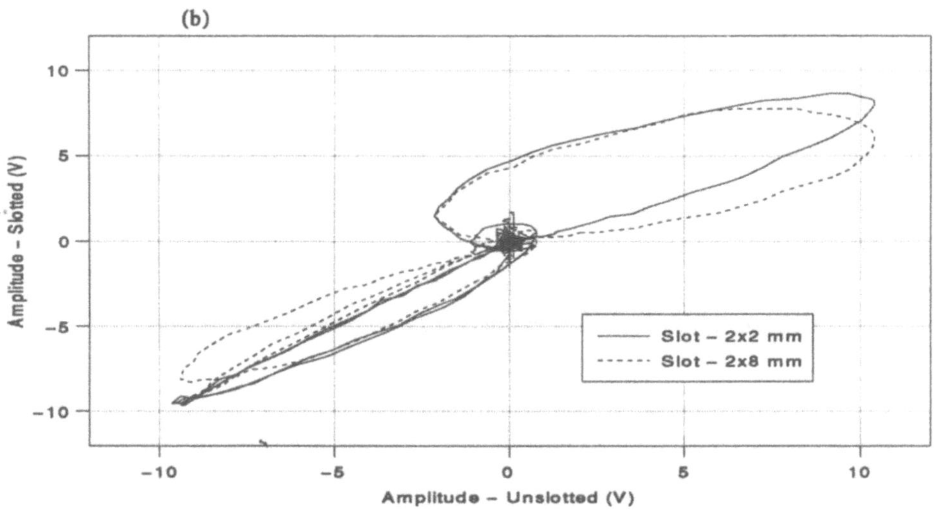

Figure 3: Regression analysis of pulses obtained for the bar with and without defects: (a) predicted pulses (b) measured pulses.

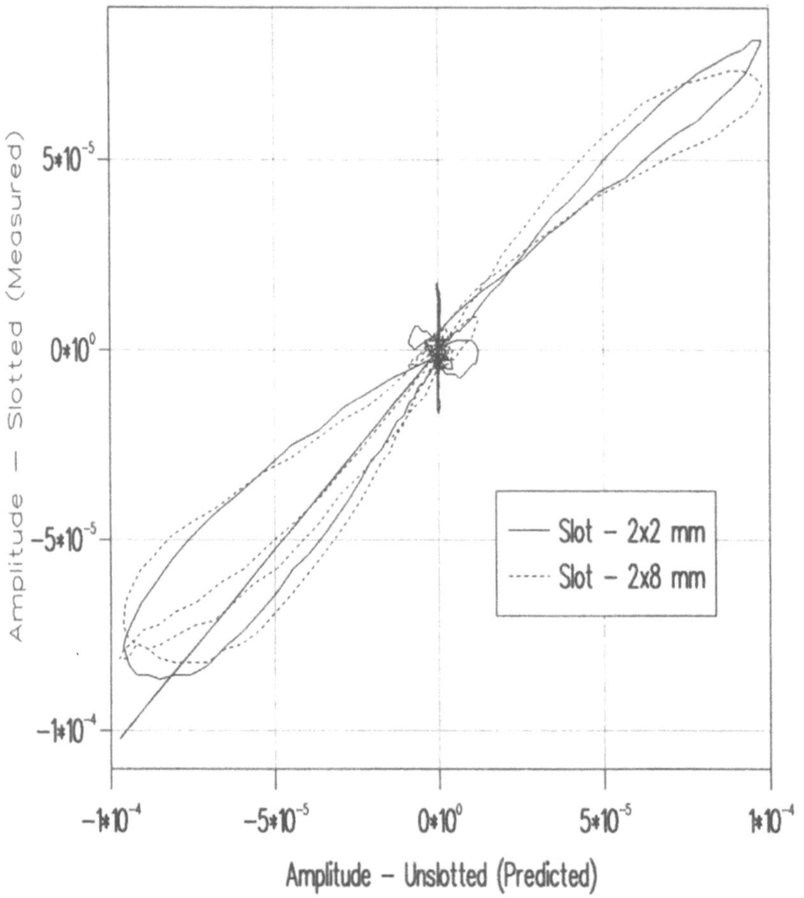

Figure 4: Defect detection using regression analysis.

Laser vibrometry for impact measurements

N.A. Halliwell[*], S.A. Kember* and S.J. Rothberg**

Since the advent of the laser in the early 1960s the field of optical metrology has provided accurate experimental data in situations in which, hitherto, it would have been considered unobtainable. The vibration engineer requires a time resolved measurement of solid surface motion and in an impact situation this can cause problems for contacting transducery and may often preclude their use. This paper examines the use of laser vibrometry to provide an in-situ measurement of vibration velocity when solid surfaces impact. After a brief review of the theory of the measurement the limitations of this technology are addressed before practical examples of successful measurements are provided and discussed.

1 Introduction

Laser Doppler Anemometry (LDA) is now a well established technique [1] which is used for non-intrusive measurements in fluid flows. Fundamentally, LDA measures the velocity of a light scattering particle which is seeded into the flow of interest and it is assumed that the particle follows the flow faithfully. For solid surface velocity measurement, light scattering particles can be considered to be replaced by light scattering surface elements and the technique is then referred to as laser vibrometry. When coherent laser light is scattered from a diffusely reflecting solid surface, such as those found in engineering practice, a laser speckle pattern [2] is formed in space in front of the target. In what follows we will show that the vibration velocity signal is detected by tracking the intensity modulation of one or more speckles when they are mixed with a coherent reference beam from the same laser source. It is the speckle pattern dynamics which limit the performance of laser vibrometers.

2 Theory of operation of laser vibrometers

The physical principle of a laser vibrometer measurement relies upon the detection of the Doppler frequency shift in coherent light which occurs when it is scattered from a moving object. Figure 1 shows a laser beam incident on solid surface target. The Doppler frequency shift f_D in light backscattered by the target moving with velocity U is given by [3]

$$f_D = 2U/\lambda \qquad (1)$$

where λ is the laser wavelength and it has been assumed that the refractive index of the scattering media (air) is unity. In general the Doppler frequency shift measured corresponds to the surface velocity component in the direction of the incident laser beam. In this way, by measuring and tracking the change

[*] Department of Mechanical Engineering, Loughborough University, Loughborough, Leics, UK. Fax: +44 (0)1509 223934.

in f_D, which is typically of the order of MHz, a time resolved measurement of the solid surface velocity U can be made.

The scattered light shown in Figure 1 has a frequency which is of the order of 10^{15} Hz and cannot be demodulated directly. The Doppler shift f_D is measured electronically by mixing the scattered light with a reference beam derived from the same coherent source, onto the surface of a photodetector. The latter responds to the *intensity* from the total light collected and this non-linear detection produces a heterodyne or "beat" in the current output, the frequency of which is equal to the difference in frequency between the two collected beams. This is shown schematically in Figure 2, where a beam splitter has been used to mix the two beams. The output of the photodetector is referred to as the Doppler signal but when this has non-zero frequency it is not possible to distinguish whether the surface is moving away from or towards the detector. For a vibration measurement we require detection of both amplitude and direction of the target motion and this is provided by frequency pre-shifting the reference beam.

Figure 3 shows an equivalent optical geometry in which the reference beam has been frequency pre-shifted by an amount f_R. With reference to this figure it is now clear that when considering the frequency spectrum of the Doppler signal the frequency pre-shift provides a carrier frequency which the frequency shift due to the target surface velocity modulates. When the Doppler signal is demodulated it is now possible to produce a time-resolved analogue of the solid surface velocity in both amplitude and direction as shown schematically in the figure.

A complete description of the various means of frequency pre-shifting the reference beam and of demodulating a Doppler signal is beyond the scope of this paper and readers are referred to the text by Durst et al [1]. All demodulation techniques produce a time-resolved voltage analogue of the Doppler signal with an associated frequency response limit.

3 Practical considerations

3.1 Speckle dynamics and limitations of a measurement

It is the dynamics and characteristics of the laser speckle pattern [2] collected by the photodetector which dictate the noise floor of a laser vibrometer measurement.

An average speckle size can be calculated and attributed to the distribution observed on the screen [4]. Within this correlation region the amplitude and phase of the light may be assumed to be constant. In laser vibrometry, the photodetector active area in the instrument is sampling a speckle pattern. One or more speckles may be sampled, depending upon the optical geometry used in a given measurement situation and also the collecting optics used (if any) for the scattered light. If the detector area collects several speckles, the current output is proportional to the instantaneous sum of the intensity distribution which has been sampled.

When a frequency pre-shifted reference beam is superposed onto the photodetector a heterodyne beat occurs within each speckle area. It is useful to think of each speckle area on the detector as a separate interferometer producing a distinct value of time varying intensity and associated phase. In practice, the photodetector output is the resultant phasor addition of the outputs from all the separate interferometers corresponding to each speckle collected. In this way, for a stationary target the output from the photodetector (i.e. the Doppler signal) has a constant amplitude and phase.

When the target moves normal to the surface and parallel to the incident laser beam, the Doppler effect produces a uniform rate of change of phase across the detector area and frequency modulation of the detector output occurs with negligible amplitude modulation.

If, however, the target tilts or moves other than normal to the surface, the characteristics of the speckle pattern sampled by the detector will change spatially and/or temporally. This effect produces a phase modulation of the detector output, which is demodulated and is indistinguishable from the change in Doppler frequency associated with normal-to-surface movement. This is a major source of noise in laser vibrometers and is referred to as pseudo-vibration [5].

The worst case scenario for speckle noise problems is when laser vibrometer measurements are taken on rotating surfaces. In this situation the speckle noise is periodic and is distributed in spectral peaks at the fundamental rotation frequency and higher order harmonics. It is not unusual to detect levels as high as 10^{-3} m/s in these vibration spectra. Fortunately in impact measurements these levels would normally be negligible but care should be taken in interpreting vibro-impact spectra where the vibration signal itself has been periodic. For a discussion of speckle effects the reader is referred to the papers by Rothberg et al [5-7].

3.2 Response to transient vibration

In addition to speckle noise considerations referred to in Section 3.1, there are two further considerations of importance in a vibro-impact situation. The first is to ensure that the velocity to be measured is within the dynamic range of the instrument as in many cases the velocity amplitudes measured will be well in excess of those normally encountered in routine vibration measurements.

The second and subtler limitation concerns the ability of the instrument to follow a sudden change in velocity of the target i.e. a sudden acceleration. Commercial instrumentation is supplied with a defined frequency response but in a transient situation it is the 'slew rate' of the electronic processing which is the limiting factor. In what follows we suggest a simple means of assessing the required slew rate for an accurate measurement to be taken and how this can be related directly to the frequency response of the laser vibrometer.

A typical velocity impulse v(t) from a vibro-impact situation is shown in Figure 4. With reference to this figure Meier-Dornberg [8] has provided a simple method of obtaining the Fourier transform for any transient time function whose characteristics consist of a "pulse time" to a maximum value $v(t)_{max}$ and a maximum time derivative $v'(t)_{max}$. The expression for the velocity spectrum V(f) can be represented by three straight lines hinged at frequencies f_1 and f_2 defined by $f_1 \cong 1/\pi t_1$ where t_1 is the effective pulse time and $f_2 \cong 1/\pi t_2$ where t_2 is a notional time over which the large value of acceleration $v'(t)_{max}$ is recorded. This spectrum is also shown in Figure 4

A useful guide, therefore, is to ensure that the frequency f_2 as defined by the notional time t_2 over which the acceleration is large is below the upper frequency limit of the laser vibrometer. In practice an upper frequency limit of 150 kHz corresponds to a notional time t_2 of 2 μsec.

4 Applications

4.1 Hammer drill

Measurements were taken on a hammer drill striker as part of a study of the hammer action. The hammer action is contained within a tightly fitting tube, precluding the use of contacting transducers. To allow optical access, part of the hammer drill was replaced by a tube with internal dimensions equal to the original. This replacement tube had two perspex windows and contained the original hammer action, see Figure 5(a). With reference to this figure the striker was covered in retro-reflective tape and the laser beam was incident on the point marked with an **x**. The laser vibrometer measures motion in the plane of the beam and the beam was set at a known angle to the motion of the striker to measure a component of that motion, see Figure 5(b).

Figure 6(a) and (b) show the time resolved velocity signature and spectrum respectively for the striker.

The velocity levels measured are sufficient to ensure that speckle noise effects can be considered negligible. It was estimated that the slew rate necessary for this measurement corresponded to a frequency f_2 of 46 kHz which was well below the 150 kHz frequency response of the instrument used in the measurement.

4.2 Golf ball impact

Modern designs of golf "woods" have a hollow, metal construction which causes them to "ring" with a characteristic sound on impact with the golf ball. The ball experiences gross elastic deformation, with velocities approaching 60m/s during the impact lasting approximately 450 μs. Laser vibrometers have been able to offer significantly greater insight than previous studies which have relied mainly on stroboscopic or high-speed video. The non-contact and remote nature of the measurements is well suited to study of the lightweight bodies – the whole club head mass is only 200g while the ball mass is 45g maximum – particularly when, as in this case, the bodies are projected significant distances from the impact site.

The laser vibrometer used in these studies had velocity and acceleration (slew rate) limits of approximately 15m/s and 2000 km/s^2 respectively. While the acceleration limit proved adequate, initial studies on the golf ball using the "Iron Byron" robot golfer quickly revealed whole body and deformation velocities in excess of the limit within the first third of the impact. To overcome this problem the maximum speed of delivery of the club head was reduced from 45m/s to 35m/s to permit a lateral deformation measurement to be made perpendicular to the line of flight of the ball. At the same time a measurement was made at 75° to the line of flight from which forward velocity was calculated. In both cases probe laser beams were carefully aligned to exclude "other" velocity components such as those from the backspin of the ball.

Figure 7a shows the lateral deformation velocity measurement. Its features include an initial inactive period in which wave propagation is seen through the ball cover without major deformation of the whole ball followed by deformation and recovery phases. The highest acceleration corresponds approximately to a velocity change of 7 m/s in 30 μsec giving an acceleration of 233 km/s which was well within the slew rate limit of the laser vibrometer. Figure 7b shows the forward velocity indicating clearly how significant deformation of the ball occurs in the early stages of impact without appreciable movement of the whole ball.

Of particular concern in the measurements was the issue of the movement of the probe laser beam across the target surface due to the inherent whole body translations of club and ball. This is not from the point of view of speckle noise levels which are low relative to the measured velocities and also lessened in importance because the whole body translation is not oscillatory. This is simply concerned with the fact that the instantaneously probed region at the beginning of a measurement was separated from that at the end of the measurement by considerably more than the laser beam size. As a result, a "displacement compensation technique" [10] has been developed requiring measurements to be made with pairs of parallel vibrometer beams for a closer estimate of the motion of a fixed point on the body as it moves between the beams. The improvement in data quality was particularly apparent in measurements on the golf club crown (top surface) which was believed from laboratory measurements to "ring" at two closely spaced frequencies. Uncompensated data showed no evidence of the expected "beat" but compensated data, such as that shown in Figure 8, showed the effect clearly.

5 Conclusions

The use of laser vibrometers to take impact measurements has been described. In contrast with the use of laser vibrometers in standard vibration measurements, the impact measurement may challenge commercial instrumentation in terms of dynamic range, frequency response and slew rate. Typical velocities in impact situations may be orders of magnitude higher than those encountered in normal vibration situations. It is the ability of the Doppler signal processing to follow rapid changes in velocity (i.e. the slew rate of the processing) which is paramount for an accurate measurement. A simple practical method for relating the required slew rate to the frequency response of the instrument has been introduced.

References

1. Durst F., Melling A. and Whitelaw J.H. 1981. Principles and Practice of Laser Doppler Anemometry (second edition), London: Academic Press.
2. Dainty J.C. (editor), Laser Speckle and Related Phenomena, New York: Springer Verlag.
3. Watrasiewicz B.M. and Rudd M.J. 1976. Laser Doppler Measurements. London: Butterworth.
4. Francon M. 1979. Laser Speckle and Applications in Optics. London: Academic Press.
5. Rothberg S.J., Baker J.R. and Halliwell N.A. 1989. Journal of Sound and Vibration 135, 516-522. Laser vibrometry: pseudo vibrations.
6. Rothberg S.J., Baker J.R. and Halliwell N.A. 1990. Journal of Laser Applications 2(1), 29-36. On laser vibrometry of rotating targets: effect of in-plane and torsional vibration.
7. Rothberg S.J. and Halliwell N.A. 1994. Transactions of the American Society of Mechanical Engineers, Journal of Vibration and Acoustics 116(3), 326-331. Vibration measurements on rotating machinery using laser Doppler velocimetry
8. Meier-Dornberg K.E. 1969. VDI Berichte 135, 9-14. Die Beschreibung von Stossvergängen durch ihre Zeitsfunctionen, Fourier und Schockspektren.
9. Hocknell A., Jones R. and Rothberg S.J., "Experimental Analysis of Impacts with Large Elastic Deformation: 1. Linear Motion", Measurement Science and Technology, 7, 1996, 1247-1254.
10. Hocknell A., Jones R. and Rothberg S.J., "Remote Vibration Measurements: Compensation of Waveform Distortion due to Whole Body Translations", Journal of Sound and Vibration, 214(2), 1998, 285-307.

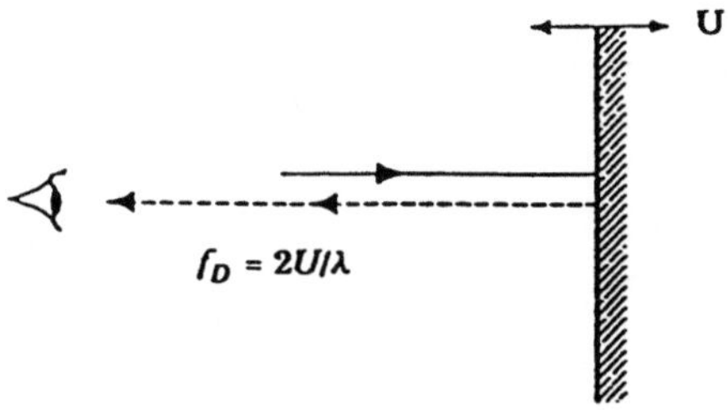

Figure 1: The optical geometry for solid surface velocity measurement

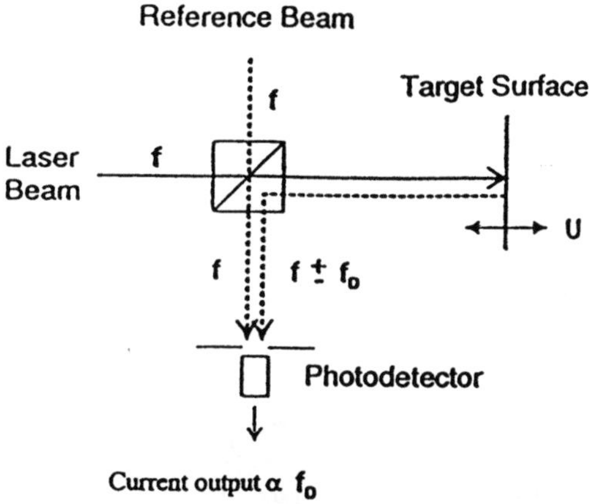

Figure 2: Reference beam heterodyning

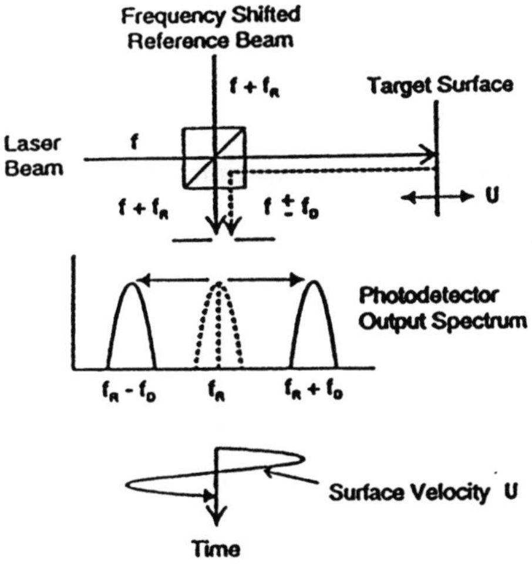

Figure 3: Frequency shifting and time resolved measurement

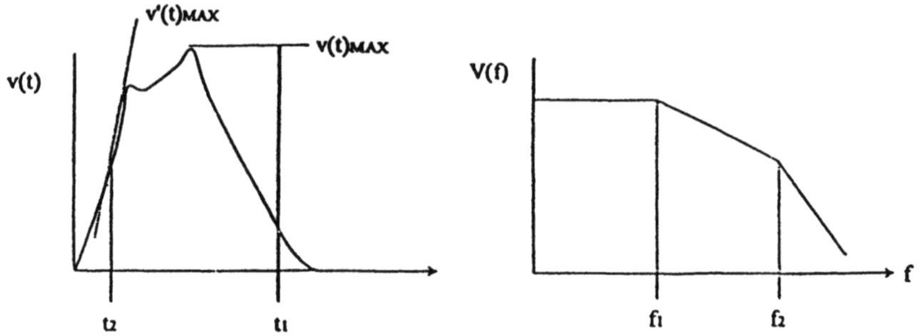

Figure 4: Velocity impulse and approximate spectrum

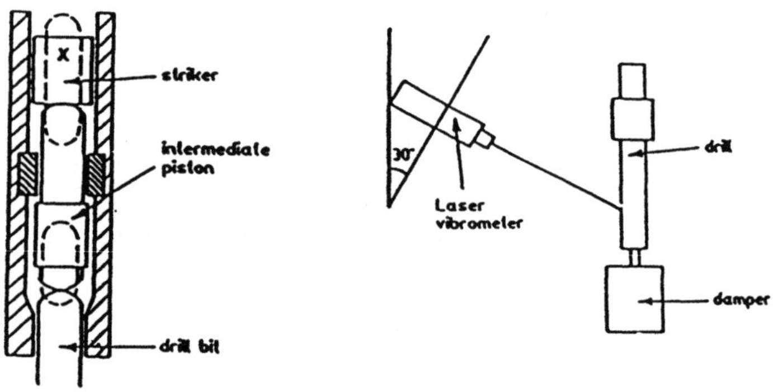

Figure 5: Experimental hammer drill

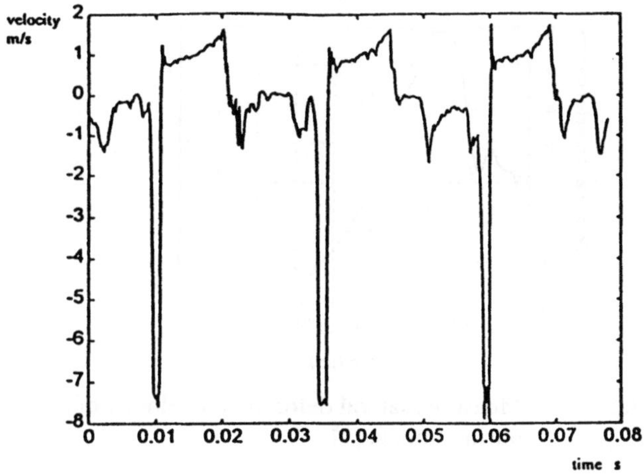

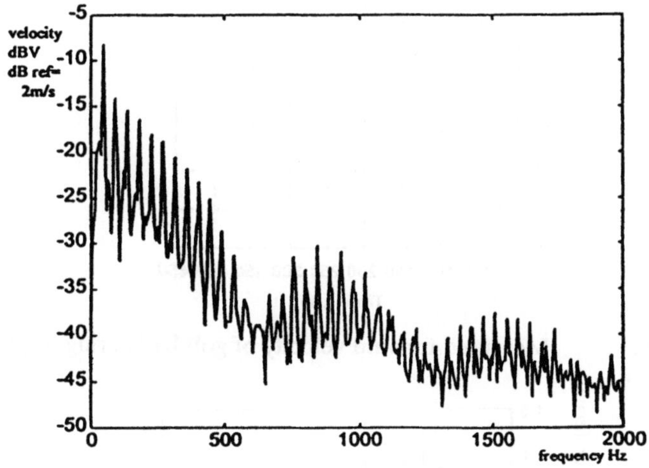

Figure 6: Measurements taken from drill
a) Time domain
b) Frequency domain

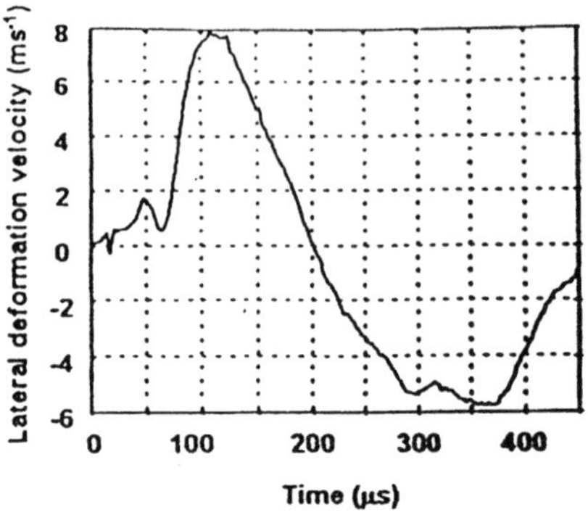

Figure 7a: Measured lateral deformation velocity of golf ball during impact [9]

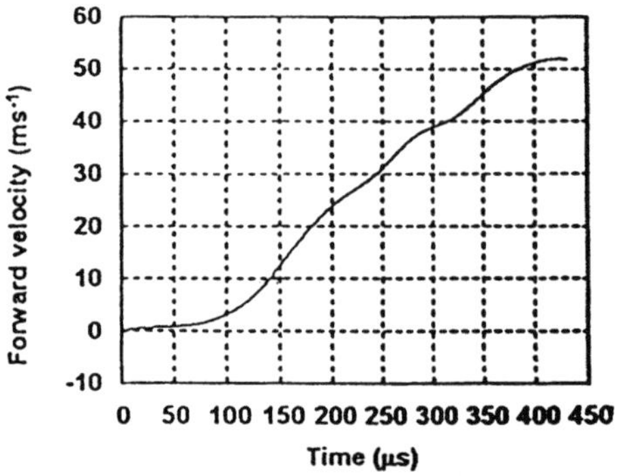

Figure 7b: Measured forward velocity of golf ball during impact [9]

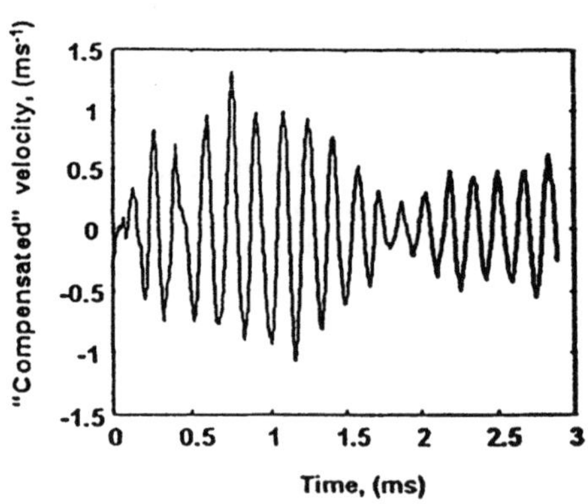

Figure 8: Vibration velocity of golf club crown during impact, re-calculated from original measurements using displacement compensation technique [10]

On the attainable performance of shock control by visco-elastic bumper

A. M.Veprik and V. I. Babitsky**

Modern visco-elastic materials, composed of energy absorbing alloys are used widely for shock control. The simplest visco-elastic bumper may provide the low-rebound and soft trim of undesirable mechanical deflections without generation of dangerous impact accelerations.
In the present paper, the simplified model of a single collision of a free lumped body with the both linear and nonlinear "Kelvin-Voigt" bumper was considered. The analytical solution, describing the dynamics of impact in the linear case was obtained. The parameters of the optimal linear shock absorber along with the attainable performance were estimated.
The enhancement of the capability of shock control was achieved by the use of nonlinear visco-elastic bumper. With the help of the numerical simulation it was shown that the use of the damped softening nonlinear bumper might improve the ability of the bumper to control the peak acceleration and restitution ratio. The approach to the problem of synthesis of the optimal nonlinear bumper was made.

1 Model of linear shock absorber

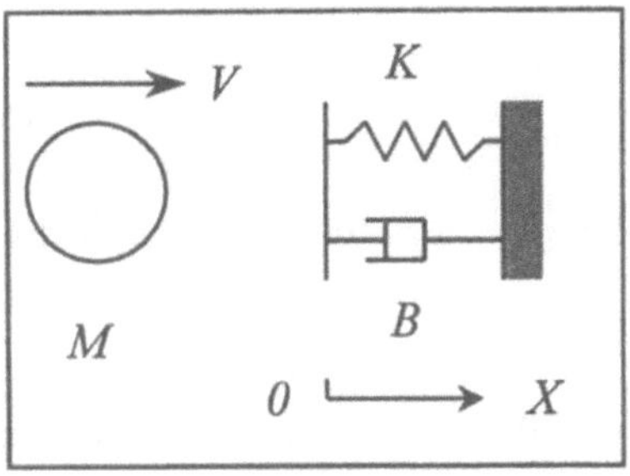

Figure 1

Figure 1 portrays the model of axial collision of a free lumped body of mass M with a linear *"Kelvin-Voigt"* bumper. The bumper was modelled schematically as a parallel combination of linear spring K and dashpot B with "grounded" right-hand side terminals. Such a model was considered in [1].
Dynamic balance between the force of bumper reaction $R = KX + B\dot{X}$ and the force of the body inertia $M\ddot{X}$ governs the development of collision. By equating the above forces, one can obtain the differential equation of the impact – the equation of the cooperative motion of the body and the free terminal of the bumper:

$$M\ddot{X} + B\dot{X} + KX = 0, \tag{1}$$
$$X(0) = 0,\ \dot{X}(0) = V,\ t \in [0, \alpha]$$

where α denotes the moment of separation.

*Department of Mechanical Engineering, Loughborough University, Loughborough, Leicestershire LE11 3TU UK.

The condition of the body separation in time instant $t=\alpha$ is the change of the sign of the force of bumper reaction or of the body acceleration [1]:

$$R(\alpha)=0 \quad \text{or} \quad \ddot{X}(\alpha)=0 \tag{2}$$

By making use of the following notations

$$h=\frac{B}{2M}, \quad A=\frac{V}{\Omega}, \quad \Omega^2=\omega^2-h^2, \quad \omega^2=\frac{K}{M}, \quad \xi=\frac{h}{\omega},$$

the well-known solution of (1) may be expressed as following:

$$X(t)=Ae^{-ht}\sin\Omega t\,, \quad t\in[0,\alpha], \tag{3}$$

It is worth noting that damped and undamped natural frequencies Ω and ω along with the loss factor ξ are the apparent quantities related to the mass of the body.

The corresponding derivatives with respect to time were obtained from (3) as follows:

$$\begin{aligned}
\dot{X}&=Ae^{-ht}(\Omega\cos\Omega\,\mathrm{t}-h\sin\Omega\,t)\\
\ddot{X}&=-Ae^{-ht}\left[(\Omega^2-h^2)\sin\Omega\,t+2h\Omega\cos\Omega\,t\right]\\
\dddot{X}&=-Ae^{-ht}\left((3h\Omega^2-h^3)\sin\Omega\,t+(3h^2\Omega-\Omega^3)\cos\Omega\,t\right)\\
&t\in[0,\alpha]
\end{aligned} \tag{4}$$

By solving (2) for α, the time at which rebound occurs was found out:

$$\alpha=-\frac{1}{\omega\sqrt{1-\xi^2}}\operatorname{atan}\frac{2\xi\sqrt{1-\xi^2}}{1-2\xi^2} \tag{5}$$

The peak value of acceleration may occur at the very beginning of the collision due to the immediate step-wise reaction force $R(0)=VB$ or at the time instant when $\dddot{X}(t)=0$.

The value of the initial acceleration may be calculated by using the statement $M\ddot{X}_0=R=VB$ from which it follows:

$$\ddot{X}_0=\frac{VB}{M}=2V\xi\omega \tag{6}$$

The maximum value of acceleration occurs in the time instant when $\dddot{X}(\vartheta)=0$ ($\vartheta<\alpha$), therefore, with the use of results of (4) for the third-order derivative:

$$\vartheta=-\frac{1}{\omega\sqrt{1-\xi^2}}\operatorname{atan}\frac{(4\xi^2-1)\sqrt{1-\xi^2}}{\xi(3-4\xi^2)} \tag{7}$$

By substituting the corresponding values of the time instant ϑ from (7) into the expression for the acceleration from (4) and after rearrangements the following maximum value has been obtained:

$$\ddot{X}_{max} = V\omega\gamma(\xi) \tag{8}$$

where

$$\gamma(\xi) = \frac{1}{\xi(3-4\xi^2)} \cdot \exp\left[-\frac{\xi}{\sqrt{1-\xi^2}} \operatorname{atan} \frac{(4\xi^2-1)\sqrt{1-\xi^2}}{\xi(3-4\xi^2)}\right] \times$$

$$\times \cos\left[\operatorname{atan} \frac{(4\xi^2-1)\sqrt{1-\xi^2}}{\xi(3-4\xi^2)}\right]$$

Finally, combining (6) with (8), we obtain for the maximum maximorum of acceleration:

$$\hat{\ddot{X}} = V\omega\Gamma(\xi), \tag{9}$$

where

$$\Gamma(\xi) = \begin{cases} \gamma(\xi) & if \quad \gamma(\xi) \geq 2\xi \\ 2\xi & if \quad \gamma(\xi) < 2\xi \end{cases} \tag{10}$$

The value of $\xi = 0.5$ is the exact solution for the equation $\gamma(\xi) = 2\xi$. That means, that for the visco-elastic bumpers with $\xi > 0.5$ the peak of acceleration occurs in the very beginning of collision.

The peak deflection is occurred when $\dot{X}(\tau) = 0$ ($\tau < \alpha$), therefore by the help of the expression for the first-order derivative from (4), the instant of this event occurring may be obtained as follows:

$$\tau = \frac{\operatorname{atan} \frac{\sqrt{1-\xi^2}}{\xi}}{\omega\sqrt{1-\xi^2}} \tag{11}$$

By substituting the corresponding values of the time instant τ from (11) into the expression for the deflection from (4) and after rearrangements, the following peak value was obtained:

$$\hat{X} = \frac{V}{\omega} \cdot \Phi(\xi) \tag{12}$$

where

$$\Phi(\xi) = \frac{1}{\sqrt{1-\xi^2}} \cdot \exp\left[-\frac{\xi}{\sqrt{1-\xi^2}} \operatorname{atan} \frac{\sqrt{1-\xi^2}}{\xi}\right] \cdot \sin\left[\operatorname{atan} \frac{\sqrt{1-\xi^2}}{\xi}\right]$$

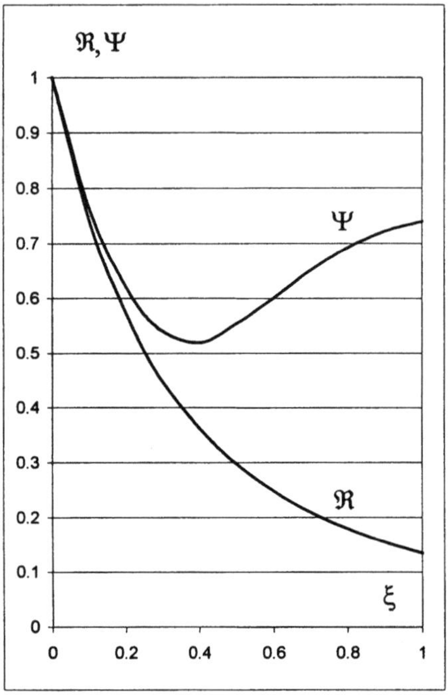

Figure 2

2 Optimisation of linear shock absorber

As it can be seen from (9) and (12), the peak values of deflection and acceleration depend upon the approaching velocity V, apparent natural frequency and loss factor of the bumper. The value $\Pi = \hat{\ddot{X}}\hat{X} = V^2\Psi(\xi)$ where $\Psi(\xi) = \Gamma(\xi)\cdot\Phi(\xi)$, does not contain the factor ω. Figure 2 depicts the function $\Psi(\xi)$. Considering further the factor Ψ as a dynamic quality of shock absorber [2], we conclude that optimal loss factor equals 0.4 independently of the bumper stiffness and approaching velocity. For the optimal bumper $\Psi_{opt} = 0.52$ as shown. In comparison with the undamped bumper, the bonus will be about 50% decrease of the chosen criterion. This result correlates with the result of analysis [2] as applied for the visco-elastic vibration isolator.

By making use of the expressions for the velocity from (4) and for the moment of rebound (5), we can calculate the restitution ratio:

$$\Re = -\frac{\dot{X}(\alpha)}{V} = -\exp\left[-\frac{\xi}{\sqrt{1-\xi^2}}\operatorname{atan}\left(-\frac{2\xi\sqrt{1-\xi^2}}{1-2\xi^2}\right)\right]\times \\ \times\sin\left[\operatorname{atan}\left(-\frac{2\xi\sqrt{1-\xi^2}}{1-2\xi^2}\right)\right] \tag{13}$$

It appears that at visco-elastic impact the restitution ratio is the function of the bumper loss factor ξ only (see Figure 2). This result correlates with [1].

The above analysis higlights the following features of the shock control by the linear visco-elastic bumper:

- The linear visco-elastic shock absorber may be optimised by a proper use of viscous damping. The quality factor enriches the "optimal" value of 0.52 as the loss factor goes to 0.4; that is close to the maximum attainable value of 0.5, as established in [2].
- The peak acceleration and deflections at impact are linear functions of the approaching velocity.
- The restitution ratio perfoming by the optimised bumper equals 0.36.

For particular applications, such as bumpered vibration isolation systems [3,4,5], hand-held percussion machines [6], manipulator's stops, cushioning, etc. it appears necessary to improve the quality of shock absorbing. Such features of a shock control as the peak acceleration limiting and restitution ratio reduction with the growth of approaching velocity may be achieved by making use of the damped nonlinear shock absorber with softening elastic characteristic.

3 Model of nonlinear shock absorber

Figure 3 represents the model of axial collision of a free lumped body of mass M with the nonlinear *"Kelvin-Voigt"* bumper. The bumper was represented schematically as a parallel combination of nonlinear spring and nonlinear dashpot performing nonlinear reaction with elastic and damping components $F(X)$ and $\Phi(\dot{X})$ respectively. Dynamic balance between the nonlinear force of bumper reaction $R(X,\dot{X}) = F(X) + \Phi(\dot{X})$ and the force of the body inertia $M\ddot{X}$ governs the development of collision.

By equating the above forces, the differential equation of the impact – the equation of the cooperative motion of the body and the free terminal of the bumper may be obtained:

$$M\ddot{X} + \Phi(\dot{X}) + F(X) = 0 \quad t \in [0,\alpha] \quad X(0) = 0,\ \dot{X}(0) = V \qquad (14)$$

where α denotes the moment of the rebound governing by the equation (2).

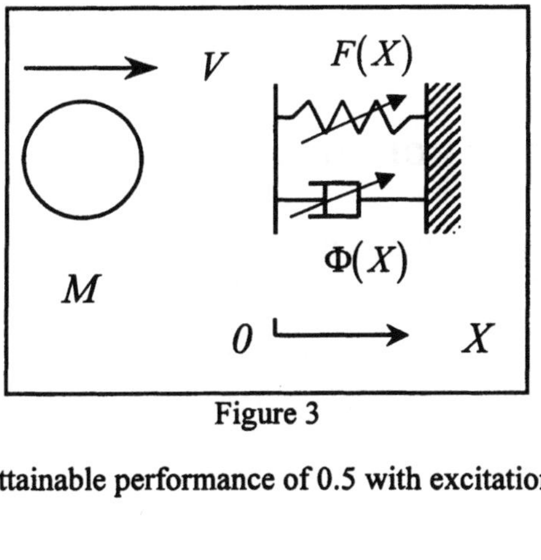

Figure 3

As Mindlin showed in [6], the conservative softening bumper (hyperbolic tangent elasticity) as applied for the design of vibration isolator might perform the function of saturation of the peak acceleration levels with the growth of the excitation energy. It was established also that such vibration isolator has a poor quality factor (close to 1) when the excitation is relatively low-powered, yet it achieves apparently maximal attainable performance of 0.5 with excitation power growth [2].

It is evident also that the restitution ratio of such a conservative bumper equals unity. It makes it inapplicable in a number of cases, such as bumpered vibration protection systems due to the danger of the nonlinear resonance arising as a result of an accidental environmental disturbance.

In order to improve the quality of shock control in a full range of pre-impact velocities, it appears necessary to implement the damping component into the bumper embodiment (as shown in the Figure 3). In the present paper both elastic and damping components were considered as nonlinear elements performing a *hyperbolic tangent type of nonlinearity* as follows:

$$F(X) = KD\tanh\left(\frac{X}{D}\right), \qquad \Phi(\dot{X}) = BU\tanh\left(\frac{\dot{X}}{U}\right) \tag{15}$$

where K, B denote the initial slopes of the curves "elastic force vs. deflection" and "damping force vs. velocity", respectively; D, U denote the characteristic values of deflection and velocity governing the character of saturation of the above curves and, therefore, the full development of the collision. The meaning of these values is evident from the following reasoning: the maximal possible values of the elastic and viscous component of the bumper reaction are limited with the values KD and BU, respectively. Considering (15), equation (14) may be rewritten for the unit mass as follows

$$\ddot{X} + 2\omega\xi U\tanh\left(\frac{\dot{X}}{U}\right) + \omega^2 D\tanh\left(\frac{X}{D}\right) = 0, \qquad t \in [0, \alpha] \tag{16}$$

$$X(0) = 0, \; \dot{X}(0) = V$$

The analytical solution of (16) is not possible. The Simulink model was used for numerical simulations.

As compared with the case of the linear bumper, now we have four variables for the optimisation, namely ω, ξ, D, U. In the framework of the present paper the attainable performance of the nonlinear bumper will be discussed and approach to the optimisation will be made.

4 Atainable perfomance of nonlinear shock absorber

For the present analysis the following parameters of the nonlinear bumper were considered: $\omega = 2\pi \cdot 100 \approx 628(rad/s)$, $\xi = 0.4$, $D = 0.0005m$. The value of the approaching velocity V (expressed in the units of *m/s*) was varied within the interval $[0, 5]$ and the value of the parameter U (expressed in the units of *m/s*) was varied within the interval $[0, 2.5]$.

Figure 4 portrays the bumper quality parameter Ψ as a function of the approaching velocity V obtained at different values of the parameter U. It can be seen that a decrease in the parameter U results in the decline of the bumper quality at the low pre-impact velocities. For example, at $U = 0.01\ m/s$ and $V = 0.1\ m/s$, the above factor enriches the value of $\Psi_{max} = 0.9$ (as shown in Figure 4). The further growth of the pre-impact velocity results in the asymptotic improvement of the bumper quality factor Ψ up to the maximal attainable value of 0.5, thus over-performing the linear visco-elastic bumper (see curve labeled LIN in Figure 4).

The growth of the parameter U (up to the 0.6) results firstly in the decrease of the value Ψ_{max} and does not affect the values of the parameter Ψ in the range of high pre-impact velocities. The further growth of the above parameter results in the growth of both values Ψ_{max} and Ψ in the range of high pre-impact velocities.

Figure 5 represent the behaviour of the parameter Ψ_{max} with the change of the parameter U. It can be seen that there exists the optimal value of the parameter $U_{opt} \approx 0.6$.

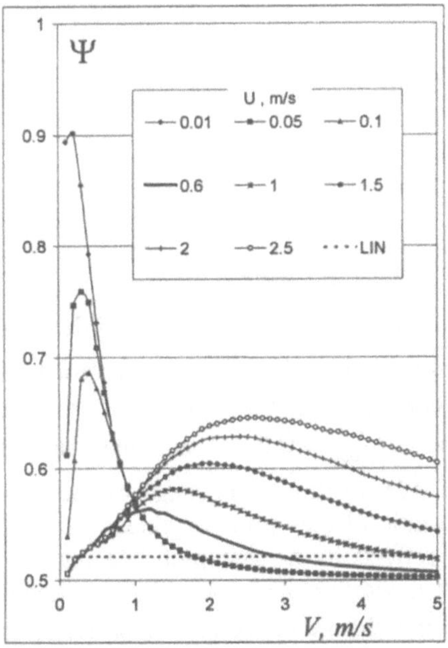

Figure 4

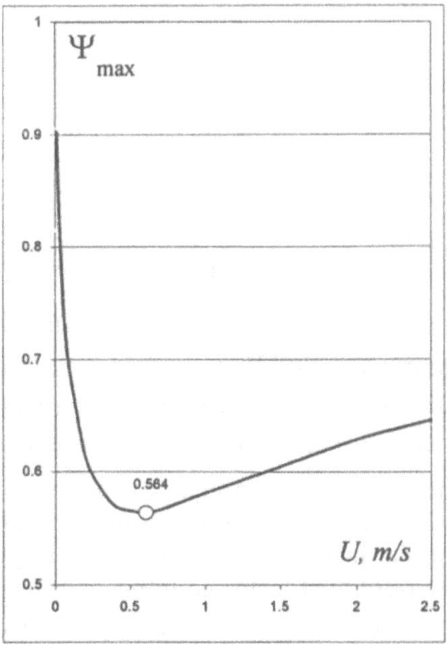

Figure 5

Returning to Figure 4 (solid line without labeling), we conclude that "optimised" bumper behaves in an optimal way as follows:

- Parameter Ψ_{max} is minimised over the full range of the pre-impact velocities. The correspondent minimax value is of about 0.56; that is only 8% more than the maximal attainable value of 0.5.
- Bumper quality is close to the maximal attainable value of 0.5 in the range of high pre-impact velocities.

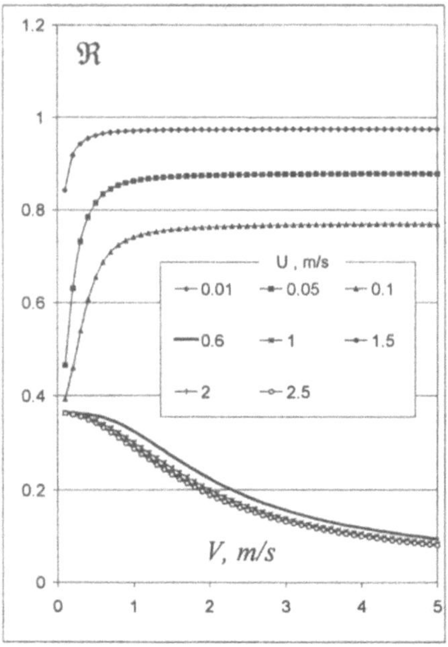

Figure 6

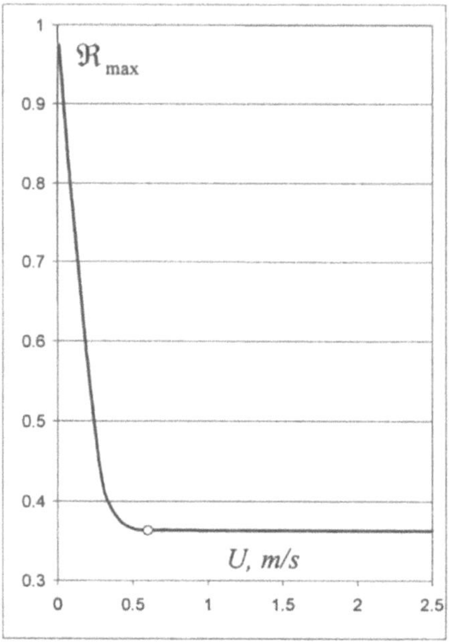

Figure 7

Figure 6 portrays the restitution ratio $\Re_{max}$ of the bumper as a function of the pre-impact velocity V at different values of the parameter U. The lowest values of the parameter U (quasi-conservative case) correspond to a high value of $\Re$. Yet it appears that from the particular optimal value $U_{opt} \approx 0.6$ the maximal value of the restitution ratio equals to the value of 0.36 in the whole range of the pre-impact velocities, thus over-performing the ratio of the optimised linear bumper. Further increase of the pre-impact velocity reflects in the asymptotic decrease of the restitution factor up to the zero level. It is worthwhile to note that additional increase in the factor U is not reflected in the change of the restitution ratio, as shown in Figure 7.

The duty of the limitation of the growth of the peak accelerations with the growth of the pre-impact velocity performed by the nonlinear bumper is illustrated in Figure 8. It is evident how the nonlinear bumper over-performs the linear one over the whole range of the pre-impact velocities. Figure 9 represents the simulated time histories obtained for the optimal value of parameter U at different values of the pre-impact velocity V.

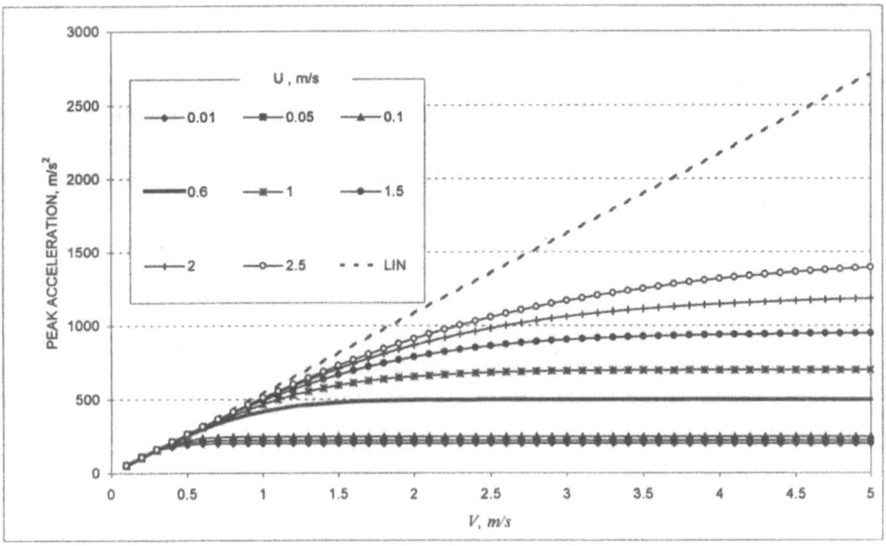

Figure 8

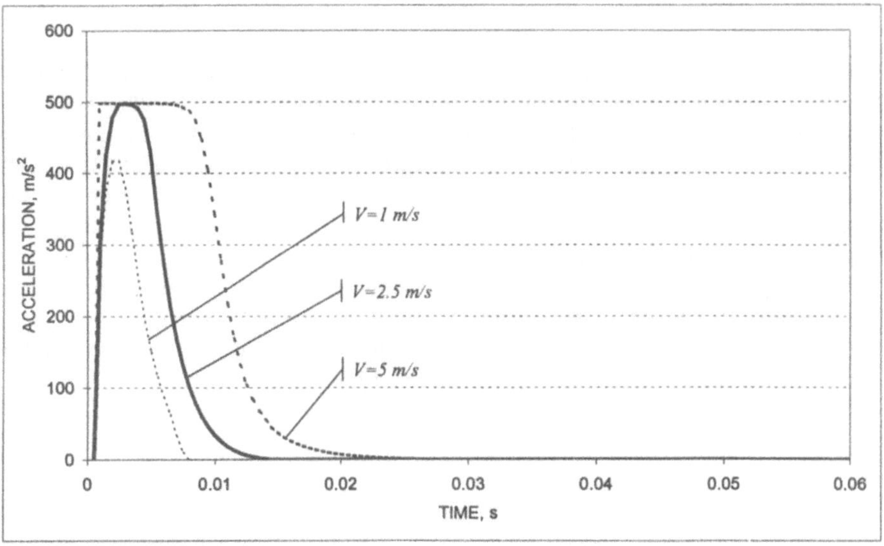

Figure 9

5 Conclusions

With the help of numerical simulations it was shown that a damped shock absorber with softening nonlinear characteristics performs the following features:

- Quality factor of shock control may to be maintained at the optimum level of about 0.5 in the whole range of the pre-impact velocities by the proper use of the damping strategy.
- Peak values of impact acceleration may be limited at any desirable level by the matching the nonlinear elastic and damping properties of the bumper in the whole range of the pre-impact velocities.
- Restitution ratio may be maintained at significantly lower level (as compared with linear visco-elastic bumper and undamped softening bumper) in the whole range of the pre-impact velocities by the proper use of the damping strategy.

The preliminary approach to the problem of optimal synthesis of a damped nonlinear shock absorber was made.

References

1 Panovko JG, Introduction to the theory of the mechanical impact. Moscow, Nauka, 1977 (in Russian).

2 Shock and Vibration Handbook , Harris CM and Crede CE (Eds.). New York: McGraw-Hill Book Company, 1961.

3 Babitsky VI, Veprik AM, 5th International Congress on Sound and Vibration, Australia, Smart nonlinear structure of vibration isolation, 5: 2971-2980, 1997.

4 Veprik AM, Meromi A, Leshecz A, Novel technique of vibration control for split Stirling cryocooler with linear compressor, Proceedings of SPIE's 11th Annual International Symposium on Aerospace/Defense Sensing, Simulation and Controls “AeroSense”, Orlando Florida, Vol. 3061 Infrared Technology and Applications XXIII: 640-651, 1997.

5 Babitsky VI, Veprik AM, Universal bumpered vibration isolator for severe environment, Journal of Sound &Vibration 218(2): 269-292, 1998

6 Babitsky VI, Hand-held percussion machine as discrete nonlinear converter, Journal of Sound &Vibration 214(1): 165-182,1998.

Interaction analysis of mechanical system and hydraulic impact buffer

Vladimír ČECH[†]

1 Introduction

Hydraulic impact buffers (brakes and shock absorbers) are used to damp the impulse forces afflicting the mechanical systems.

In the mentioned meaning we understand damping as a **special control process** for which a priori is given **the ideal law of control** i.e. variation of the ideal hydraulic buffer force F_b with respect to time or path afflicting the parts of the mechanical system and transmitted by hydraulic buffer (brake, shock absorber) (Fig. 1). The law of control is selected on the basis of knowledge of dynamic characteristics of the mechanical system (the natural frequencies proper and mode-shapes, specific damping, etc.), nominal characteristics of the hydraulic buffer (brake) and limiting - variation - conditions (law of conservation of impulses, energy, maintaining the initial force of prestressed elastic elements, hydraulic buffer stroke, etc.). It is respected that the hydraulic buffer force [3,7,8,10]

$$F_b = K_b \left(\frac{x_r}{S_c\left(x_r, \Delta p\right)} \right)^2 \tag{1}$$

in the first approximation is proportional to the power of relative velocity of its piston against the cylinder x_r (Fig. 1) („constant of proportionality" K_b) and variable control area of the throttling (regulation) element of buffer S_c (x_r, Δp), which is a predefined function of the piston motion relative to the cylinder x_r (**a priori control program**) or pressure drop Δp, too.

In actual operational conditions, **the actual law of control F_b** (hydraulic buffer force) can significantly differ from the ideal one as a result of certain systematic and random deviations of system parameters from its nominal values. Thus, the actual forces in the mechanical system can be significantly higher than their projected value.

The most frequent are the changes in the value of exciting impulse and in the volume of dispersing air together with the temperature of liquid of hydraulic buffer. Moreover, in offering **standard buffers** [14], the designer, after all, cannot know the dynamic characteristics of mechanical systems in which the buffers will be used, i.e. he cannot „tune" them for the specific conditions. Similar problem arises when the hydraulic buffers (brakes, shock absorbers) are mounted

[†] Brno Military Academy Department of Weapon Systems, 65 Kounicova St., 612 00 BRNO, CZECH REPUBLIC, Tel. No. 00420-5-4118-2281;, Fax. No. 00420-5-4121-1269, E-mail: cech.v@vabo.cz

to **the test stand** [15, 3, 7] separately or with a part of the original mechanical system.

2 Calculation model

For modelling of the above mentioned effects, the calculation model with at least two degrees of freedom (Fig. 1) should be used while one degree of freedom is needed to model the motion of the hydraulic buffer proper (x_r, F_b) together with the motion of incident to ($t=t_o$: $x_{r\,o}$, $x_r = x_{r\,o}$) or accelerated mass m_r by external force F_e ($t \leq t_o$: $F_e \equiv 0$; $x_r = x_{r\,o}$, $x_r = x_{r\,o}$) and the second degree of freedom provides for modelling, at least approximately, the dynamic qualities of the mechanical system proper (reduced mass m_s, reduced deformation x_s and nonlinear stiffness resulting in reaction R_s by which the mechanical system is contractually loaded).

Nonlinear elastic element (F_r) is intended to maintain the initial configuration of the system ($t< t_0$) and to return it into this position in the final state [10, 14, 15] while negotiating especially the friction forces F_r and the element of gravity g . m_r. In general, the effect of gravity of bodies must be respected as well, i.e. we assume that the axis of the system includes angle φ with the horizontal plane (Fig. 1).

During the process the mechanical system proper is loaded with overall braking resistance (force) R_b

$$R_b = F_b + F_s; \qquad F_s = F_r - F_f \cdot sign\ x_r - g \cdot m_r \cdot \sin\varphi \tag{2}$$

During the process $F_b >> F_s$ (Fig. 4,6)

The system motion is described by two equations of motion

$$m_r \cdot \left(x_r + x_s\right) = R_b - F_e \tag{3a}$$

$$m_s \cdot x_s = R_s - R_b - m_s \cdot g \cdot \sin\varphi \tag{3b}$$

Change in working pressure p_i of any working volume of hydraulic buffer (fig. 1,2) i = 1,2,3,..... is given by differential equation (equation of continuity) [11,3,4]

$$C_i \cdot \frac{dp_i}{dt} = Q_{ci}, \qquad i = 1,2,3,\ldots. \tag{4}$$

where C_i is instantaneous capacity of the i-th volume inclusive the changes in its magnitude (x_r), wall compliance and inclusive the effect of air dispersed with vapours of liquid. Q_{ci} is total instantaneous „retained" (nonexecuted) volume discharge of liquid and the air dispersed with the vapours of liquid of the i-th volume. If air has penetrated into the system, the magnitude of C_i (analogy to mechanical compliance) changes during the process 100 - 150 times (sudden solidification, softening [7]!

Mass outflow of liquid m_j or discharge velocity of liquid v_j by arbitrary hydraulic resistance j = 1,2,.... i.e. by variable resistance, too (Fig. 1, 2) is described by simplified Euler equation (equation of motion) [11, 3, 4]

$$m_j = \rho_k\left(p_k\right) \cdot S_{cj} \cdot \left(v_j \pm \dot{x}_r\right) \quad j = 1,2,\ldots\ldots, i = 1,2,\ldots\ldots \tag{5a}$$

$$\frac{dv_j}{dt} = \frac{1}{2 \cdot l_{Ej}}\left[v_{TB}^2 - v_{Tj}^2 + 2 \cdot l_{oj} \cdot \ddot{x}_s\right] \tag{5b}$$

$$v_{TB} = \sqrt{\frac{2 \cdot \Delta p}{\rho_k\left(p_k\right)}} \qquad \Delta p = p_i - p_k \geq 0 \qquad i,k = 1,2,\ldots;\ i \neq k\,, \tag{5c}$$

where ρ_k (p_k) is the density of liquid at pressure p_k, k = 1,2,...; v_{TB} is the stationary theoretical discharge velocity of liquid by the j-th resistance; v_{Tj} is the nonstationary theoretical discharge velocity of liquid by the j-th resistance; $v_j = \varphi_j \cdot v_{Tj}$ is the nonstationary actual discharge velocity of liquid by the j-th resistance; $\varphi_j = \varphi_j$ (Re_j, L_j, ε_j) is the velocity coefficient [3, 5]; l_{oj} and l_{Ej} are reduced lengths of the accelerated column of liquid [3, 5].

System of differential equations (3, 4, 5) has variable (nonstationary) parameters and it is pulse excited and that is why it can be only numerically calculated.

3 Results obtained

A 100 mm tank gun D10 mounted in T-55 tank (Fig. 3) or mounted to the ballistic range test stand (Fig. 4, 5, 6) has been used for experiments. The decisive parameter for identification is a pressure of liquid in the first working volume of hydraulic brake p_1 (Fig. 1, 2, 3, 4, 5) since during the working stroke (recoil) $p_1 >> \Delta p_{13} = p_1 - p_3$ and $p_1 >> p_2$ (Fig. 2).

The system is designed for an 100 HE-high explosive cartridge, or an 100 APT-armour piercing tracer cartridge ($F_e = F_H$ is a force of shot [1], impulse of shot $I_{H1} \cong 22\ 000$ Ns, $F_{Hmax} \cong 2.8$MN) fired from T-55 tank. The accuracy of identification of parameters obtained is evident from Fig. 3 ($x \equiv x_r$, λ=max x_r). The corresponding curve of $\dot{x}_r$ and R_b shown in Fig. 4 is denoted as „standard". Mentioned here, at the same time are the curve of ideal „R_b - ideal" and the curve of R_b designed by the gun designer „R_b - project" (assumption: noncompressible liquid, air-free system, liquid stationary discharge, constant discharge or velocity coefficients φ_j). Controlled peak output by hydraulic brake is approx. 3.5 MW (Fig. 4: $F_{bmax} \cong 320$ kN, $\dot{x}_a \cong \dot{x}_r \cong -11\text{ms}^{-1}$). The most important natural frequencies of T-55 tank are given by oscillations of the spring-mounted parts and are of approx. 1 Hz, i.e. they do not affect the course of braking process [3] (Fig. 1: $x_r \cong x_a$ or $|x_r| >> |x_s|$).

If the 100 mm tank gun is mounted in a **test stand** [3], the system has its natural frequencies 9; 40.5; 58.3; 82; 108; 147.6;...Hz. In this case, the effect of spring mounting (x_s) of the mechanical system upon the hydraulic brake function proper should not be neglected (formula (1), x_r=x_a-x_s, Fig. 4,6), i.e. the system behaves like a „**bound hydro-mechanical system**". The identification revealed that the decisive natural frequency of the system is approx. 40.5 or 58.3 Hz [3].

The system was excited by the shot of an APDSFST cartridge (Fig. 4, F_{Hmax}≅2.8MN and impulse of shot I_{H2}≅14 000 Ns< I_{H1}) with standard air penetration into the hydraulic circuit of approx. 5.5%. The difference between velocities x_a and x_r („spring-mounting") results in initial „oscillations" of pressure p_1 and consequently of the whole braking resistance R_b which produces the development of „a peak" in the curve of force R_S loading the mechanical system proper, i.e. the system is not sufficiently „tuned".

The behaviour of the system in **a critical situation** (volume of air of 23.5%, Fig. 5,6) was examined by experiments. The coincidence of the calculated pressure curve p_1 (Fig. 5) and applying the theory of neglecting compressibility of liquid and gases and considering the quasi-stationary discharge of the liquid (formula 1)) fails to yield a sound agreement with the experiment (Fig. 5, curve „computed - 2"). The „spring mounting" effect of the system is evident from the curve of velocities x_r and x_a (Fig. 6) which directly affects the curve of pressure p_1 and, consequently the overall braking resistance R_b - 1 (Fig. 6, R_b - 2 - curve of Fig. 4). In this case the system is loaded by maximum value of force Rbmax approx. 510 kN, i.e. a value 1.4 times higher than that of Rbmax - standard ≅ 360 kN.

References:

1. ALLSOP, D. - POPELÍNSKÝ, L. et al: Brassey's Essential Guide To Military Small Arms. Design Principles and Operating Methods. Brassey's, London - Washington 1997, p.361.
2. BREPTA,R. - PŮST,L. - TUREK,F.: Mechanical vibrations. Technical Guide 71, Praha, Sobotáles 1994, p. 590.
3. ČECH, V.: Dynamic characteristics of the hydraulic damper - brake of Gun. MA Brno 1993, p. 173 + 218 = 391.
4. ČECH,V.: Modelling of the nonstacionary liquid discharge through the throttling hole. In: Proceedings of 15th International conference on Fluid Power 95 - 28. ÷ 29.11.1995, Olomouc, p. 122 ÷ 132.
5. ČECH,V.: Discharge coefficient of the throttling hole and dependence on its Reynolds number and on its relative length. In: Proceedings of 30th Spring International Conference Modelling and System Simulation MOSIS '96 - Volume 2, 23. ÷ 25.4.1996, Krnov, p. 99 ÷ 104.
6. ČECH,V.: Basic characteristic of hydraulic Impact Buffers and modelling of their operation. In: Proceedings of National Conference with International Participation Engineering Mechanics ′96 - Volume 2, 13. ÷ 16.5.1996, Svratka, p. 45 ÷ 50.

7. ČECH,V.: Analysis of Cannon Mechanisms Operation. In: Proceedings of the 2nd European Nonlinear Oscillations Conference, Volume 1, 9 ÷ 13.9.1996, Praha, p. 125 ÷ 128.
8. ČECH,V.: The quality modelling of the hydraulic dampers and brakes as the elements for the control of the loads in the dynamic systems. In: Proceedings of National Conference with International Participation Engineering Mechanics '97 - Volume 2, 12. ÷ 15.5.1997, Svratka, p. 23 ÷ 28.
9. ČECH,V.: Hydraulic Circuits of Gun Recoil Mechanisms. In: Proceedings of 2nd Conference on Barrel Weapon Systems. MA Brno 21. ÷ 23.10. 1997, p. 163.
10. ČECH,V.: Recoil mechanisms II. Basics of design and operation. MA Brno 1992, p. 303.
11. HENKE,R.W.: Fluid Power Systems and Circuits. Published by Hydraulics and Pneumatics Magazine 1983, p. 374.
12. JULIŠ,K. et al.: Mechanics. Volume 2, Dynamics. Technical Guide 66, Praha, SNTL 1987, p. 685.
13. NOSKIEVIÈ, J. et al.: Kavitacion in hydraulic machines and devices. Praha, SNTL 1990, p. 334.
14. PROSPECTUS: Stossdämpfer für die Industrie. International firm ENIDINE USA, Orchard Park, NY, USA.
15. REIMPELL,J.: Fahrwerktechnik: Stossdämpfer. Stoss - und Schwingungsdämpfer. Feder - und Dämpferbeine. Konstruktion und Einbau. Würzburg, Vogel - Buchverlag 1983.

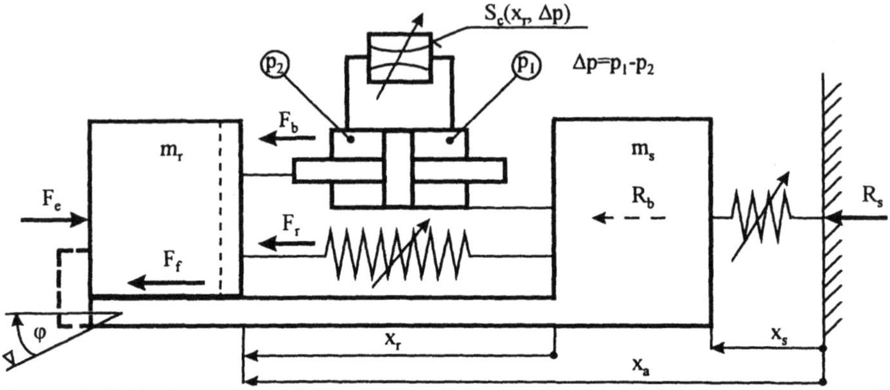

Fig. 1 Calculation model of the system with two degrees of freedom and hydraulic buffer (shock absorber, brake) [3, 7, 9].

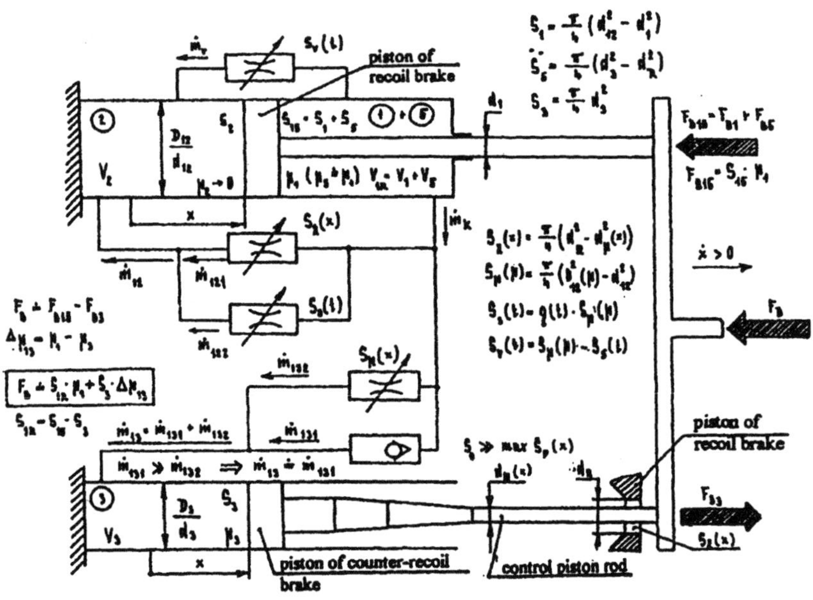

Fig. 2 Simplified diagram of hydraulic circuit of combined brake suitable for modelling the processes during the basic working travel (recoil) of pistons which is used also to model the processes during the reverse working travel (counter-recoil) of pistons.

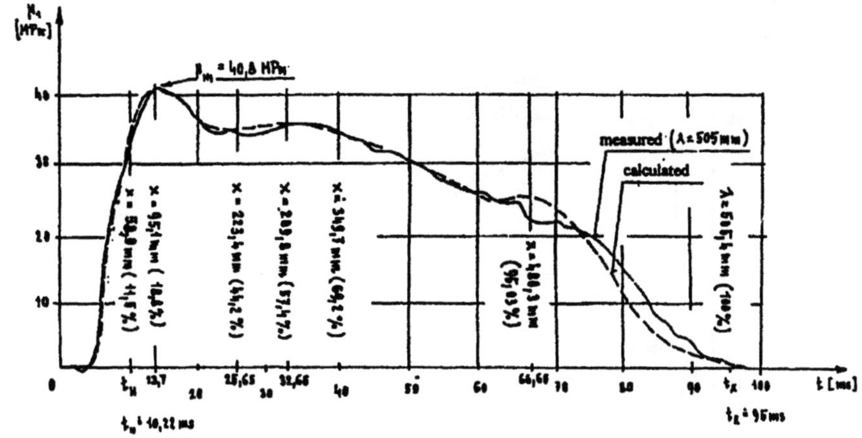

Fig. 3 Comparison of the result of mathematical modelling of the pressure curve p_1 in working volumes 1 and 5 during recoil (Fig.2) of combined brake of 100 mm tank gun of T-55 tank with a measured curve (100 HETP-high explosive target practice cartridge, volume of air approx. 1.8%). The measured curve is filtered by lowpass filter – Butterworth filter of the 10th order with $f_0 = 500$ Hz.

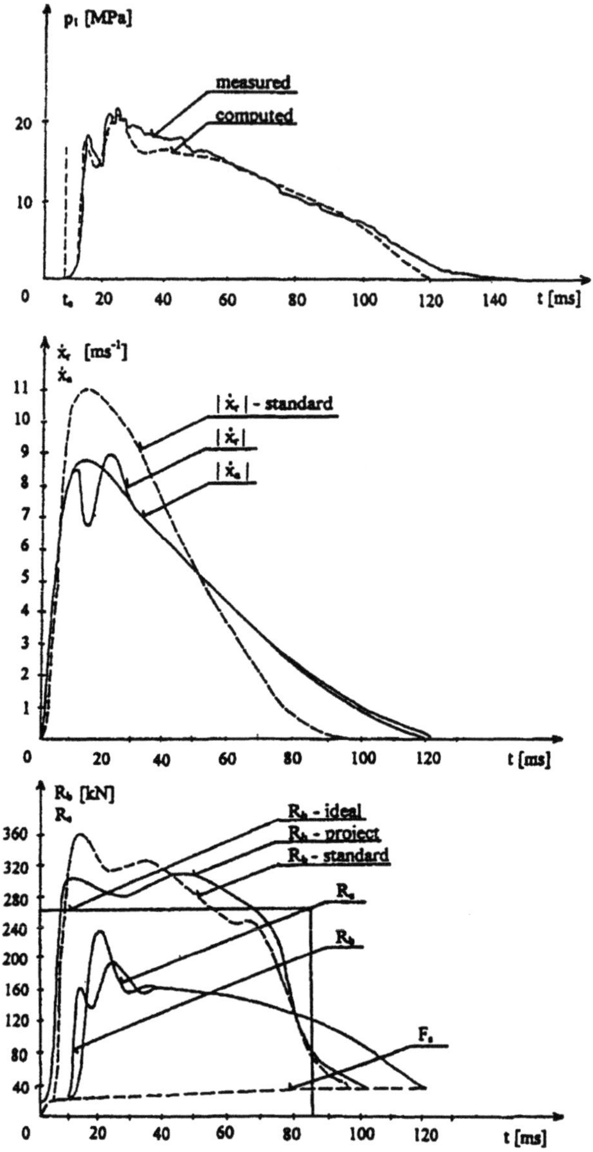

Fig. 4 Comparison of the result of mathematical modelling of the pressure curve p_1 in working volumes 1 and 5 during recoil (Fig.2) of combined brake of 100 mm tank gun of T-55 tank with a measured curve (100 APDSFST-armour piercing discarding sabot – fin stabilised with tracer BM-20; volume of air approx. 23.5%) and corresponding velocity curves $|\dot{x}_a|, |\dot{x}_r|$ and forces R_b, R_s.

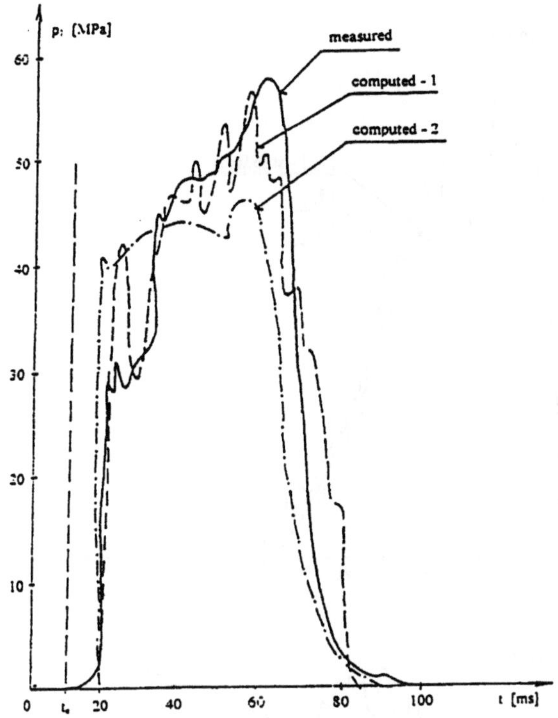

Fig. 5 Comparison of the result of mathematical modelling of the pressure curve p_1 in working volumes 1 and 5 during recoil (Fig.2) of combined brake of 100 mm tank gun of T-55 tank with a measured curve (100 HETP-high explosive target practice cartridge, volume of air approx. 23.5%). The measured curve is filtered by lowpass filter – Butterworth filter of the 10th order with $f_0 = 500$ Hz.

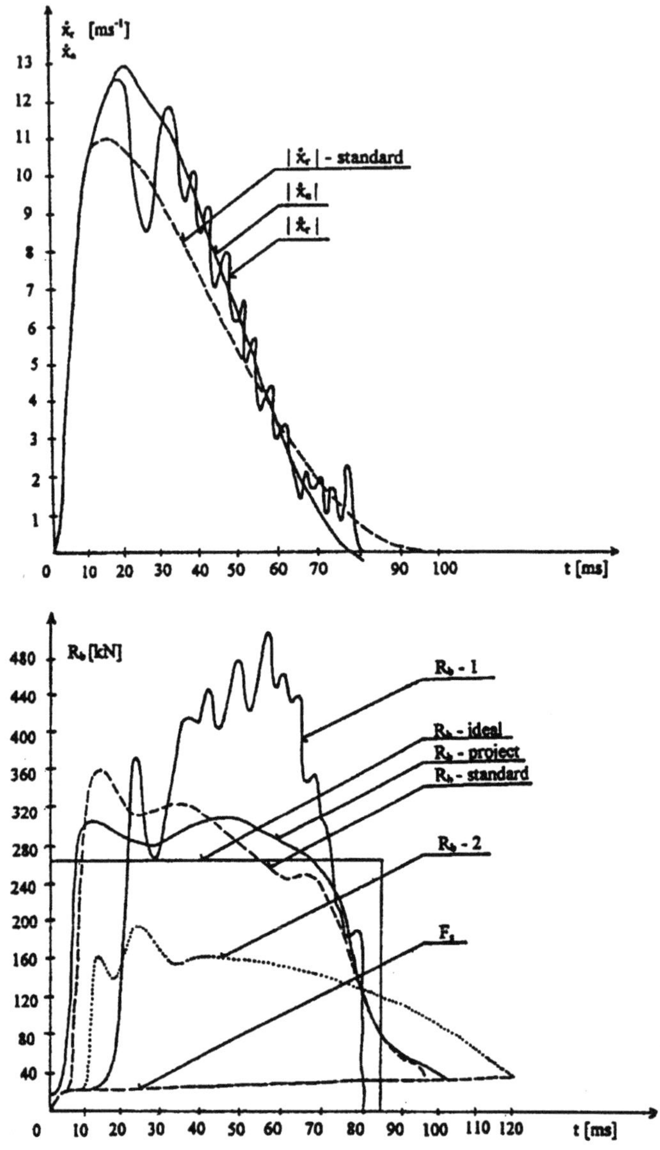

Fig. 6 Calculated velocities $|x_a|$, $|x_r|$ and forces R_b, R_s for the volume of air of the hydraulic circuit of the combined brake of approx. 23.5% and the pressure curve p_l in accordance with Fig. 5 (computed - 1).

Dynamics of the impact force generator

***Barbara Błażejczyk-Okolewska* (✓) *and Krzysztof Czolczynski* (✓)**

During the last years the interest of scientists in multibody mechanical systems in which the phenomenon of impact occurs has been still growing. Numerous investigations are carried out on the possibility of employing this phenomenon to increase the efficiency of operation of industrial machines or to eliminate those phenomena which are undesirable during machine operation. In the present paper a principle of operation of the impact force generator being an element of the rotor of the heat exchanger has been presented. Step disturbances of the rotational velocity of the rotor caused by the generator are aimed at intensification of the heat exchange process.

1 Introduction

In numerous industrial machines the impact of their movable parts is either the basic principle of their operation or the effect which improves their operating efficiency. The classic examples of such machines or devices are: a pneumatic hammer, impact dampers, or heat exchangers. One of many research projects conducted at the Technical University of Łódź deals with a construction of a new heat exchanger characterised by a higher coefficient of efficiency than that of heat exchangers being now produced [1]. One of the factors that contribute to intensification of the heat exchange process are disturbances in the rotational velocity of the rotor of the heat exchanger. These disturbances can have a character of forced torsional vibrations of the rotor, which causes that the angular velocity of the rotor oscillates around a constant average value. These oscillations can have (depending on the assumed parameters of the generator) a harmonical or even chaotic character [2].
Another type of disturbances which can result in intensification of the heat exchange are step disturbances of the rotational velocity. The simplest way to generate such disturbances is to employ the phenomenon of impact which causes, according to the Newton's hypothesis, step variations of the velocity of the bodies impacting on each other.
Below a physical model of the generator of step disturbances of the angular velocity has been presented and a principle of its operation has been discussed. Two selected aspects of the generator operation have been shown: the self-synchronisation phenomenon and the effect of multiplication of the number of fenders on the generator operation.

2. Physical model of the generator

The object of considerations is a system composed of two main parts (Figure 1):

- a rotor equipped with one or more fenders, driven by an electric engine,

---

(✓) Technical University of Łódź, Stefanowskiego 1/15, 90-924 Łódź, Poland

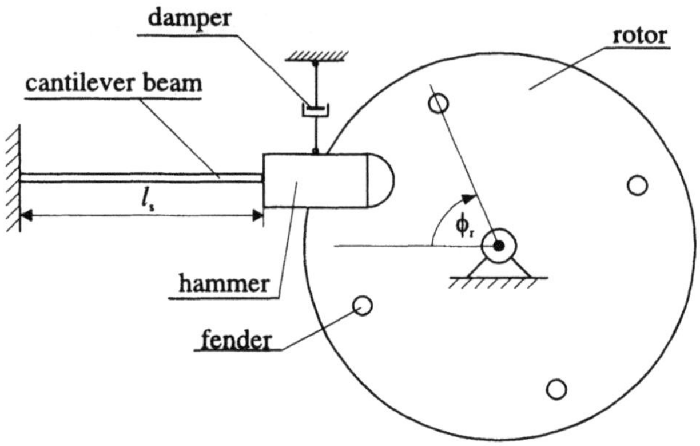

Figure 1. Schematic diagram of the generator.

- a hammer in the form of a cylinder with a semicircular end, mounted on the end of a cantilever beam.

During the operation of the generator, the rotor fenders impact on the hammer, which causes vibrations of the hammer on one hand and the desired step variations of the rotational velocity of the rotor on the other hand. The hammer vibrations are suppressed by a viscous damper. The mass of the cantilever beam and the damper has been neglected. As the geometry of the generator is rather sophisticated, two various kinds of impacts occur during its operation:

I - the fender collides with the cylindrical part of the hammer, and the line of impact is perpendicular to the hammer axis,

II - the fender collides with the spherical part of the hammer, and the line of impact goes through the centre of the fender and the centre of the basis of the spherical part of the hammer.

The numerical simulation of the free motion of both the hammer and the rotor has been made by employing the Runge-Kutta procedure.

The drive engine is connected to the rotor by means of a massless elastic shaft. The aim of this shaft is to minimise undesirable variations of the engine rotational velocity. Equations of motion of both the hammer and the rotor and eqiations of impacts (based on the Newton law) have been presented in details in [3].

3 Data of the system

The next part of this paper presents the results of the numerical simulation of the motion of the system, whose basic data are as follows:

- the hammer:

inertial moment B=12.14×10^{-6} kgm^2, mass m=89.83×10^{-3} kg, radius R=0.01 m, length of the cylindrical part l_h=0.03 m, inertial moment of the cross-section of the spring I=63.6×10^{-12} m^4, logarithmic decrement of damping Δ=ln(1.5), ln(3.5) or ln(7).

- the rotor:
inertial moment B_r=0.01 kgm^2, inertial moment of the engine B_e=0.015 kgm^2, distance between fenders and the axis of rotation r=0.1 m, stiffness coefficient of the shaft connecting the rotor and the engine k_s=0.25×10^3 Nm, coefficient of restitution k=0.9.

4 Influence of the length of the cantilever beam on the system operation

Figure 2 presents maps of impacts for many generators which differ in the length of the cantilever beam l_s (see Figure 1). On these maps we may observe:

- the values of the angular velocity of the rotor before (ω_{bi} ▼) and after (ω_{ai} ▲) impacts,
- the values of the angle ϕ_r describing position of the selected fender at the impact time (see Figure 1),
- the curves showing the value of the basic frequency of free vibrations of the hammer α_1 and its fractions $1/2\alpha_1$, $1/3\alpha_1$ and $1/4\alpha_1$.

Figure 2a shows a map for a well damped hammer (Δ=ln(7)). As can be seen, in the whole range of l_s under investigation the motion of the rotor is regular: during each rotor rotation one impact of the fender on the hammer occurs, and the velocities ω_{bi} and ω_{ai} are the same at each impact. The values of these velocities are only to a slight degree dependent on the length of the cantilever beam l_s. The cause of this phenomenon lies in the fact that, owing to extensive damping of the hammer motion, the hammer is always close to its static equilibrium position at the moment of impact. Significant changes (especially in ω_{ai}) are observable only for $l_s > 0.27$, when the frequency of impact is close to the natural frequency of hammer vibrations α_1. In Figure 2b an impact map for a generator in which the damping coefficient of hammer vibrations is two times smaller (Δ = ln(3.5)) is presented. On this map we may observe that the values of the rotational velocity ω_{bi} and ω_{ai} start to depend on the cantilever beam length l_s, especially in the neighbourhood of the curve 1/2 α_1, when the natural frequency of hammer vibrations α_1 is nearly two times higher than the frequency of impacts. This means that the diminishing of the damping Δ leads to an increase in the possibility of controlling the intensity of impacts (ω_{bi} - ω_{ai}) by means of changes in the cantilever beam length. On the other hand, the diminishing of the damping causes an increase in the vibration amplitude of the hammer. An increase in the amplitude of the hammer vibrations is followed by a loss of motion regularity when the frequency of impacts is close to the natural frequency α_1 ($0.3 < l_s < 0.35$). Then impacts occur at different values of the angle ϕ_r, and, which is even worse, impacts of the second kind occur when the fender collides with the spherical part of the hammer. Such impacts may lead to excitation of significant axial forces acting on the cantilever beam, and, as a result, to damages of the generator. Figure 2c presents a map of impacts prepared for the minimal value (Δ=ln(1.5)) of the damping coefficient. In Figure 2c, a close relation between the variations of both the velocities ω_{ai} and ω_{bi} and the function of α_1, 1/2 α_1 and 1/3 α_1 can be easily observed. For instance: when l_s = 0.12 m, the rotor velocity ω_{bi} (close to its average angular

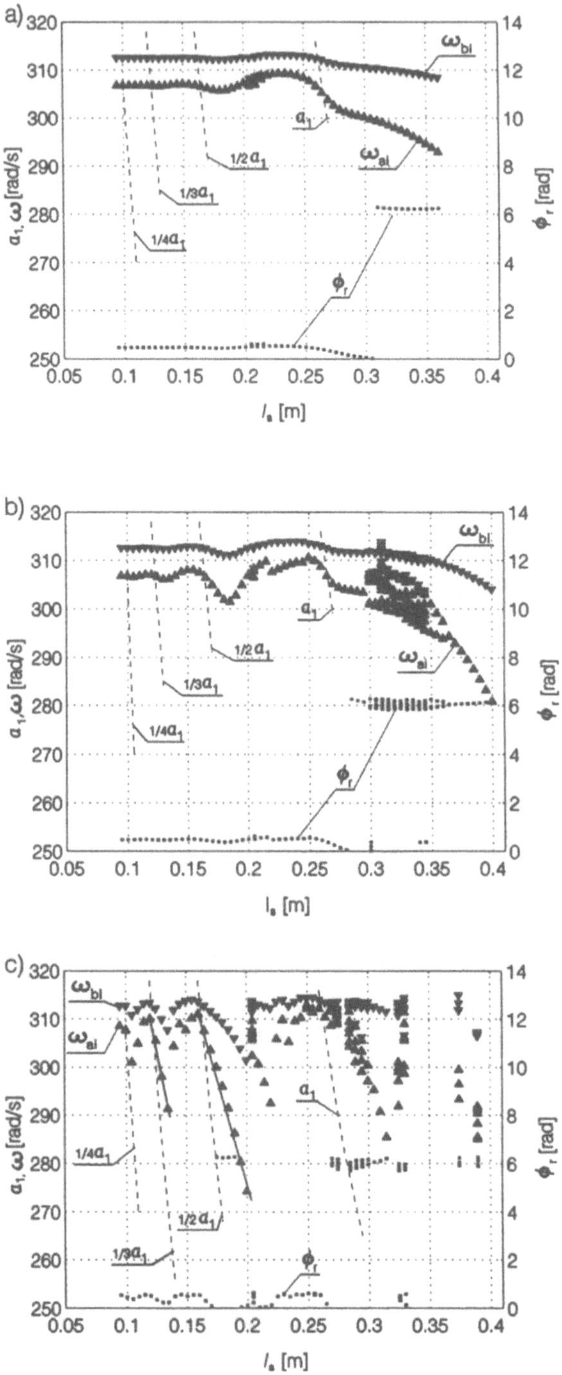

Figure 2. Map of impacts - Generator with one fender; a) $\Delta=\ln(7)$, b) $\Delta=\ln(3.5)$, c) $\Delta=\ln(1.5)$.

velocity) is equal to approximately 1/3 of the value of the basic frequency of free vibrations of the hammer α_1. It means that the impact forcing of these vibrations has a subharmonic frequency. Under such conditions the generation of the hammer vibrations occurs at minimal values of the impact impulse. As a consequence, the values of ω_{bi} and ω_{ai} are high, while the difference between them - minimal.
An increase in the cantilever beam length causes of course a decrease in the value of $1/3\alpha_1$. Despite it, however, the hammer "wants" the impact forcing of its free vibrations to have a subharmonic character, so, as a consequence of the variation of the geometrical configuration of the system at the impact time, the impacts are stronger and stronger, which, as a result, causes the diminishing of both ω_{bi} and ω_{ai}. This way of affecting the angular velocity by the hammer has been called the self-synchronisation of the system and described for the first time in [4]. When l_s exceeds the value 0.16 m, the conditions for easy generation of the hammer vibrations with the next subharmonic frequency $1/2\alpha_1$ arise and the situation repeats.
As can be seen, a choice of the length of the cantilever beam makes it possible to control both the average velocity of the rotor and the intensity of the impacts (the difference ω_{bi}-ω_{ai}).
As a result of the growing amplitude of hammer vibrations, an irregular motion of the hammer is observed in the neighbourhood of the harmonic resonance ($l_s > 0.2$): impacts of both the I and II kind occur, whereas in the neighbourhood of $l_s = 0.35$, the motion of the system is not possible at all: impacts of the II kind cause its jamming. In the range $0.36 < l_s < 0.4$ an irregular motion of the rotor is observed, whereas impacts occur its every 1-3 rotations.
To sum up, control of the intensity of impacts in a generator with one fender is possible for a small value of the damping coefficient of hammer vibrations, owing to the phenomenon of self-synchronisation. The frequency of excitation of hammer vibrations must have a subharmonic character, whereas the machine operation in the neighbourhood of the harmonic resonance of the hammer is not possible at all.

5 Operation of the generator with two or four fenders

Figure 3 shows a map of impacts made for a generator with two fenders mounted on the rotor. Simulation of the generator motion has been performed for the logarithmic decrement of the damping of hammer vibrations $\Delta = \ln(7)$. Despite the fact that the value of Δ is so high, large sensitivity of the system (significant changes in ω_{bi} and ω_{ai}) to changes of the cantilever beam length is observed in the range $0.17 \text{ m} < l_s < 0.25$ m, i.e. in the range between the first subharmonic and harmonic excitation. Both the fenders impact here on the hammer, whereas the velocities connected with these impacts are the same: we observe single lines of ω_{bi} (▼) and ω_{ai} (▲), and two lines of the markers (■) denoting the angular position ϕ_r of one of the fenders, distant by π radians from each other. Thus, a generator with two fenders allows for generation of two impacts during one rotation of the rotor. Moreover, unlike in the generator with one fender (cf. Figure 2a), control of the impact intensity and average angular velocity of the rotor is possible by means of changes in the cantilever beam length. For $l_s > 0.25$ m, a disturbance in the regularity of the motion takes place and next, for

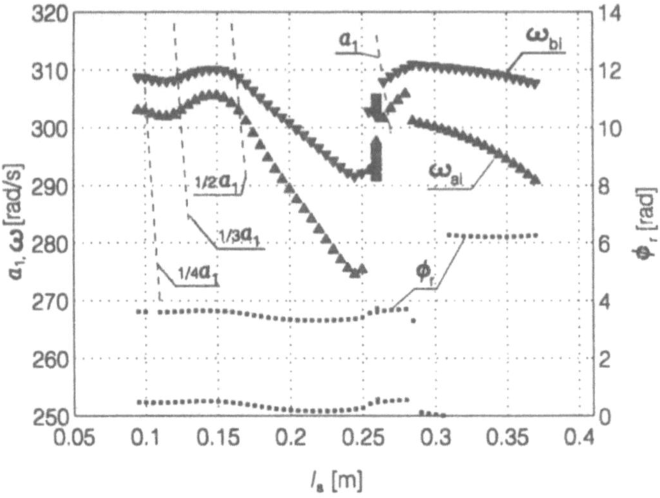

Figure 3. Map of impacts - generator with two fenders.

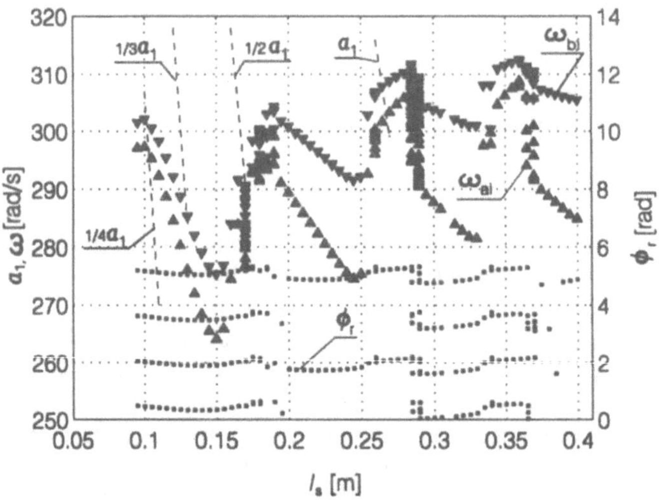

Figure 4. Map of impacts - generator with four fenders.

$l_s > 0.29$ m, an impact on one of the fenders only occurs: in this range the operation of the generator is identical to the operation of a generator with one fender.

Figure 4 presents an analogous map of impacts prepared for a generator with four fenders ($\Delta = \ln(7)$). A basic, regular motion of the system, during which identical (with respect to the value of ω_{bi} and ω_{ai}) impacts of each fender on the hammer occur, is observed in the range 0.1 m $< l_s < 0.17$ m.

For $l_s = 0.1$ m the angular velocity of the rotor is equal to 1/4 of the basic frequency

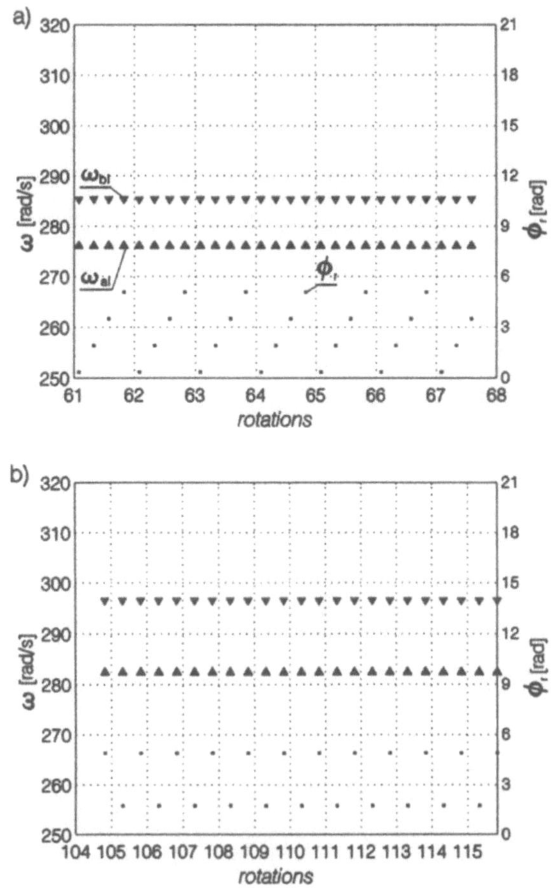

Figure 5ab. Map of impacts - generator with four fenders; a) l_s=0.13 m, b) l_s=0.22 m;

of free vibrations of the hammer, and thus, as there are four fenders, the hammer operates under the conditions of the harmonic resonance. Owing to the phenomenon of self-synchronisation, there is a strong correlation between the function of 1/4 α_1 (dotted line) and the functions of ω_{bi} and ω_{ai} in the range under discussion.

The confirmation of the fact that there are four impacts of the fenders on the hammer during one rotation of the rotor is found in Figure 5a. We can see there the velocities of the rotor before and after the impact and the values of the angular position of one of the fenders as a function of the number of rotor rotations (l_s = 0.13 m).

In the range 0.17 m < l_s < 0.2 m we observe a disturbance of the motion regularity, whereas for larger values, i.e. in the range 0.2 m < l_s < 0.25 m, regular, identical collisions of every second fender with the hammer occur: in this range of l_s the generator operates like a generator with two fenders. A map of impacts showing only

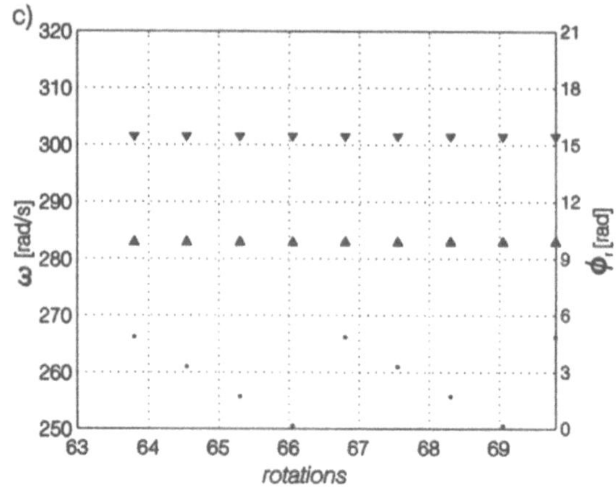

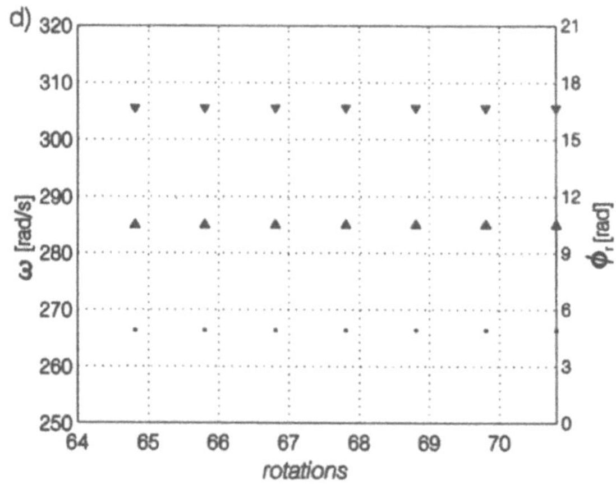

Figure 5cd. Map of impacts - generator with four fenders; c) l_s=0.32 m, d) l_s=0.4 m;

two impacts per one rotation of the rotor for l_s = 0.22 m is shown in Figure 5b.
The behaviour of the generator for 0.3 m < l_s < 0.36 m is highly interesting. Observing single lines of the markers of ω_{bi} and ω_{ai}, and four lines of the markers of ϕ_r, we can have an impression that, similarly as in the range 0.1m < l_s < 0.17 m, four impacts of the fenders on the hammer occur. We are disabused of this false conviction by a map presented in Figure 5c (l_s = 0.32 m). It is clearly seen on it that one impact of the fender on the hammer occurs per one rotation of the rotor, and each fender collides with the hammer every four rotations of the rotor; the rotor velocity before and after the impact is the same at each impact. When l_s exceeds the value 0.37 m only one of the fenders collides with the hammer (cf. Figure 5d for l_s=0.4 m), and the generator operates like a generator with one fender.
In principle a change in the rhythm of impacts is accompanied by a loss of their

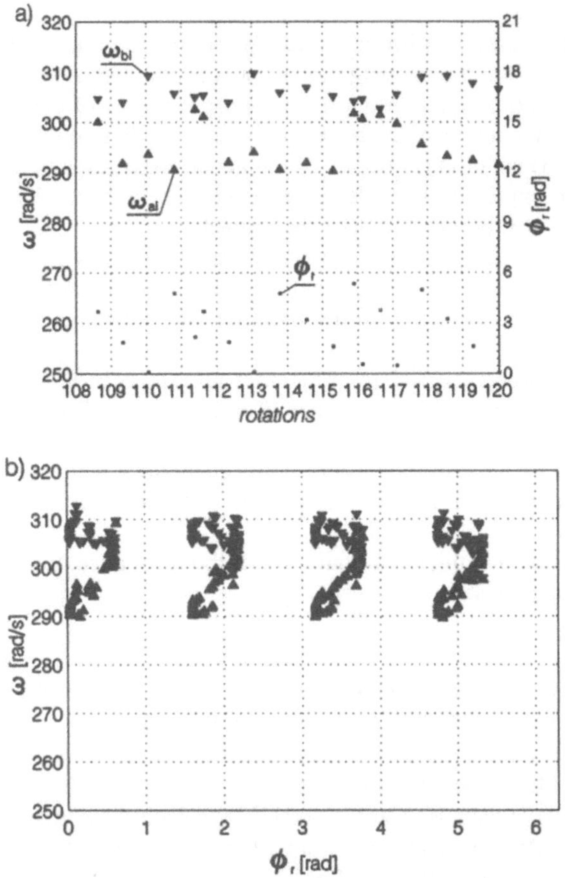

Figure 6. Generator with four fenders, l_s=0.29 m; a) map of impacts, b) Poincare map.

regularity. It takes place, for instance, for l_s = 0.29 m: the blurred columns of the velocity markers (▼, ▲) point out to the motion irregularity, which is confirmed by a map shown in Figure 6a. It can be read from it that one or two impacts of the fender on the hammer occur per one rotation of the rotor. A further analysis of this phenomenon is presented in Figure 6b depicting the velocities ω_{bi} and ω_{ai} as a function of the angular position of one of the fenders. Data from 200 rotations of the rotor have been collected on it. This map is a kind of the Poincare map. It can be seen that the velocity markers form 4 regions which the more densely are filled in, the longer the simulation of the motion is. This fact points out that these regions are attractors and the motion of the generator has a chaotic character.

6 Conclusions

The object of the numerical investigations presented here is a mechanical impact force generator. During the investigations it was found that in the majority of cases the system exhibited a regular motion. By a proper choice of the value of the basic frequency of free vibrations of the hammer it was possible to obtain desired changes both in the average rotational velocity of the generator rotor, and in the intensity of impacts, which is expressed as a difference between the value of the angular velocity of the generator rotor before and after the impact.

In the system under investigation, the phenomenon of self-synchronisation, which consists in the fact that the generator hammer forces such an angular velocity of the rotor at which the motion of both the hammer and the rotor is regular, has been observed.

A value of the damping coefficient of hammer vibrations is of great importance for safe operation of a generator, which has been especially observed for a generator with one fender.

At a high value of the damping coefficient, generators with two or four fenders provide a stronger possibility to control the average rotor velocity and impact intensity.

In the regions situated in the neighbourhood of the resonance, the phenomenon of a chaotic motion can be seen.

References

1 Wawszczak W.: Wysokoobrotowy wymiennik ciepła. Scientific Bulletin of Łódź Technical University, **705**, 207, Łódź 1994 (in Polish)

2 Wawszczak W. and Jagiełło B.: Generation and analysis of torsional vibration. Mechanics and Mechanical Engineering **1** (1), 1997, p. 43-60.

3 Błażejczyk-Okolewska B., Czołczyński K. and Jach A.: Dynamics of the impact force generator. Mechanics and Mechanical Engineering **1** (2), 1997, 103-109.

4 Błażejczyk-Okolewska B. and Czołczyński K.: Some aspects of dynamical behaviour of the impact force generator. Chaos, Solitons and Fractals **9** (8), 1998, 1307-1320.

Driver seat suspension design and identification for commercial vehicles

I Szepessy[*] *and I Wahl*[*]

In commercial vehicles the secondary suspension (seat, driver's cabin) is responsible for the driver's comfort. The fatigue caused by long termed vibrations to which the truck drivers are exposed, can be very severe: spinal hernia, dislocation of the spinal disks, etc. This paper reports on the results of an industrial contract work, optimizing the parameters of a pneumatically suspended seat suspension system. The design problems of the measurement system will be described. The paper introduces the parametric identification process of the seat suspension, and analyzes the measured results from point of view of the human discomfort.

1 Introduction

Designing driver seat suspension the most important thing is the ride comfort. The condition of the comfort is to absorb the vibrations acting from the road and not to allow such a frequency on the seat pad as the eigenfrequency of the driver. The main goal is to develop such a driver seat which can operate in the range of frequency not including the eigenfrequencies of the human driver and the gain of the seat be the least in range of frequency appearing on the floor of the vehicle. Thus it can be acceptable if the eigenfrequency is situated between 1-4 Hz-s.

It is glowingly actual issue to design a controlled driver seat, which assumes the task of establishment of vibration comfort from the body suspension, and it is able to assure better vibration comfort in passive suspensions. It is necessary to have a discrete or continuous model and its parameters of the driver seat to control it (model based control).

More and more commercial vehicles are equipped with air spring seat suspension, which is only statically controlled. The investigated driver seat has the same structure as well. However, it causes same problem that the characteristic of the air spring is linear only in case of small displacements. The shock absorber built into the seat is generally asymmetrical, hence this is can also cause some problems estimating linear parameters.

---

[*]Department of Automobiles, Technical University of Budapest, 1521 Budapest, Hungary

2 Mathematical Model

Fig. 1. shows the 2 DOF mechanical model of the seat suspension with the notations:

m_1: mass of seat pad and frame
c_1: stiffness of the spring
k_1: damping ratio of shock absorber

m_2: mass of human driver
c_2: stiffness of seat pad
k_2: damping ratio of seat pad

h: excitation from the floor of the vehicle.

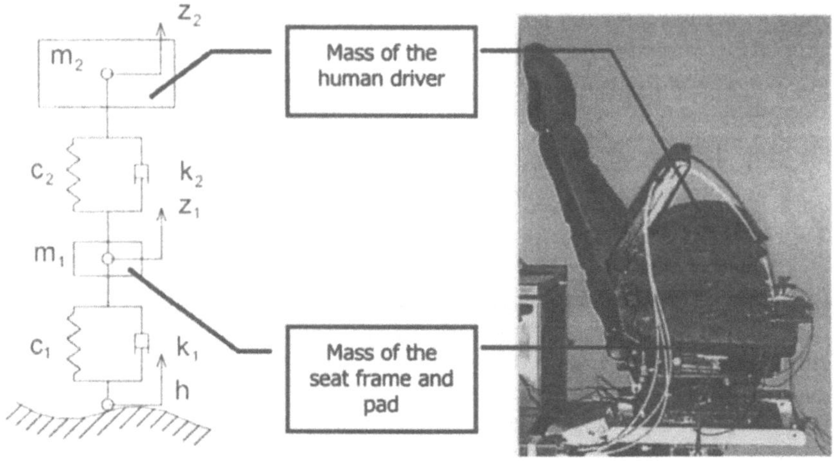

Figure 1. 2 DOF model of the seat suspension

The linear Newton-equations of the dynamic motion can be written as follows:

$$m_1\ddot{z}_1 = c_1(h - z_1) + k_1(\dot{h}_1 - \dot{z}_1) - c_2(z_1 - z_2) - k_2(\dot{z}_1 - \dot{z}_2) \quad (1)$$

$$m_2\ddot{z}_2 = c_2(z_1 - z_2) + k_2(\dot{z}_1 - \dot{z}_2) \quad (2)$$

In spite of the fact that the real seat suspension is nonlinear, we applied linear model as in generally used investigations of dynamical structures. Naturally, this is a good approximation of the real system. After Laplace-transformation (s: Laplace operator, capital letters are Laplace-transformed forms):

$$m_1 s^2 Z_1 = c_1 H - c_1 Z_1 + k_1 sH - k_1 sZ_1 - c_2 Z_1 + c_2 Z_2 - k_2 sZ_1 + k_2 sZ_2 \quad (3)$$

$$m_2 s^2 Z_2 = c_2 Z_1 - c_2 Z_A + k_2 sZ_1 - k_2 sZ_2 \quad (4)$$

From Eq. 4. we can express the displacement of the motion of the seat pad:

$$Z_2(s) = \frac{k_2 s Z_1 + c_2 Z_1}{m_2 s^2 + k_2 s + c_2} = \frac{k_2 s + c_2}{m_2 s^2 + k_2 s + c_2} Z_1(s) \tag{5}$$

The transfer function between the displacement of the seat pad (output) and the frame (input) can be expressed from the Eq. 5.:

$$G(s) = \frac{Z_{output}(s)}{Z_{input}(s)}$$

$$G_{21} = \frac{Z_2}{Z_1} = \frac{k_2 s + c_2}{m_2 s^2 + k_2 s + c_2} \tag{6}$$

Getting the transfer function between the displacement of the frame and the floor it has to write the follows from Eq. 3.:

$$G_{1h} = \frac{Z_1}{H} = \frac{k_1 s + c_1 + k_2 s Z_2 + c_2 Z_2}{m_1 s^2 + k_1 s + c_1 + c_2} \tag{7}$$

The transfer function between the seat pad and the floor can be expressed as the product of the transfer function between the displacement of the seat pad and the frame as well as the transfer function between the motion of the frame and the floor.

$$G_{2h} = \frac{Z_2}{H} = G_{21} \cdot G_{1h} = \frac{k_2 s + c_2}{m_2 s^2 + k_2 s + c_2} \cdot \frac{k_1 s + c_1 + k_2 s Z_2 + c_2 Z_2}{m_1 s^2 + k_1 s + c_1 + c_2} \tag{8}$$

Eq. 7. and 8. include the motion of the seat pad "Z_2". If we want to express the transfer function between the seat pad and the frame without using "Z_2" we have to substitute Eq. 5. for Eq. 3. Expressing G_{1h} we get as follows:

$$\frac{Z_1}{H} = \frac{k_1 s + c_1}{m_1 s^2 + k_1 s + k_2 s + c_1 + c_2 - \frac{(k_2 s + c_2)^2}{m_2 s^2 + k_2 s + c_2}} \tag{9}$$

From this equation we can write the polynomial form of the transfer function between the seat pad and the frame:

$$G_{1h} = \frac{Z_1(s)}{H_1(s)} = \frac{b_3 \cdot s^3 + b_2 \cdot s^2 + b_1 \cdot s + b_0}{s^4 + a_3 \cdot s^3 + a_2 \cdot s^2 + a_1 \cdot s + a_0} \tag{10}$$

where the parameters are the follows:

$$b_3 = \frac{k_1}{m_1} \qquad a_3 = \frac{m_2(k_1 + k_2) + m_1 k_2}{m_2 m_1}$$

$$b_2 = \frac{m_2 c_1 + k_1 k_2}{m_2 m_1} \qquad a_2 = \frac{m_2(c_1 + c_2) + m_1 c_2 + k_1 k_2}{m_2 m_1}$$

$$b_1 = \frac{c_1 k_2 + k_1 c_2}{m_2 m_1} \qquad a_1 = \frac{c_1 k_2 + k_1 c_2}{m_2 m_1}$$

$$b_0 = \frac{c_1 c_2}{m_2 m_1} \qquad a_0 = \frac{c_1 c_2}{m_2 m_1}$$

As we have mentioned the transfer function between the seat pad and the floor can be expressed as the product of the transfer function between the motion of the seat pad and the frame as well as the transfer function between the motion of the frame and the floor:

$$G_{2h} = G_{21} \cdot G_{1h} = \frac{Z_2}{Z_1} \cdot G_{1h} = \frac{k_2 s + c_2}{m_2 s^2 + k_2 s + c_2} \cdot G_{1h} \tag{11}$$

The polynomial form of the transfer function is:

$$G_{2h} = \frac{Z_2}{H} = \frac{b_2 \cdot s^2 + b_1 \cdot s + b_0}{s^4 + a_3 \cdot s^3 + a_2 \cdot s^2 + a_1 \cdot s + a_0} \tag{12}$$

where the parameters are the follows:

$$a_3 = \frac{m_2(k_1 + k_2) + m_1 k_2}{m_2 m_1}$$

$$b_2 = \frac{k_1 k_2}{m_2 m_1} \qquad a_2 = \frac{m_2(c_1 + c_2) + m_1 c_2 + k_1 k_2}{m_2 m_1}$$

$$b_1 = \frac{c_1 k_2 + k_1 c_2}{m_2 m_1} \qquad a_1 = \frac{c_1 k_2 + k_1 c_2}{m_2 m_1}$$

$$b_0 = \frac{c_1 c_2}{m_2 m_1} \qquad a_0 = \frac{c_1 c_2}{m_2 m_1}$$

3 Measurement

The dynamic investigation of the seat suspension is carried out using a test bench including hydropulsator. The schematic structure of the measuring system can be found in Fig. 2. The hydropulsator is controlled by the same computer, which processes the measured signals getting from an AD/DA measuring card.

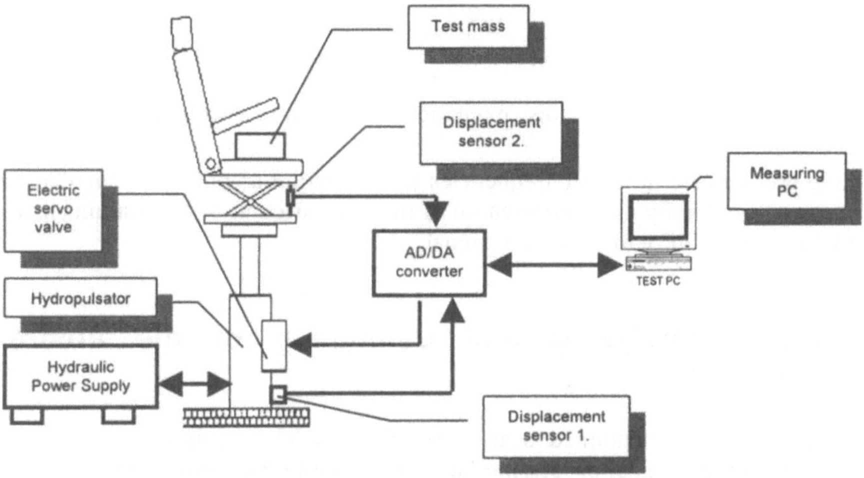

Figure 2. Schematic structure of the measuring system

The measuring PC measures the displacement of the floor (z_1) generated by the hydropulsator as well as the relative displacement between the seat frame and the floor (z_2-z_1). The former is measured by the displacement sensor 1., and the latter is measured by the displacement sensor 2.

The amplitude and frequency of the producible excitation signals by the hydropulsator are limited. The maximal available frequency was 30 Hz-s in case of 1-mm amplitude.

On the other hand, it must be regarded that the seat can be simulated by a two-mass, 2DOF vibrational system, where the sprung parts are the spring and damper between the scissors mechanism and the springy seat pad. The second eigenfrequency of the seat is over 20 Hz-s because of stiffness and damping rates.

It had to create an excitation signal, which covers the 0-30 Hz frequency range to estimate the parameters of the seat. The 0-30 Hz frequency generated white noise is fulfilled the expectations (Fig. 3.).

As we have mentioned above the real seat suspension system is nonlinear. Since the nonlinear system had been modeled by a linear system it was expedient to keep the amplitude of the excitation signal in small range, in which the linear approximation is acceptable.

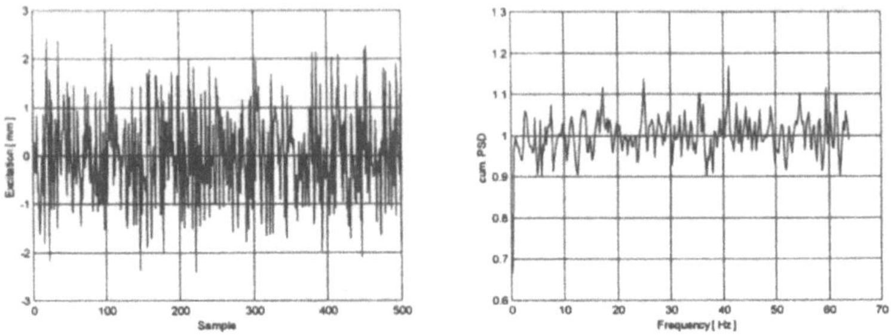

Figure 3. White noise excitation signal and its averaged PSD

In the present case at higher frequencies the excitation amplitude is smaller, thus this range can hardly estimate because of the inner friction of the seat mechanism and the measurability of the output signal.

4 Signal analyzing and estimation of the abstract parameters

In chapter 2 the mentioned seat suspension model is continuous. Since the measurement and signal processing has been made by PC, thus it was obvious to use discrete estimation method. Rapidity was also an important factor, which was the reason to use a discrete **Autoregressive Moving Average** structure and **Least Squares** parameter estimation.

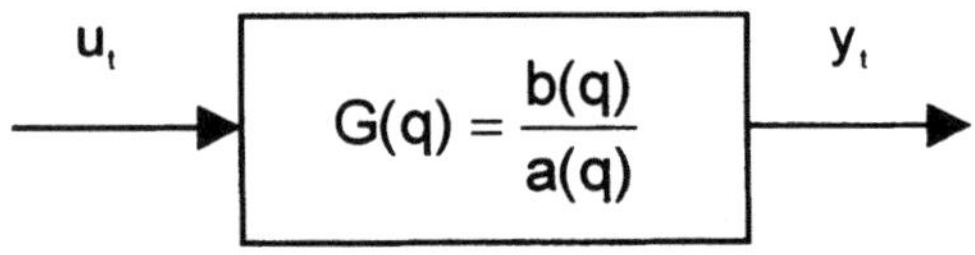

y_t : system output
u_t : excitation signal
q: shift operator

Figure 4. Polynomial discrete transfer function model

Concerning the above model, the equation between the input and output signals is as follows:

$$y_t + a_1 \cdot y_{t-1} + ... + a_p \cdot y_{t-p} = u_t + b_1 \cdot u_{t-1} + ... + b_q \cdot u_{t-q}$$

$$u_{t-1} = q \cdot u_t \qquad y_{t-1} = q \cdot y_t$$

$$(1 + a_1 \cdot q + ... + a_p \cdot q^p) \cdot y_t = (1 + b_1 \cdot q + ... + b_q \cdot q^q) \cdot u_t$$

The most frequently used model structure can be seen below (Fig. 5.). At this model the noise model is adverted by H(q) transfer function as well.

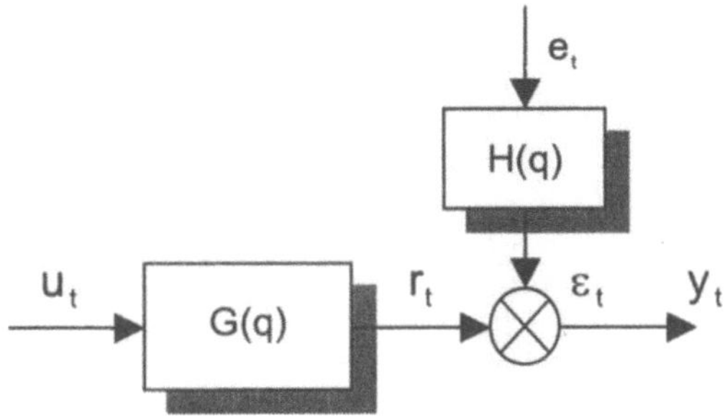

Figure 5. Model structure respecting output noise

$$y_t = \frac{b(q)}{a(q)} \cdot u_t + \frac{1}{a(q)} \cdot e_t \qquad \text{LS structure}$$

The parameters of the model are identified by Least Squares estimation method. The goal of LS estimation is to minimize the residual (ε) between the measured and estimated output (Fig. 6.).

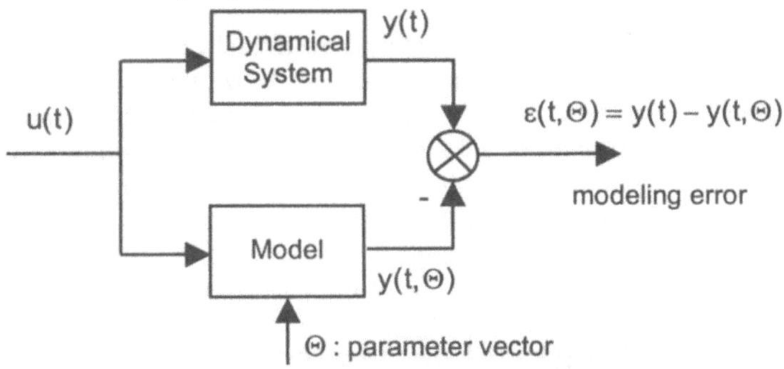

Figure 6. Least Squares parametric estimation

The LS estimation and parameter computation from discrete values to continuous time has fulfilled by an own MATLAB program.

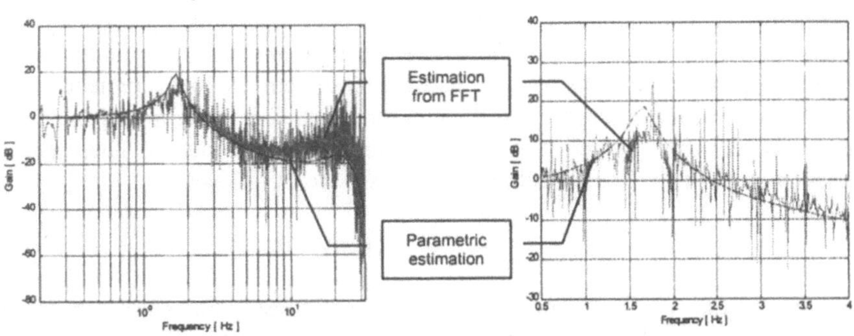

Figure 7. Parametric and non-parametric estimation of transfer function of seat suspension

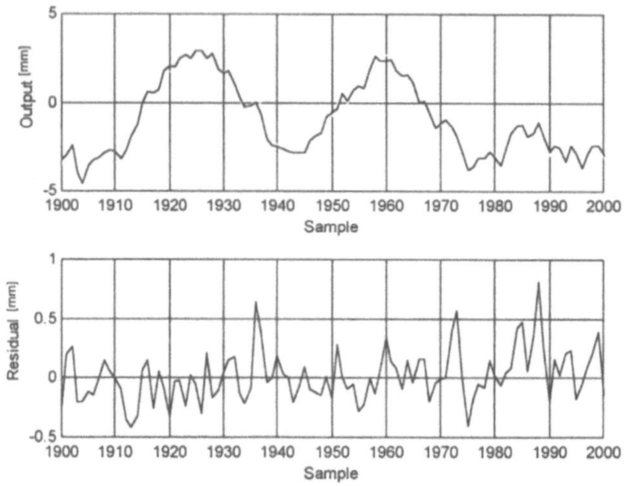

Figure 8. Measured (light) and estimated (dark) output (above) and the residual of the estimation (below) with 64 Hz sample frequency

Fig. 7. and 8. shows the results of the LS estimation. It is clearly distinguishable that the first eigenfrequency of the seat situated at 1.7-1.8 Hz, which is fairly good from point of view of vibration comfort. At this eigenfrequency the parametric estimation gives higher gain than the FFT estimation and the real value as well, which can cause error at computing the physical parameters (damping). However, from the approximate value of the parameter we can produce more exact estimation.

5 Determination of physical parameters

In chapter 4. only the continuous abstract parameters were determined by LS parameter estimation. Naturally, the goal was to define the physical parameters adequacy. These can be determined from the parameters of the transfer function between the seat pad and the frame (Eq. 10.).
The number of the parameters are 8, where some parameters are equal ($b_1=a_1$ and $b_0=a_0$). It is obvious that the parameters obtaining from the measurement are not equal, because of the inaccuracy of the measurement. The linear model includes 6 real parameters, from which it is easy to determine 2 parameters. These are the masses (m_1 and m_2). The predetermined masses are as follows:

$$m_1 = 25 \text{ kg} \qquad m_2 = 60 \text{ kg (driver mass)}$$

The other physical parameters are computable if these parameters are utilized from the abstract b_3, b_2, a_3, a_2 parameters. At this calculation it has to advert that the discrete-continuous recalculation can be solved only numerically (Zero Order

Holder) or approximately (Tustin-transformation), and the equation system describing the physical parameters is overdetermined as well. The approximated parameters are as follows:

- parameters of the seat : $c_1 = 1.35 \cdot 10^4$ N/m $k_1 = 140$ Ns/m
- parameters of the seat pad: $c_2 = 2 \cdot 10^5$ N/m $k_2 = 500$ Ns/m

Fig. 9. shows the difference between the FFT estimated, parametric estimated discrete transfer functions and the physical parameters based on continuous transfer function.

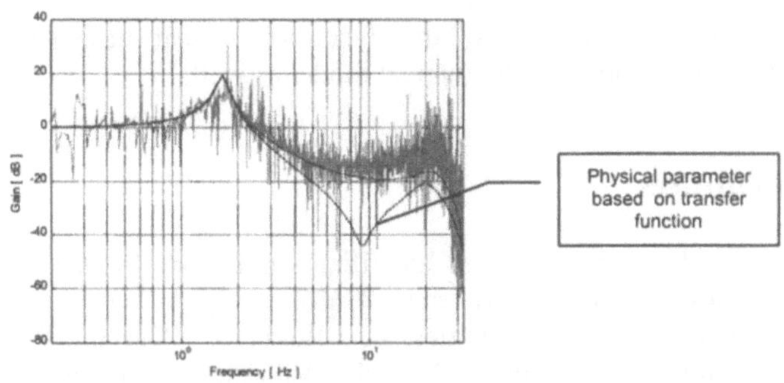

Figure 9. The parameter based on transfer function in comparison to the estimated

6 Parameter sensibility investigation based on the identified parameters

After determining the real parameters it is possible to make a quick computational parameter sensitivity analysis. Changing the real parameters the trend of transfer function is very visual. During the investigation the parameters of lower part (c_1 and k_1) have been changed, using 1/2, 3/4, 4/3, 2/1-times of the nominal value. Fig. 10. shows the result of the parameter changing.

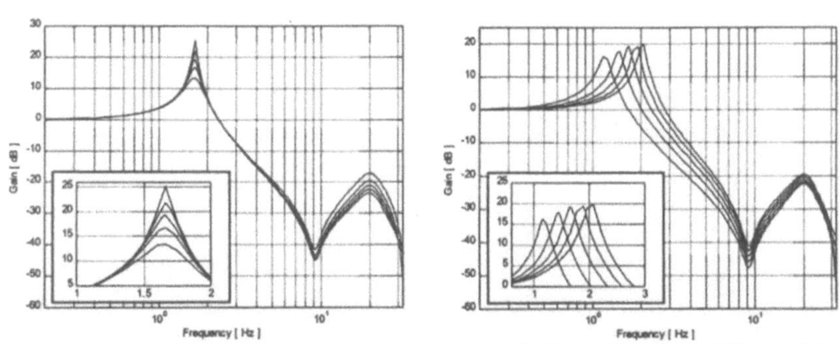

Figure 10. Result of changing damping ratio (left) and spring stiffness (right)

7 Conclusions

The linear model provides adequate result to estimation of the physical parameters, however this model is accurate only in case of excitation signals with limited amplitude.
The seat suspension can be identified in wider range of operation with respecting to the nonlinearities. To solve this problem such an excitation instrument is needed, which can excite up to 32 Hz-s as high amplitude that gives well measurable output signals even in case of -40 dB gain as well.
Measuring similar mechanical systems it has to advert the effect of unwished neglected degrees of freedom.
The further investigations of the dynamics of seat suspension system will be included the nonlinear mathematical model as well.

References

1 Ljung, L. (1987): System Identification Theory for the User, PTR Prentice Hall, Englewood Cliffs, New Jersey 07632
2 Michelberger, P. - Keresztes, A. - Bokor, J. - Várlaki, P. (1991): Statistical Identification of Nonlinear Vehicle Vibrating Structures., Proceedings of Identification and System Parameter Estimation, Pergamon Press, Vol. 1, pp. 358-363.
3 Griffin, M. J. (1990): Handbook of Human Vibrations, Academic Press, Harcourt Brace & Company, Publishers
4 Michelberger, P. - Keresztes, A. - Bokor, J. - Várlaki, P. (1988): Modelling and Identification of Bus Axle and Frame Structure Dynamics., International Journal of Vehicle Design, Vol. 9, No. 1.
5 Amirouche, F.L., Tong, R.T., Palkovics, L. (1996): Human Body Vibration Control and Ride Comfort - A Two-State Semi-Active Suspension Design for Cabs and Seats., ASME Winter Annual Meeting, Atlanta, Georgia, November.

Two-grooved automatic nail machine

Sergei Vasilishin and Vilen Roizman***

An original technological two-grooved nail production scheme has been developed, researched, and introduced to industry.
On the basis of this technology a range of original designs for new generation of nail machines have been developed that are unique and outperform previously existing models as to amounts of power consumption and specific metal content.
Let us consider as specific metal content as the following value:

$$SMC = \frac{M}{nF}\left[\frac{kg.\min}{mm^2}\right]$$

where M is weight of the machine (kg); n - quantity of nails per minute. F - nail cross-section area.
Actually, in denominator there is section area of all nails made per time unit per minute.
Formula of specific power consumption (SPC) can be shown in the following way:

$$SPC = \frac{N}{nF}\left[\frac{kW.\min}{mm^2}\right]$$

where N is power of a motor of the machine.
Before 1989 according to indexes of power consumption and metal content firms producing nails were rated as follows:
1st place - MORONI (Italy)~
2nd place - TANISAKAS (Japan):
3rd place - WAFIOS (Germany);
4th place - PRIGMA-PRESS (Ukraine) whose automatic machines were the most heavy and economically inefficient.
In Figure 1 the mechanical diagram of AB4 115 having been produced before 1989 is shown:

---

* "Prigma-Press'
** Technological University of Podilia.I I. Institutskaia Str., 280016 Khmelnitskv, Ukraine. Fax +(0038)+(03822) 23265.

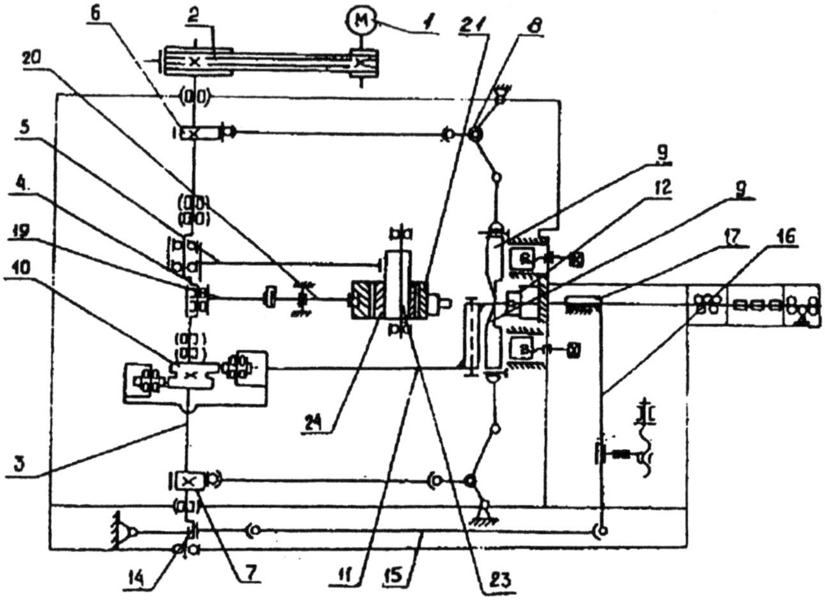

Figure 1. Mechanical diagram of nail automatic machine AB4115.

The machine is actuated from electric motor 1 through flywheel 2 to drive shaft 3; the motion is transmitted by means of cranks 4 and 5 to heading mechanism slide; by means of two eccentrics 6 and 7 to toggle 8 and further to cut slides 9; by means of two-unit knuckle 10 to three-arm lever 11 and further to slide 12, clamping dies 13; by means of crank 14 through link 15, lever 16 to carriage 17 clamping chisel 18 of the feeding mechanism. Crank 4 puts with the help of connecting rod beam 19 rocker 20 into motion. The connecting rod transmits horizontal motion to heading slide 21, simultaneously crank 5 through connecting rod 22 swings eccentric 23, which through stone 24 gives vertical motion to the slide. As the result of mixing the two motions the slide travels in closed trajectory and goes on straight line during heading.

The automatic machine was a one-shaft unit. All driven elements were located on the shaft. The elements consequently, according to sequence diagram, controlled the motors of heading, cutting, clamping and feeding mechanisms.

Complicated trajectory of heading tool (figure 2a) was performed from mechanism with plenty of mechanical ties, and that leaded to quick wear of friction influenced parts (hinges, stones, eccentrics) and loss of mechanical accuracy of the mechanism. One had to stop the machine in this case for change of worn parts.

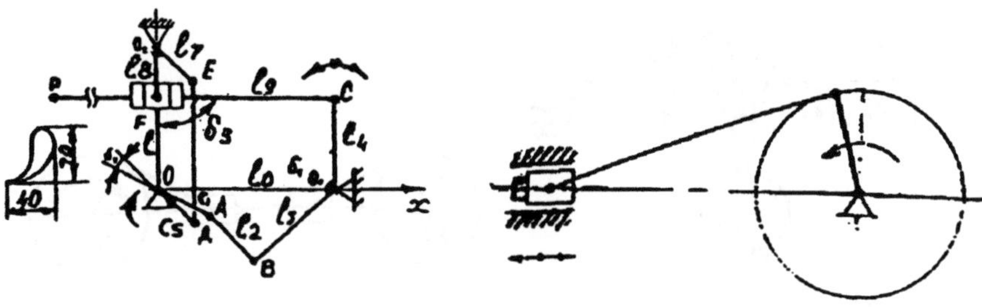

a) b)

Figure 2. Heading mechanism.

The second drawback of the above machines is that force of the clamping mechanism (figure 3a) is transmitted to dies in vertical direction that causes taking up of the clamping force by the frame, by its cantilever part. As a result it is impossible to reach absolutely stiff clamp of the nail, because the cantilever flexes under the load. As the result a nail with extended head, with more rigid tolerances for heads in the clamp places, changes dimensions that influences quality of clamping and, subsequently, shape of nail head.

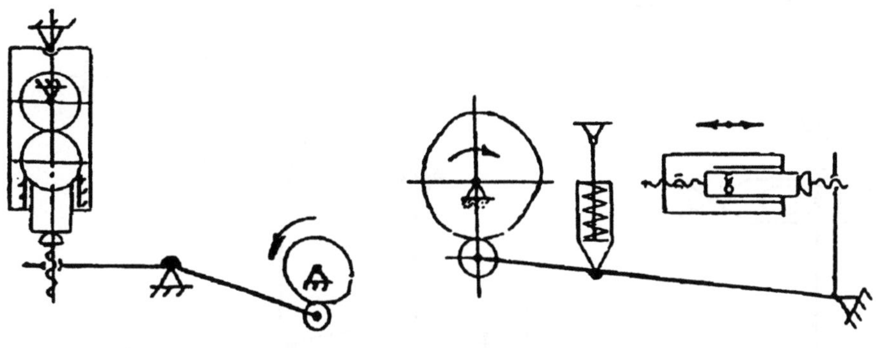

a) b)

Figure 3. Clamping mechanism.

The other drawback of AB series machines is present at the cutting mechanism(figure 4a). Cutting mechanism of peen type involves 8 kinematic ties. During the cutting hinges in kinematic ties are loaded and energy of elastic deformation is reserved.

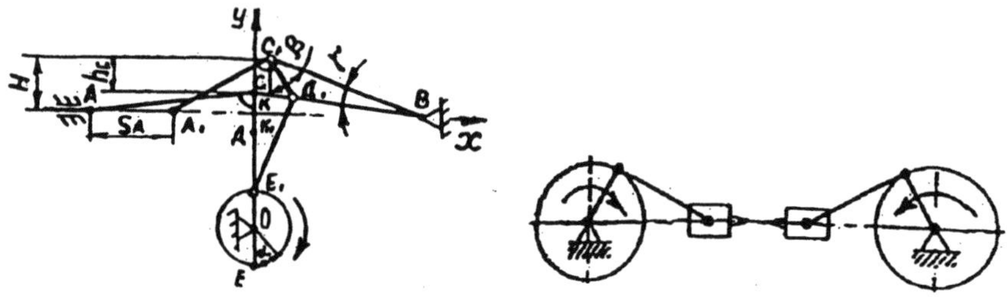

a) b)

Figure 4. Cutting mechanism.

Knives, as a rule, due to material properties do not cut wire efficiently and wire at the end of cutting is broken off from the metal and in the result of what the entire loaded system springs and knives shock one another by cutting edges. Intensive wear of knives comes and durability diminishes.

One more drawback of the machine is in feeding mechanism (figure 5). As a result of wire bringing on its surface various inclusions (metal dust, rust, scale etc.),during travel through feeding mechanism these waste elements are accumulated on the guiding column. It is followed by intensive wear of friction pairs: guiding column and bushing. In time in the pairs with mutual friction a clearance (gap) appears, play of feeding carriage increases and it results in increase of nail curvature.

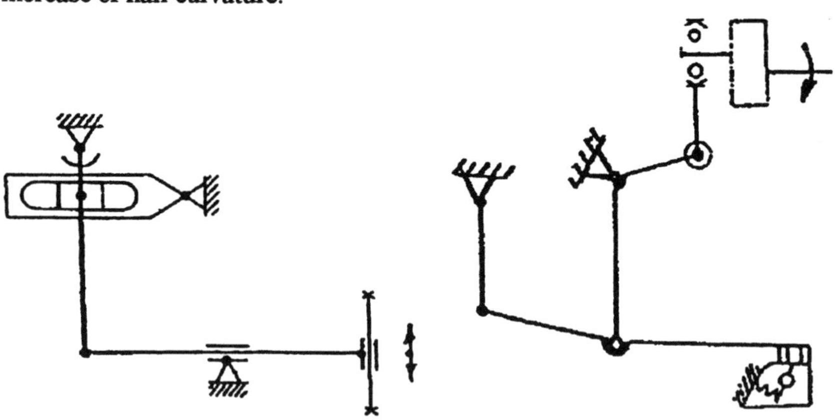

a) b)

Figure 5. Feeding mechanism.

A serious shortcoming of AB type automatic machines is high vibration the cause of which is big quantity of spatial mechanisms which are very difficult to balance under conditions of limited space of a nail machine.

At high speed unbalanced forces of the second order are so considerable that it is necessary to use special mechanisms to balance them (diametric balancing), but that is impossible under conditions of overall dimensions. This phenomenon suppressed growth of capacity. The machines worked steadily only at low speed. But to justify somewhat this system we should note that it has its advantages, providing for simple calculation of cyclogram of the entire technological process that is quite complex and strict. And here we must say that nail machines can be compared with fast shooting weapon, where a shot is changed for a stroke of the heading slide forming a nail head (up to 600 pieces per min.), and this "shot" must be extended in time in order to have plastic (without cracking) deformation of a portion of wire into a head, that is why increase of capacity must be gained only within idle and back stroke of the slide. And this indispensable non-uniformity was previously reached by using multi-link mechanisms with all the resulting drawbacks.

As a matter of fact it could be possible to reason the old scheme:
to draw up a mathematical model of a complex vibro-impact system and to try to find such values of its parameters that would eliminate defects. And for theory such a model would probably constitute a matter of interest. But such an approach would hardly give a positive result in practice, not only for the reason of complicated vibro-impact system, but because it is impossible to find among the drawings accurate ones, agreed to mathematical model, mass coefficients, damping, rigidity and the rest of values.

That is why we did not take and solve such a task, but have found an entirely new productive idea: multigroove production of nails at one machine, and for the first stage we have developed a two grooves scheme, i.e. production of two nails simultaneously from two lines of wire fed.

As the basis kinematic scheme of mutually perpendicular shafts was taken which makes it possible to place all actuating and driven mechanisms in one plane that makes all these mechanisms quite simple and allows to process two lines of wire.

The automatic machine operates as follows:
from electric motor 1 through V-belts through pulleys 2 and 3 rotation is conveyed to crankshaft I which gives reciprocating motion to the heading slide and conveys rotational motion to the side shafts II and III through two pairs of conical gear wheels 5 and 6. From the side shafts, through an eccentric, slides 7 of sharpening and cutting-off mechanism are set in motion, and from the right side shaft II from clamping knuckle 8 through lever 9 slide 10 of clamping mechanism is set in motion.

On the left cantilever of crankshaft I faceplate 11 is placed which gives through tie-rod 12 and lever 13 reciprocating motion to the feeding carriage 14. During the feed wire is gripped by chisel 15 and pulled through straightening rollers 16 positioned before feeding mechanism. The carriage moves along guides 17. Thus, travel is performed along a straight line. On the right cantilever of the crankshaft a brake coupling 20 is positioned, it conveys motion from pulleys through flywheel to crankshaft and serves for disengagement of the crankshaft from the operating motor and hard braking of the machine moving parts. At models AB4411A, AV113 instead of brake coupling a flywheel is installed which is connected with the crankshaft with friction spacer.

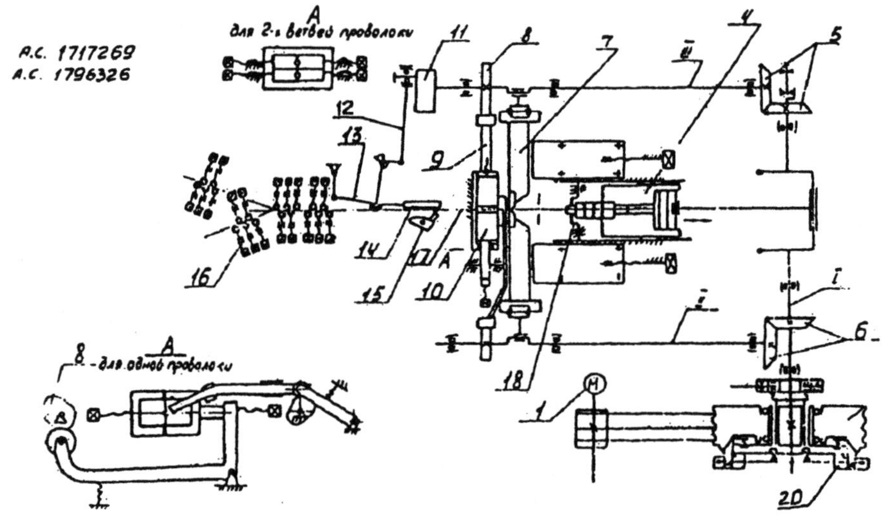

Figure 6. Mechanical diagram of AB4115 automatic machine.

Advantages of the new kinematic scheme in comparison to the previous one are obvious. First of all they consist in simplification of kinematic schemes of the entire automatic machine and its basic units by change of complicated multi-link mechanisms for simple, e.g. crank-and-slide ones. It became possible, because the kinematic scheme now is based upon drive system of three shafts which lie in one plane, two of them are perpendicular to the main drive shaft, that allowed to place all actuating mechanisms in the plane of maximum rigidity of the frame.

The main difference of the design between the machines of AV and AB are shown in mechanical diagrams (figures 2...5). The feeding mechanism of the machine (figure 5b) is designed as a four-link unit driven from a crank, installed on the left side. Automatic machine AB4118 is supplied with microdrive for low-speed nail production. Such a kinematic scheme provides a range of advantages in comparison to AB series automatic machines.

The heading mechanism (figure 2b) is designed as crank-and-slide one. As against to AB series the slide is fixed in round guides that provides for rigid straight travel. The bearing in the crank is performed as shells that provides for a proper rigidity, minimum overall dimensions.

This design allows to balance the mechanism with rated load on the crankshaft. The straight rigid travel provides for a straight rigid stroke that influences favorably upon shaping of nail head.

The clamping mechanism (figure 3b) at AV automatic machines in comparison to those of AB is turned by 90 degrees. It allows to close the clamping force load inside the frame body that increases rigidity of the mechanism. With such a construction and comparatively small dimensions a necessary force is created for clamping during nail head forming. Besides, assembly of clamping lever by horizontal position of the die is much more convenient. The lever does not hinder outlet of nails out of the automatic machine, because it is placed under the die unit, and length of lever shoulders allows to reduce the force of wire clamping on the knuckle by four or five times, when at AB machines this force was reduced only by 2 to 2.5 times. In general this increases longevity and durability of the mechanism. The cutting mechanism (figure 4b) is designed as crank-and-shaft one. The bearings on the crank are made as shells. In comparison to AB mechanisms the number of kinematic ties is decreased by three times, length of the slide guides. Rigidity of the frame is increased. Elastic deformations by such a design are diminished considerably. Because of the fact that AV machines possess only 2 joints (instead of 8 at AB type) the total size of clearances (gaps) at joints is by 4 times less. The result is that there is no mutual shocking of knives while cutting during partition of nail from wire, i.e. there is much less springing phenomena in this design. Testing has proved that durability of nails is by two times higher at the automatic machines of the new type.

The feeding mechanism (figure 5b) at the machines of the new type is designed on two columns. This allowed to increase rigidity of the mechanism, to diminish wear of guides, to make the motion of the carriage along the straight line more steady. The quality of nail as to curviture has become higher. The durability of the feeding mechanism has increased by 4 times.

To improve ergonomic properties at the automatic machines of large standard sizes microdrive is supplied. This allowed to better arrangement of the machine, to make operating safer.

Driven actuating mechanisms at the machines of the new type are located on three mutually perpendicular shafts, positioned in one plane, while at AB machines they are placed on one shafts in the rear. The new arrangement allowed to distribute the load from technological efforts more uniformly over the entire perimeter of the frame, that in its part allowed to design a more rigid bed frame of the frame type with longitudinal reinforcing ribs along of it, which take up the heading force. Bed frame deformations are decreased from 0.4 to 0.1 mm that improved considerably quality of formed nail head.

Finally, we would like to stress it once again, that the new automatic machine possesses, on the contrary to the old one, two grooves, that increased greatly its capacity with the same or less speed of nail head deformation.

In the new scheme the number of hinges has been decreased approximately by 2 times and the problem of backlash wear is settled; now balancing of mechanisms has become possible, it provided for decrease of vibrations and transition to hard alloy covering of knife edges that has been impossible before for the reason of mutual shocking during "giving back" of elastic energy by the machine spent by it during "loading" of the automatic machine for the nail formation and released after cutting of the nail.

The design has made it possible to form a frame design with longitude reinforcing ribs which take up the load better. This has made it possible to close the load during nail heading inside the frame and diminish deformations of it by 4 times. Finally it has resulted in considerable improvement of quality of the nails produced and increase of output possibilities.

While designing the new construction all necessary calculations: kinematic, energetic, strength are programmed in computer and made the basis for automatic designing of nail machines that makes it possible to organize in a shorter time production of a range of the automatic machines of this new type, differing in wire diameter to be processed. The result of this work is creation of the new generation of automatic machines which as against to those of previous and foreign ones are two-grooved, straighten wire in two mutually perpendicular directions, possess the highest capacity and the lowest power consumption for nail production. This makes it possible to gain the first place in the world within the sphere of nail machines.

During operational development of the new design the problem of securing serviceability and reliability of the machine units, first of all exposed to wear okand friction, units of feeding, heading and cutting (about 30 parts in total) has been solved. But description of he researches having been conducted cannot be presented within the limits of this report and is the matter of a special article.

CONCLUSIONS.

An original technological two-groove scheme of nail production has been developed and introduced.

A range of nail machine constructions has been created which are noted for more reliable design of heading, cutting, clamping and feeding mechanisms.

Analytical and numerical methods for the analysis of the stitch forming devices of blindstitch sewing machine.

Dr. Deltchev[*] *and Z. Tcherneva-Popova*[*]

This paper presents a new approach in making models of stitch forming devices in a blindstitch sewing machine. A mathematical model of stitch forming devices is proposed. In this first part of the work a kinematical analysis using Denavit and Hartemberg's parameters is done. This model is not unique for the sewing machine but with slight modifications could be used to make analysis of multibody system with the same number of bodies. An optimal model for the impact of the needle with the fabric is proposed. To solve the problem, MATLAB computational software is used.

1 Introduction

The most important of stitch forming devices for this type of machine namely the mechanism of the looper and the mechanism of the needle are represented as systems of interconnected bodies, the interconnections being the kinematical joints of fifth class. The restriction of using only kinematical joints of fifth class is not essential because spherical joints could be described by a sequence of three rotations after including two additional bodies without inertial parameters.

In first approximation all the bodies except for the carrier of the needle are regarded as rigid bodies.

The aim of this paper is to present models of the stitch forming devices, namely mechanism of the needle and mechanism of the looper.

2 Description of the system

To prove existence of vibro-impact processes one should determine whether or not there are periodically repeated motions combined with the impacts. Lets make some further analysis of the systems being investigated.

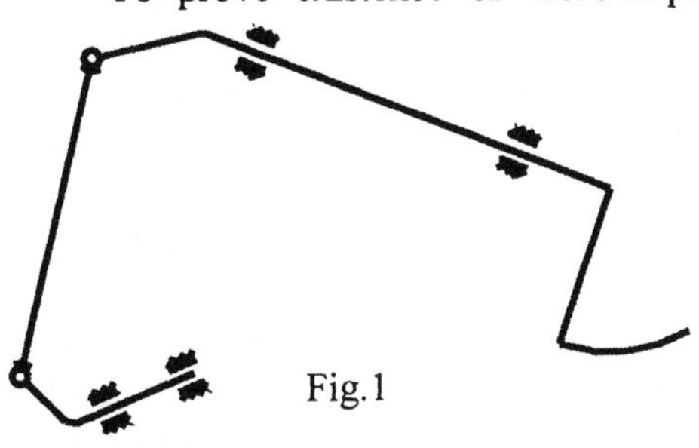

Fig.1

Both mechanisms (Fig. 1 and Fig. 2) could be represented with six bodies after replacing spherical joint with two additional bodies.

[*] Department of Mechanics, Technical University of Sofia,
Sofia – 1797, "Kl. Ohridski" N8, bl. I, Bulgaria

The first problem which has to be solved is the determination of the orientation of the bodies with respect to a given reference configuration (inertial coordinate system) or, equivalently the determination of the relative orientation between the connected bodies.

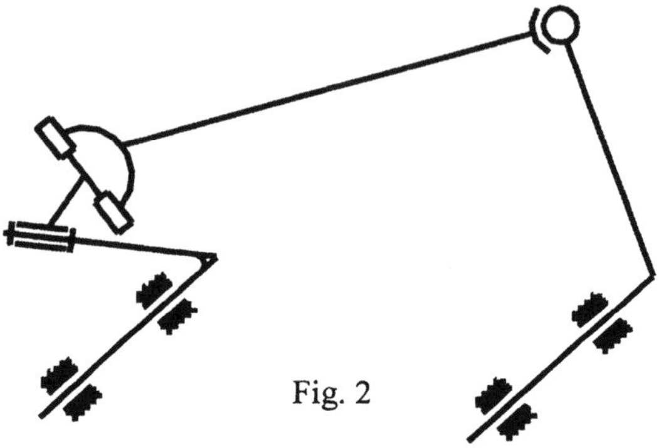

Fig. 2

The optimal (in terms of simplicity and adequateness) model for the impact of the needle with the fabric is a heavy point mass attached to a linear spring that enters in contact with a massless spring. The heavy mass point symbolizes the head of the needle and the massless spring symbolizes the fabric being sewn. After impact, the appropriate model should include an element that comprises the effects of dry friction following the penetrating of the needle in the fabric. After making these models and solving them it will be possible to outline the ways this machine could improve its performance

The most convenient way for investigating mechanisms containing only kinematical joints of fifth class is proved to be using Denavit and Hartemberg's parameters.

To each body in the system coordinate system is linked in the following way (Fig. 1):

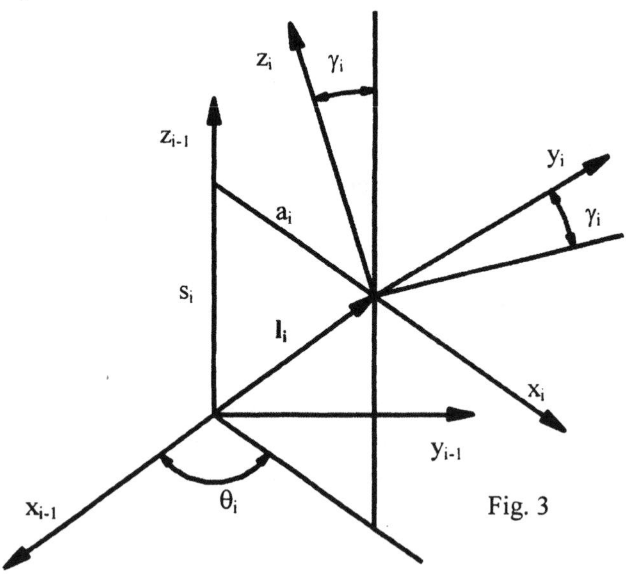

Fig. 3

Axes $O_i z_i$ are along the axis of the kinematical joint of the body with this number. Axis $O_i x_i$ is along the common perpendicular between $O_{i-1} z_{i-1}$ and $O_i z_i$ and has the direction from $O_{i-1} z_{i-1}$ to $O_i z_i$. And y_i is placed in such way that $O_i x_i y_i z_i$ form a Cartesian coordinate system. The corresponding parameters are defined as follow:

$$\theta_i = \angle(x_{i-1}, x_i),\ \gamma_i = \angle(y_{i-1}, y_i),\ a_i = dist(z_{i-1}, z_i),\ \ s_i = dist(x_{i-1}, x_i)$$

Let us consider now a joint connecting the (i-1)$^{-th}$ and i^{-th} bodies in the system and relative orientation of a base $x^{(i)}$ associated to i^{-th} body with respect to the base $x^{(i-1)}$ associated with (i-1)$^{-th}$ body. This relative orientation can be described by the transformation matrice $\tau_{i-1,i}$, given by the relation:

$$x^{(i)} = \tau_{i-1,i} \cdot x^{(i-1)} \tag{2.1}$$

The (3 x 3) matrices $\tau_{i-1,i}$ are orthogonal. To fully describe position of the i^{-th} body with respect to the to the (i-1)$^{-th}$ body we should know the position of O_{i-1} (i. e. vector quantity $\mathbf{O_{i-1}O_i}$). With the given parameters and orientation of the systems (see fig.1) it could be written in the form:

$$\mathbf{l_i} = \left[a_i.cos(\theta_i)\ \ \ a_i.sin(\theta_i)\ \ \ s_i\right]' \tag{2.2}$$

When using homogeneous coordinates these expressions could be combined:

$$T = \begin{pmatrix} \tau & l \\ 0 & 1 \end{pmatrix} \tag{2.3),$$

here 0 is (1 x 3) - zero matrice and T is the homogeneous matrice of passage from the (i-1)$^{-th}$ base to the i^{-th}

3 Analytical description

After replacing expression (2.1) and (2.2) in (2.3) we can obtain the following expression for the homogeneous matrice of passage from (i-1)$^{-th}$ body to the i^{-th}

$$T_{i-1,i} = \begin{pmatrix} \cos(\theta_i) & -\sin(\theta_i).\cos(\gamma_i) & \sin(\theta_i).\sin(\gamma_i) & a_i.\cos(\theta_i) \\ \sin(\theta_i) & \cos(\theta_i).\cos(\gamma_i) & -\cos(\theta_i).\sin(\gamma_i) & a_i.\sin(\theta_i) \\ 0 & \sin(\gamma_i) & \cos(\gamma_i) & s_i \\ 0 & 0 & 0 & 1 \end{pmatrix} \tag{3.1}$$

It should be noted that given systems are closed system of bodies in other words the last body in the system coincide with the first which could be expressed in terms of a matrix multiplication:

$$T_{01}.T_{12}.T_{23}.T_{34}.T_{45}.T_{56}.T_{67} = E \tag{3.2),$$

where E is the (4 x 4) - identity matrix.

4 Numerical solution

To solve this matrix equation each of the matrix is developed in Taylor series in the following way:

$$T_{i-1,i}(\theta_i) \cong \overline{T}_{i-1,i} + \left[\left. \frac{\partial T_{i-1,i}(\theta_i)}{\partial \theta_i} \right|_{\theta_i = \overline{\theta}_i} \right] . \partial \theta_i \qquad (4.1)$$

This partial derivative could be expressed in terms of a matrix multiplication:

$$\frac{\partial T_{i-1,i}(\overline{\theta}_i)}{\partial \theta_i} = Q.T_{i-1,i} \qquad (4.2),$$

$$\text{where } Q = \begin{pmatrix} 0 & -1 & 0 & 0 \\ 1 & 0 & 0 & 0 \\ 0 & 0 & 0 & 0 \\ 0 & 0 & 0 & 0 \end{pmatrix}$$

Equation (4.2) then could be rewritten:

$$T_{i-1,i}(\theta_i) \cong \overline{T}_{i-1,i} + Q.\overline{T}_{i-1,i} d\theta_i = (E + Q.d\theta_i)\overline{T}_{i-1,i} \qquad (4.3)$$

After substitution of (4.3) in (3.2) and neglecting the infinitesimal in order higher than 2 from (3.2) the following expression is obtained:

$$\begin{aligned} &T_{01} \cdot Q.\overline{T}_{12} \ldots \overline{T}_{67} . d\theta_2 + T_{01} \cdot \overline{T}_{12} \cdot Q.\overline{T}_{23} \ldots \overline{T}_{67} . d\theta_3 + \ldots + T_{01} \ldots \overline{T}_{56} \cdot Q. d\theta_7 = \\ &= E - T_{01} \cdot \overline{T}_{12} \ldots \overline{T}_{67} \end{aligned} \qquad (4.4)$$

This equation is equivalent to linear system for the unknowns:

$d\theta_2, d\theta_3, \ldots, d\theta_7$. After solving this system (that may be done iteratively) the relative rotations in the kinematical joints could be found. To find relative velocities ω_i^r, accelerations ε_i^r (3.2) is differentiated one respectively two times.

Equation for the relative velocities is:

$$T_{01} \cdot Q.T_{17} . \omega_2^r + T_{02} \cdot Q.T_{27} . \omega_3^r + \ldots + T_{06} \cdot Q.T_{67} . \omega_7^r = -\omega_1^r . Q \qquad (4.5),$$

here ω_1^r is the angular velocity of the main shaft of the machine. This equation is equivalent to a linear system of 16 equations for the unknowns. Four of the equations are trivial statements 0=0 and 1=1. From the rest of 12 equation at most 6 are linear independent, so the solution could be found after choosing 6 linear independent equations or by using least square method. The same problems could be encountered when solving the system for the relative accelerations.

The expressions for the relative rotations are too large to be placed in this paper but there are no significant difficulties finding ε_i^r after differentiating (4.5).

Having find $\theta_i^r, \omega_i^r, \varepsilon_i^r$ it is not a problem to find absolute angular velocities and accelerations with the formula:

$$\omega_i^a = \begin{pmatrix} 0 & -\omega_{iz} & \omega_{iy} \\ \omega_{iz} & 0 & -\omega_{ix} \\ -\omega_{iy} & \omega_{ix} & 0 \end{pmatrix} = \dot{\tau}_{0i}.\dot{\tau}_{0i}' \qquad (4.6),$$

$$\varepsilon_i^a = \begin{pmatrix} 0 & -\varepsilon_{iz} & \varepsilon_{iy} \\ \varepsilon_{iz} & 0 & -\varepsilon_{ix} \\ -\varepsilon_{iy} & \varepsilon_{ix} & 0 \end{pmatrix} = \dot{\tau}_{0i}.\dot{\tau}_{0i}' + \ddot{\tau}_{0i}.\tau_{0i}' \qquad (4.7),$$

here ' stands for transpose and τ_{0i} is the orientation of the i^{-th} body in the system to the inertial coordinate system. The computation of $\tau_{0i}, \dot{\tau}_{0i}, \ddot{\tau}_{0i}$ is done with the recursion:

$$\tau_{0,i} = \tau_{0,i-1}.\tau_{i-1,i}, \qquad \dot{\tau}_{0,i} = \dot{\tau}_{0,i-1}.\tau_{i-1,i} + \tau_{0,i-1}.\dot{\tau}_{i,i-1},$$
$$\ddot{\tau}_{0,i} = \ddot{\tau}_{0,i-1}.\tau_{i-1,i} + 2.\dot{\tau}_{0,i-1}.\dot{\tau}_{i,i-1} + \tau_{0,i-1}.\ddot{\tau}_{i-1,i} \qquad (4.8)$$

The results of kinematical analysis are illustrated in the next figures in the following way: in the first four rows the kinematical parameters of mechanism of the needle are placed. In the first column relative rotations in the joints are presented, and in the second column the corresponding relative angular velocities. The same method is used for illustrating the kinematical parameters of the looper's mechanism in the rows from 5 to 9. All the parameters are for one period of rotation of the main shaft of machine. On the y-axes dimension are rad for the relative rotations and s^{-1} for the relative angular velocities.

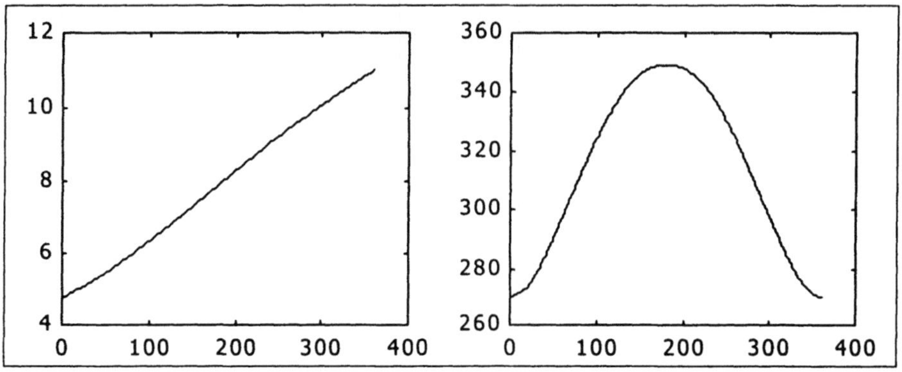

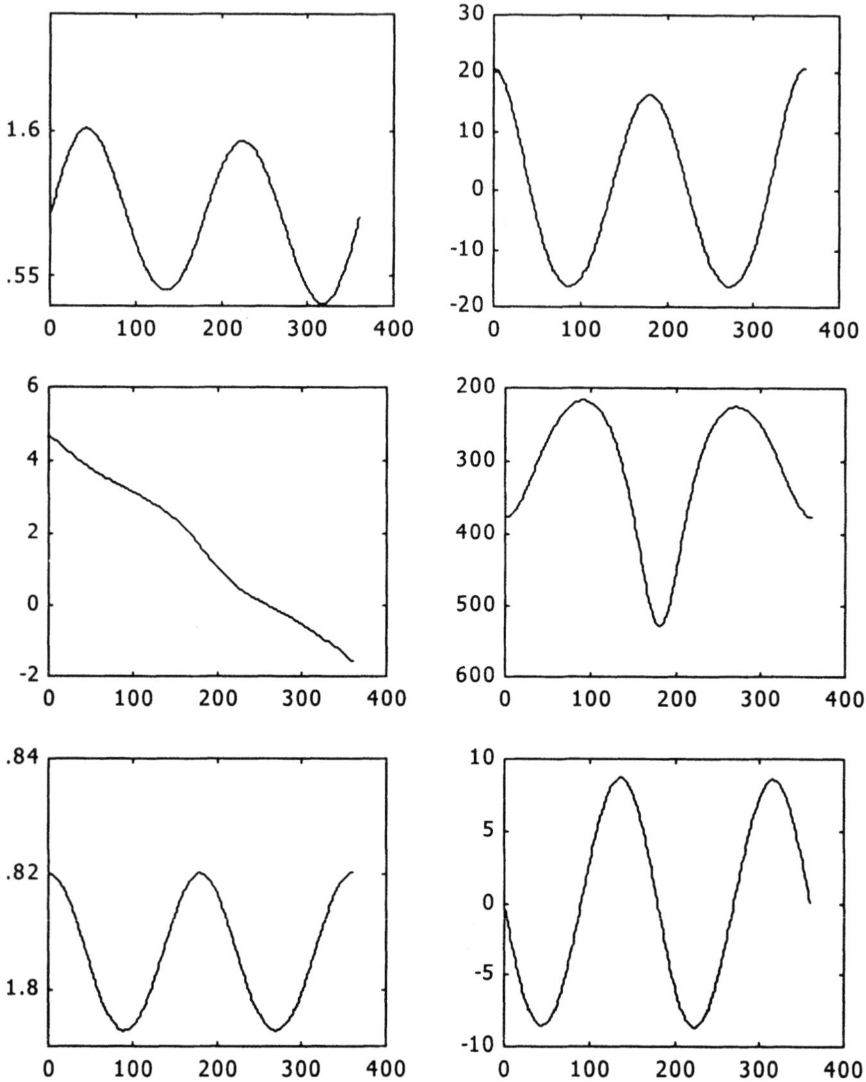
1.6
.55
0
100
200
300
400
30
20
10
0
-10
-20
0
100
200
300
400
6
4
2
0
-2
0
100
200
300
400
200
300
400
500
600
0
100
200
300
400
.84
.82
1.8
0
100
200
300
400
10
5
0
-5
-10
0
100
200
300
400

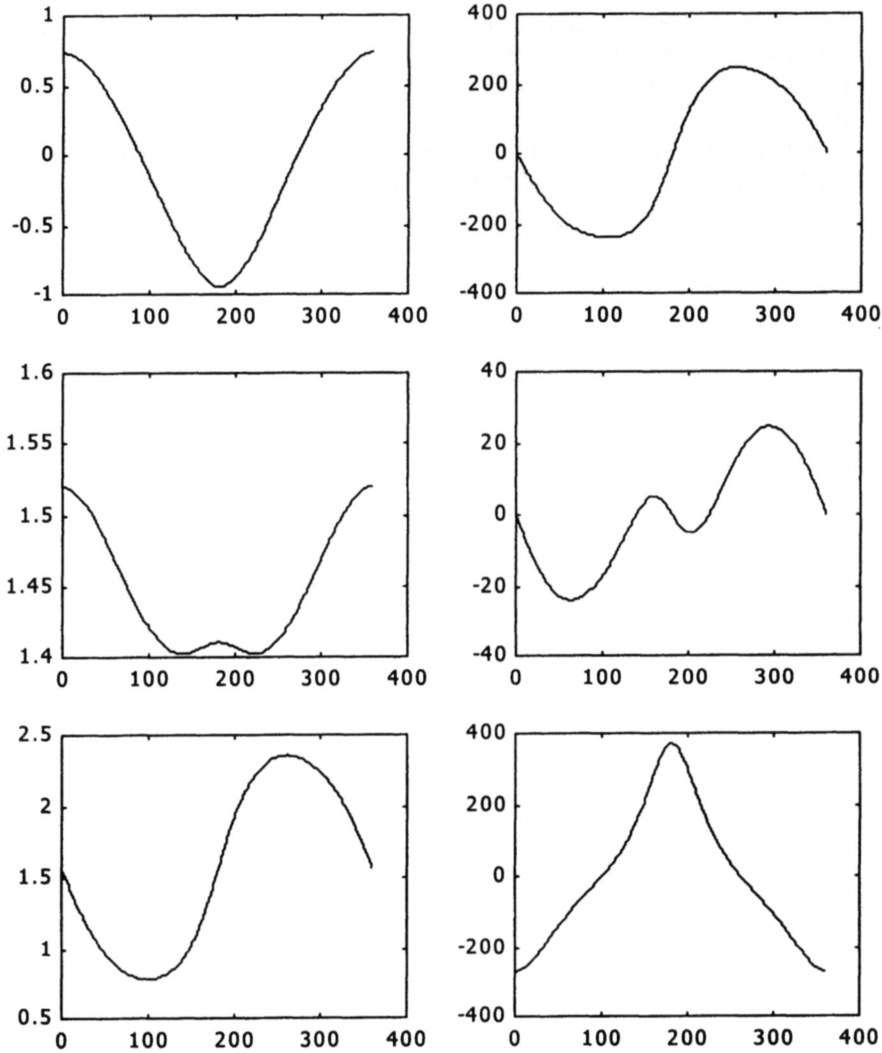

5 Conclusion

The existence of periodically components of motion which lead to vibro-impact processes is proven by means of kinematic analysis of the models of the mechanisms.

References

1. Babitsky V. I., Kolovsky M. Z. To the theory of vibro-impact systems. Machine Studies, No1, 1970 (in Russian)

2. Dubowsky S. Freudenstein F. Dynamic Analysis of Mechanical Systems with clearances. "Trans. ASME" B 93, No 1, 1971
3. Harris C.M., Crede C.E. Shock and Vibration Handbook. Mc. Graw- Hill Book 1961
4. Deltchev D. D.,Kinematical model of the mechanism of the looper in a blind stitch sewing machine, Second Symposium Tempus Jep 3310, Varna 29-31.07.1995

AUTHOR INDEX

Dynamics of Vibro-Impact Systems

MIX
Papier aus verantwortungsvollen Quellen
Paper from responsible sources
FSC® C105338

If you have any concerns about our products,
you can contact us on
ProductSafety@springernature.com

In case Publisher is established outside the EU,
the EU authorized representative is:
Springer Nature Customer Service Center GmbH
Europaplatz 3, 69115 Heidelberg, Germany

Printed by Libri Plureos GmbH
in Hamburg, Germany